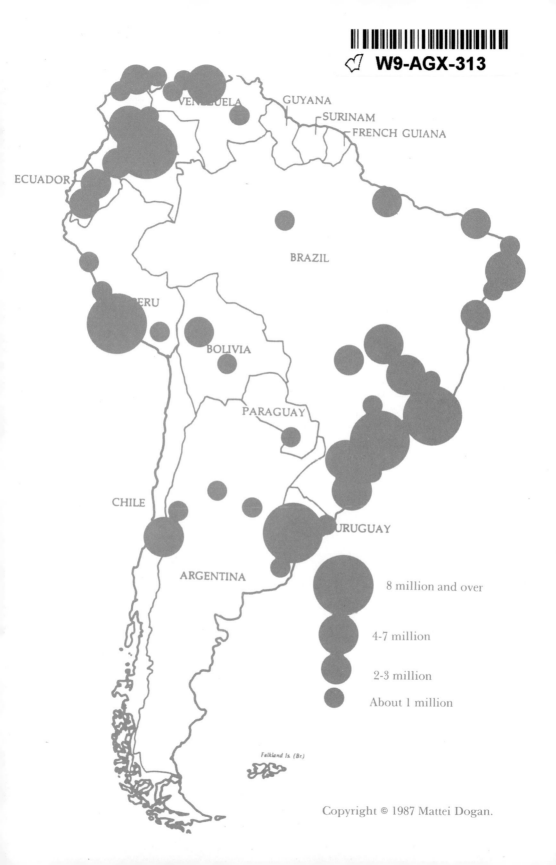

GUYANA

VENEZUELA

SURINAM

FRENCH GUIANA

ECUADOR

BRAZIL

PERU

BOLIVIA

PARAGUAY

CHILE

URUGUAY

ARGENTINA

8 million and over

4-7 million

2-3 million

About 1 million

Falkland Is. (Br.)

# Mega-Cities

# The Metropolis Era

## Volume 2

# Mega-Cities

EDITORS

## Mattei Dogan
## John D. Kasarda

**SAGE** PUBLICATIONS
The Publishers of Professional Social Science
Newbury Park   Beverly Hills   London   New Delhi

*This book is dedicated to the beautiful metropolis*
*of Barcelona*

Copyright © 1988 by Sage Publications, Inc.

*For information address:*

SAGE Publications, Inc.
2111 West Hillcrest Drive
Newbury Park, California 91320

SAGE Publications Inc.
275 South Beverly Drive
Beverly Hills
California 90212

SAGE Publications Ltd.
28 Banner Street
London EC1Y 8QE
England

SAGE PUBLICATIONS India Pvt. Ltd.
M-32 Market
Greater Kailash I
New Delhi 110 048 India

Printed in the United States of America

Library of Congress Cataloging-in-Publication Data

The metropolis era.

Contents: v. 1 A world of giant cities—v. 2.
Mega-cities.
Includes bibliographies and indexes.
1. Metropolitan areas. 2. Cities and towns—
Growth. I. Dogan, Mattei. II. Kasarda, John D.
III. Title.: Mega-cities.
HT330.M45 1987        307.7'64        87-23247
ISBN 0-8039-2602-2 (v. 1)
ISBN 0-8039-2603-0 (v. 2)

# Contents

# Introduction:

## Comparing Giant Cities

### Mattei Dogan and John D. Kasarda

Volume I (*The Metropolis Era: A World of Giant Cities*) described contemporary demographic, economic, and technological forces around the globe that have dramatically altered the location, scope, and pace of giant city growth. A number of chapters highlighted how large urban agglomerations in developing nations have expanded beyond their absorptive capacities, in terms of formal sector employment, housing, infrastructure, and the provision of essential public services. These chapters further described the innovative ways by which migrants to these cities have adapted to conditions confronting them.

An overarching conclusion was that despite the hardships greeting new urban arrivals, they consider themselves better off in the city than remaining in rural areas where chances of economic success are slim. Not only do cities offer better employment prospects, but they also provide cultural amenities, stimuli, and some very basic services lacking in most rural regions. As a result, the flood of urban migrants continues, typically to a demographically exploding primate city whose transportation and communication linkages and corresponding accessibility to domestic and foreign markets give it a substantial economic edge over smaller urban centers.

The largest metropolitan areas of developed nations are also continuing to expand, albeit at a much slower pace. Indeed, in contrast to those in developing nations, it was shown that many of their central cities are actually declining in total population size and jobs. For example, between 1970 and 1980 London lost 750 thousand residents and 430 thousand jobs, while New York City lost 820 thousand residents and nearly half a million jobs (though both cities have since rebounded). Nevertheless, as George Sternlieb and James W. Hughes demonstrate in this volume, large blue-collar job losses are still occurring in cities like New York, while their lower-income minority populations are growing. London, as Emrys Jones shows, likewise experienced a growth of over 150 thousand immigrant minorities between 1970 and 1980, despite the city's overall population and job loss.

Along with selective demographic decline, many U.S. and Western European cities that expanded rapidly during the 19th and first half of the 20th century as industrial giants, are functionally transforming from centers of producing, storing, and transporting material goods to centers of producing, storing, and transmitting information. This change in function is altering the historic roles these cities performed as economic assimilators and springboards for social mobility for their disadvantaged residents. As white-collar information-processing jobs are replacing blue-collar and other entry-level jobs in the cities, skill requirements for employment have risen substantially. With many urban minorities lacking such skills, their unemployment rates have risen significantly since 1970 as did social problems associated with their deteriorating prospects (e.g., drug abuse, violent crime, family dissolution).

Exacerbating the problems of economically transforming industrial cities has been the movement of middle- and upper-income residents to outlying areas which has drained urban tax bases, weakened secondary labor markets, and led to increased isolation and segregation of low-income minorities in the urban cores. Other outcomes of urban deconcentration are increased commuting distances, highway congestion, air pollution, and energy use. Large cities in advanced capitalist nations are particularly prone to such problems where technological innovations, especially in transportation and communication, have allowed industry and middle-income people to become increasingly footloose. The selective redistribution of industry and population has led to a fundamental spatial restructuring of these metropolises from monocentric to polycentric, as outward sprawl continues and core densities decline. We will return to this issue shortly.

## Comparative Case Studies

Assessing demographic, environmental, economic, and social conditions of giant cities around the world reveals numerous similarities and differences. In this volume, these similarities and differences are highlighted through case studies of ten giant cities: Mexico City, Tokyo, São Paulo, New York, Shanghai, Los Angeles, London, Cairo, Delhi, and Lagos. The first five cities, respectively, represent the five largest urban agglomerations in the world in 1985 (United Nations, 1985). We include Los Angeles because it is the exemplary motorized metropolis. London, tenth largest in 1985, was selected because of its historical (pre-World War II) position as the world's largest urban area. Cairo, Delhi, and Lagos are included because of their phenomenal growth and interesting contrasts. Of course, many other important giant cities could have been selected, but space constraints precluded their inclusion. A number of them, such as Moscow, Calcutta, Seoul, Jakarta, and Beijing are treated in volume 1.

A wealth of data on giant cities is diffused throughout monographs, government reports, and independent studies. It is underexploited for comparisons except in some publications of the World Bank, the Population Division of the United Nations, the World Health Organization, and the Food and Agriculture Organization. Yet, these institutions do not collect the data themselves. They usually rely on what national governments are able to gather and willing to release. For this reason, statistical data are too often inaccurate or noncomparable and, therefore, must be used critically. While keenly aware of these shortcomings, we have attempted to distill information from the variety of sources mentioned above and synthesize it along a number of comparative dimensions with that from the chapters contained in the two volumes of *The Metropolis Era*.

## Urban Growth and Decline

We have just noted that most giant metropolises in the advanced countries have experienced an aggregate decline of the population in their core areas since 1960. For some economically distressed core cities such as Detroit and Manchester, the annual rate of decline during the past 25 years has been between 1 and 2 percent, which means the loss of at least a quarter of their populations over this period. Loss in the central city has often, though not always, been compensated by an increase of the population living in the suburbs as is the case of New York, London, Paris, Madrid, and Rome. Similarly, for Tokyo, Hachiro Nakamura and James W. White show that residential locations of inner city workers have been spreading out (primarily because of transportation improvements and rising core land prices). As a result, suburbanization has accelerated in recent decades as residential densities in a number of its core areas have declined. Finally, Ivan Light describes how, in Los Angeles, the core has remained stable, while the suburbs have expanded dramatically.

In developing countries, the population of core city and suburbs have both increased at impressive rates: Baghdad has grown at an annual rate of 10 percent during the last quarter of a century and Lagos at 9.4 percent annually. The average rate for greater Kinshasa is over 15 percent a year; for Addis Ababa 11 percent; for Bangkok 12 percent in the core city and 10 percent in the suburbs. The annual rate of growth of the city itself has been over 5 percent for Khartoum, Dacca, Bogota, São Paulo, Delhi, Santiago, Pusan, and Seoul; the peripheries of these cities have increased, simultaneously, at more or less the same rate. In some cases, growth of the central city results from annexation, but in general the expansion results from an increasing density of the population in the central city and suburban rings.

Shanghai, assessed in this volume by Rhoads Murphey, is a special case since the Chinese government directly intervened during the 1950s and 1960s to slow this city's growth. Murphey describes government attempts at urban population redistribution in mainland China and why such redistributional policies were eventually circumvented.

## Models of City Growth and Decline

There has been considerable debate as to whether the process of urban growth in today's Third World cities is similar to or qualitatively different from that which took place in Europe and the United States during the 19th and 20th centuries. Some who stress similarities have developed sequential stage models of urban population growth and suggest it is possible to classify city growth over time and space. One of the best known is that of Peter Hall (1984).

Hall proposes a five-stage model of urban population evolution. The first stage (under conditions of limited economic and technological development) entails substantial rural-to-urban migration toward a primate city where the bulk of the nation's industrial activity is located. With the spread of transportation arteries, the second stage brings heightened industrialization throughout the region and results in the formation of secondary cities as alternative magnets for rural migrants, though the primate city continues to grow rapidly. Eventually the primate city core becomes so densely settled that "spillover" to the suburban rings begins. In the third stage, suburban spillover accelerates and the peripheral areas begin to grow faster than the urban core. In the fourth stage, the primate city core actually begins to lose population while its suburbs continue to grow. The city's degree of primacy declines as secondary cities become increasingly attractive to industry and migrants. Finally, during stage five, population loss of the primate city core accelerates and its immediate periphery suffers relative (though not absolute) population losses to secondary cities and nonmetropolitan areas, constituting an end to the urban life cycle.

Hall argues that many of the least developed countries are in stages one and two, newly industrializing countries such as Mexico, Brazil, and Korea and those of southern and eastern Europe are in stages two and three; most northern and Western European countries in stages four or five, and the United States and Great Britain clearly in stage five. His model is consistent with traditional regional growth theory in that it posits that eventual declines in primate cities translate into gains by subordinate urban agglomerations.

In like manner, Richardson (1980) proposes a process of "polarization reversal" whereby urban disparities are gradually reduced through population deconcentration which results from diseconomies of scale in the largest agglomerations and technological innovation, especially in transportation and communication. The expected outcome, according to traditional regional growth theory, is a full integration of the national economy through an evolving hierarchy of urban places and corresponding reduction in income inequalities across cities and regions (Berry and Kasarda, 1977:277-81).

Such models of urban growth and decline have been sharply criticized by neo-Marxian and other class-based theorists who contend that the evolution of Third World cities constitutes a unique pattern of urban development unparalleled in the Western experience. Colonial heritages, extreme poverty, ruling class hegemony, rapid population growth, and dependency on the economies of capitalist nations generate hugh primate cities which dominate Third World economies and discourage or prevent indigenous development of secondary cities (Castells, 1977; Portes and Walton, 1981; Wallerstein, 1974). The result, they argue, is increasingly greater spatial inequalities in Third World nations as their primate cities disproportionately grow at the expense of other parts of the country.

## Industry and Employment

Even though there are numerous ways to define *urban*, cities are most frequently defined as places where residents are primarily engaging in nonagricultural activities. Most giant cities today, as Mattei Dogan describes in volume 1, evolved through their favorable location for domestic and, especially, international trade. London, New York, Los Angeles, Tokyo, Lagos, and Shanghai are excellent examples. They are situated on or near habors which economically link them to the rest of the world. Likewise, Mexico City, Delhi, and São Paulo were important trading centers during their respective colonial eras.

As with population growth, it is possible for comparative purposes to develop a sequential typology of industrial and employment transformation in which many cities can be placed. This typology typically follows economic stages from a handicraft and lower-order service structure to a more formal commercial-industrial based structure, eventually reaching an information-processing, higher-order service structure. In the first stage, informal economic activities dominate with low costs of entry, family ownership of enterprises, and labor-intensive technologies. During this preindustrial phase, urban economic activities are confined to traditional sectors such as crafts and food distribution by small family enterprises (Beavon and Rogerson, 1986). The urban employed consist primarily of artisans, petty traders, food vendors, and other lower-order service providers.

In the second stage (where many giant cities of the Third World today are), economic activities are partially transformed from family enterprises to corporate production units, capital grows in importance relative to labor, and wage and salary employment expands. With technological advancement and capital accumulation, development of an extended trading network and industrial concentration further stimulates urban growth, often creating a primate city (Golden, 1981). In this industrialization stage, cities specializing in manufacturing activities expand rapidly. The manufacturing sector as a powerful export-base industry has multiplier effects, creating new job opportunities and attracting waves of rural migrants seeking employment.

With mechanization of industrial production and a growing capital-to-labor ratio, a substantial increase in manufacturing output can be achieved with small increments in the manufacturing labor force. Because of the reduced labor absorption capacity of more capital-intensive manufacturing activities, the informal sector becomes increasingly important in providing employment opportunities. This sector often has advantages compared to the formal sector, including (1) a higher potential for absorbing migrant labor, (2) higher real wages for unskilled workers, (3) less sex discrimination, (4) better opportunities for upward mobility through entrepreneurship with limited capital, and (5) no involuntary unemployment (Beavon and Rogerson, 1986; Hackenberg, 1980; Tailhet-Waldorf and Waldorf, 1983).

As the national economy matures and transportation networks expand, competition from lower-cost outlying sites reduce urban manufacturing employment. During this third stage, large-scale production units move to peripheral areas and smaller cities and are replaced by knowledge-intensive firms in the core employing well-educated, skilled persons. Higher-order, knowledge-based services are exported nationally and internationally as the functions of major cities gradually transform from goods-processing and lower-order services to information-processing and higher-order services. Although this sequential model represents the historical pattern of Western urban industrial and employment transformations, there is evidence from case studies in this volume that Mexico City, São Paulo, Cairo, Delhi, and Tokyo are following similar sequences.

Stage models of economic transformation, like models of population redistribution, are questioned on the basis of the differing circumstances of developed and developing nations and their consequences for a unified evolutionary scheme. It is argued that factors such as burgeoning labor pools, excessive unemployment, technological diffusion, colonial heritages, and the relatively smaller sizes of some Third World countries dictate different urban economic development patterns from those experienced by developed countries. For example, that Third World urban populations are supported by relatively small industrial bases has led many researchers to assume that cities in developing countries suffer from inflated tertiary sectors; such cities might be considered "overurbanized" because their populations are not justified on the basis of their formal economies. It has been assumed, moreover, that these large tertiary sectors consist of a disproportionate number of "dead-end" job opportunities. Such considerations would suggest that Third World cities suffer "abnormal" economic transformation and, furthermore, that this abnormal growth fundamentally differentiates urban industrial processes of developed and developing nations.

Our conclusion, however, is that the substantial growth in informal sector employment in Third World cities should not be considered "abnormal" since it provides vital opportunities to new urban arrivals and often serves as a springboard for economic success by those excluded from the formal sector.

Moreover, research by Preston (1979) suggests that where service-sector employment is substantially outgaining industrial employment in Third World cities, such service-sector growth has typically resulted from rising proportions of professional, technical, and administrative employees in these cities.

## Demographic Density and Crowding

The number of people per square mile is an indicator frequently used to compare cities, but it is often misleading, because density depends on how administrative borders of the central city are drawn. In general, European major cities are more compact than American cities. Certainly, when one reads that the density in the central city is, for instance, 2 thousand per square mile in Indianapolis against 33 thousand in Manchester, or 15 hundred in Kansas City against 27 thousand in Naples, or 3 thousand in Dallas against 57 thousand in Athens, or 23 hundred in San Diego against 47 thousand in Barcelona, one perceives behind these figures different types of cities on the two continents.

Nevertheless, with some exceptions, this indicator does not discriminate between giant cities in industrially advanced countries and in developing countries, although the word *crowded* is usually applied to Third World cities. In this regard, it is surprising that for very different types of cities the density in the central part is approximately the same. A series of pairs of cities, one from an industrially advanced country, the other from a developing country, shows the misleading character of the indicator "persons per square mile in the central city":

- Tokyo and Bombay (about 42 thousand per square mile)
- Athens and Madras (55 thousand)
- Osaka and Bangalore (33 thousand)
- Barcelona and Ahmedabad (46 thousand)
- Milan and Hyderabad (24 thousand)
- Moscow and Istanbul (21 thousand)
- Leningrad and Alexandria (16 thousand)
- Chicago and Pusan (14 thousand)
- San Francisco and Karachi (14.5 thousand)
- Madrid and Recife (13 thousand)
- Detroit and Belo Horizonte (9.5 thousand)
- Kiev and Baghdad (6 thousand)
- Bremen and Bogota (4 thousand)

"Density" does not help us to perceive the contrast between Tokyo and Bombay, even if both cities are enormous seaports, or between Athens and Madras. Thus, this indicator, like others, should not be used in isolation from other indicators in comparative urban research.

Cities, or districts in cities, may have a high density without being crowded, and vice-versa may be crowded even if the density is comparatively low. With most of its buildings having between five and eight floors, Paris has a high density but it is not now considered a crowded city, except for a few areas. Lagos is a good opposite example. According to an urban planning report, 83 percent of the population lives in "rooming houses," a kind of small building of one or two floors. On each floor there are some ten rooms along a central corridor; 72 percent of Lagos families live in a single room, and the average size of the family is eight persons! (Lengellier, 1978:61). Like many other Third World cities, Lagos is a relatively flat city.

Density of people in some cities may also be low simply because cars take up a lot of space. A person who lives in an apartment of 50 to 55 m$^2$ (about 538 to 592 square feet) in a building of seven floors, occupies about 8 to 9 m$^2$ (about 86 to 97 square feet) of the land. The car of this person also occupies 8 to 9 m$^2$, either in a parking lot or in the street (Sauvy, 1968:153). And one often needs a second parking lot for the car at the office or factory and a third lot at the market, stadium, or theater. In cities where there is a high proportion of households owning an automobile, like American cities, the density of people is necessarily reduced accordingly. That cars require a substantial amount of space at a variety of locations is an important, but frequently overlooked, basic fact in comparative studies of urban density.

## Transportation and Traffic Congestion

Related to the discussion above, one might ask if traffic congestion is different in nature or in degree in the giant cities of advanced and of developing countries? The literature on urban transportation is enormous. Several international conferences have been organized on this topic in recent years, gathering hundreds of experts and generating reports on transportation problems in a large number of major cities. Statistical data are available on number of automobiles per 100 households; number of persons per 100 cars; percentages of commuters using public transportation; and passenger journeys by subway, tramcar, railway, and bus. The statistical information is richer for American, European, and Japanese major cities than for Third World ones, but this imbalance does not prevent comparisons. The difficulty comes from the fact that the indicator "number of persons per car" or "automobiles per household" measures affluence or access to private motorized transportation, or simply the need to own a private car rather than actual traffic congestion (Fried, 1981, chap. 4).

Statistical data on commuters using public transit systems (subway, railway, or bus) do not control other variables, and so we find similar percentages for cities of very different sizes and wealth: greater Bombay and smaller Göteborg; Teheran and Bremen; Istanbul and Stuttgart; Madras and Berne. If we limit the sample to giant cities, a clear dichotomy appears: on one side, the American major cities, on the other the giant cities of the rest of the world. In American metropolitan areas, most people seem to have four wheels instead of two legs and they eschew public transit. More than 90 percent of commuters use private cars in Los Angeles, Denver, Houston, Dallas, Miami, Atlanta, Minneapolis, Detroit, St. Louis, Buffalo, and Seattle. The proportion is between 75 and 90 percent in the metro areas of San Francisco, Chicago, Washington, Boston, Philadelphia, New Orleans, Pittsburgh, Baltimore, and Cleveland.

At the opposite pole, more than 85 percent of commuters use public transportation in Moscow, Calcutta, Tokyo, Bombay, London, Cairo, and Paris. Some of the remaining 15 percent usually walk from home to place of work. In almost all non-American-Canadian-Australian giant cities, a large proportion of people living within the central city use public transportation; the most notorious exceptions are the two giant conurbations in West Germany and Holland: Rhine-Ruhr and Randstad. But these are not real exceptions: within each of these two megalopolises, people use private cars because they travel from small towns or from semiurban areas to cities, or even between neighboring cities.

There is also a significant difference between most of the European-Japanese metropolises and the Third World metropolises. In Moscow, Paris, London, Tokyo, Leningrad, Osaka, and to a lesser degree in Milan, Brussels, and other cities, the subway system carries most commuters, while buses and trolley-buses provide most transportation in Cairo, Seoul, Havana, Bombay, Delhi and most of the Indian major cities, and so on (Tokyo Metropolitan Government, 1985:76).

In Chinese cities, the bicycle is the primary transportation mode. Rhoads Murphey notes in his chapter on Shanghai that bicycles have created serious congestion problems in this crowded city, yet the demand for them still far exceeds the supply. Attempts to further develop mass transportation are proceeding in Shanghai with the construction of a subway system to alleviate the city's bicycle congestion and overcrowded buses.

Some cities have a balanced system of subway and buses: Rome, Madrid, Kiev, Budapest. In other cities, the railway plays an important role, particularly in Mexico City, Tokyo, Yokohama, Osaka, London, and Paris for commuting between central city and suburbs.

Simplifying, one might say that the predominant transportation vehicle in most American giant cities is the private car; in most European metropolises, the subway; in most Third World major cities, the bus; and in most Chinese cities, the bicycle.

## Automobiles and Cities

Everything has already been said in recent years about the advantages and disadvantages of public transportation and private cars. Suffice it here to quote an observation about America made by Ivan Illich (1973):

The average American devotes more than one thousand five hundred hours per year to his car. He works to buy it, to pay for gasoline, tires, tolls, insurance, taxes, tickets, parking rent, not to mention accidents. He devotes four hours per day to his car by using it or by earning the money to support it. This American spends one thousand five hundred hours in order to be able to travel ten thousand kilometers. That means that he spends one hour per six kilometers.

Many sociologists have a contemptuous view of the American's love affair with the automobile. For the American, the private car is a rational choice if public transportation is not efficient enough, or it brings certain personal freedom or social status. Moreover, it is impossible to conceive of a city of several million inhabitants where the traffic depended on horses. The automobile played a vital role in the horizontal expansion of cities. But after a certain level of growth, the car can play a negative role as well, to such a degree that some commentators have argued that the automobile has "destroyed" quite a number of central cities, especially in the United States.

One can, in fact, build a sequential model describing the decay of the cores of cities as a result of the multiplication of automobiles. (1) The increase in the number of private automobiles leads to congestion of the streets in the center of the city. (2) Public transportation (especially buses) functions less effectively because of the increased concentration of private cars. (3) Automobile bottlenecks, noise, and exhaust pollution lead to declines of property value in some areas, reduced maintenance of housing, and so on. (4) The initial center, which had a diameter of perhaps one, two, or three kilometers, expands so much that it is no longer possible to circulate on foot. (5) This situation favors the creation of a new center at a relatively large distance from the original one. (6) The car then becomes indispensable, and the increasing number of automobiles causes further decline of the central city as middle-class people migrate to suburbs to avoid the congested core and be closer to their deconcentrating jobs.

This model can be applied to many cities. For instance, the center of Athens, the typical old Mediterranean monocentric city, has been seriously affected by automobile growth in recent years. Until 1970, the number of private cars in Athens was limited. Transportation was provided by a network of buses radiating from the center to a distance of about 10 kilometers. The increasing number of cars has created severe congestion on the streets in the center. Many inhabitants have fled the degradation of the environment caused by traffic jams, pollution, and noise. Property values have declined; renovation has stopped. Air pollution has worsened. Decay of the historical center has proceeded. New commercial centers have been developed at a distance (Prevelakis, 1984:122-24). From a *city* in the old Greek sense of the word, metropolitan Athens has been transformed into an amorphous sprawl.

In Paris, for example, there are 2.1 million cars, half in the city and half in the suburbs, and about 200 thousand are in circulation simultaneously much of the time, except during the rush hours in the morning or evening when there are more. If there is a strike of the public transportation system, or if it rains heavily, the number of cars in circulation increases dramatically. If it increases by 5 percent—10 thousand more cars—the speed of circulation decreases by 20 percent.

The relation between the number of excess cars in circulation and the speed of circulation can be easily calculated, taking into consideration the total length of all streets and considering also the number of cars parked on the streets. According to an unpublished report by the Préfecture de Police, Paris, in 1976, the average speed in Paris within the city limits is 35 km per hour if there are 53 cars moving per km; 15 km per hour if there are 86 cars per km; 8 km per hour if there are 107 cars per km. Theoretically, the number of cars that can move simultaneously in the city is 220 thousand cars at a speed of 15 km per hour, or 280 thousand cars at a speed of 8 km per hour. After that number, the speed of traffic diminishes rapidly.

Surveys done by aerial photography show that within the Paris city limits, when the number of cars in circulation increases by 75 percent (110 thousand instead of 65 thousand, for example), the speed falls by 45 percent. This problem of traffic congestion can be described in simple figures. A 20 percent increase in the number of moving cars results in an increase of 45 percent in the time spent per kilometer. If these additional journeys were made by public transportation (subway or bus) instead of by private car, time and energy would be saved for all. It has been calculated that in such a situation, 80 thousand hours are lost by Paris drivers per day. The marginal cost of congestion is very high (Orselli, 1975:19). Such a calculation could be applied to many giant cities.

According to a city's topography and street patterns, frequently related to its age, the threshold of congestion may be a little higher or lower: lower in Rome and London, higher in New York and Los Angeles.

A knowledge of the topography of the city is essential for understanding the traffic flow and potential congestion. Traffic necessarily gives rise to different kinds of problems in Rio and in Istanbul, in Cairo and Bangkok, in Manila and Seoul. The topography limits to a great extent the significance of the various statistics on traffic.

A subway system, particularly in old cities, requires an enormous investment. One might calculate that the five million cars circulating in the Los Angeles metropolitan area represent a much smaller investment than the subway system in Paris, London, or Tokyo. However, these subway systems will undoubtedly last as long as the cities themselves, but cars have to be replaced periodically. The average commuter in Los Angeles spends three or four times more money for transportation than the average Parisian or Londoner. Which should be the model for tomorrow, Los Angeles or Paris?

To conclude this section, let us note some points around the world where we have observed the traffic—pedestrian or motorized—to be the heaviest: Sinjuku metropolitan station in Tokyo in the evening, when 90 percent of the people are males; Howrah Bridge on the Hoogly River in Calcutta; the Galata Bridge in Istanbul; the three bridges over the Nile and downtown streets in Cairo; Jumma Masgid area in Bombay; the highway dividing Jakarta into two parts; the Bazar in Teheran; the old district of Delhi; Avenues Santa Fe and Lavalle in Buenos Aires; Picadilly Circus in London; Place de l'Opera in Paris during office hours; the intersection of Broadway and 42nd Street in New

York; the Bay Bridge in San Francisco at 5 p.m.; the Loop in Chicago during working hours; the road from the railway station to the airport in Lagos. No doubt, readers can suggest numerous other world-class points of traffic congestion.

## Health Indicators

The extreme diversity of the world's giant cities is paralleled by the diversity of nations (Dogan and Pelassy, 1984). The ranking of giant cities on various indicators follows more or less the ranking of their nations. Thus, only by comparing conditions of life in the giant city and in its respective country can we really know if life is better or worse in the giant city than in the rest of the country. In this section, we select two indicators to compare the health conditions of large cities across nations: (1) running water and sewer facilities and (2) infant mortality. In the section that follows these comparisons, we will focus on a key social indicator: crime.

### Running Water and Sewerage Facilities

"Few laymen quite appreciate the quantities of water, sewage and air pollutants involved in the metabolism of a modern city" (Wolman, 1965:160). Water pipelines and sewerage are basic infrastructures of any giant city. Knowing, for example, what is the proportion of dwellings with inside running water one can usually tell, from this information alone, if this particular city belongs to an advanced country, to a developing country, or to a country in the middle rank.

As Robert Fried (1981) notes, water distribution lags behind electricity distribution in the Third World's cities, since stringing power lines is much cheaper than laying water pipes. In some Third World cities fewer than 10 percent of dwellings have inside running water, for instance: Jakarta, Kabul, Dacca, Rangoon, Mogadishu. In other cities between 10 and 20 percent are so equipped: Calcutta, Brazzaville, Ulan Bator, Asunción, Dar es Salaam, Tenerife. Some of the largest metropolises are in the middle rank: Cairo (40 percent), Istanbul (41 percent), Teheran (52 percent), Rio (55 percent), São Paulo (59 percent), Mexico City (64 percent). Obviously this diversity can be explained in part by the spatial ecology of the city. The exceptions are rare; for instance Addis Ababa (74 percent), where distribution of water pipelines is better than expected according to other indicators, or Manila (95 percent). An anomaly in the opposite direction is Kyoto (only 81 percent).

The sewage system is also a discriminating variable—in the statistical sense—between giant cities in developed and in less-developed countries. The percentage of dwellings attached to public sewers varies from nearly 100 percent in almost all American and Western European major cities to fewer than 10 percent in Jakarta, Baghdad, Manila, Teheran, Addis Ababa,

Rangoon, Karachi, Kinshasa, Bangkok, Bangui, Bamako, Accra, Conakry, and to less than 20 percent in Dacca, Kuala Lumpur, Kampala, or Asunción. Privies built over water are one of the main alternative methods of human waste disposal in many Third World countries, with obvious health consequences. In São Paulo, as Vilmar Faria points out in this volume, the city's two main rivers are essentially "dead" as a result of the continuous and direct flow of raw sewerage and waste.

## Infant Mortality

The infant mortality rate, probably one of the best social indicators, discriminates clearly between the giant cities in advanced countries and in developing countries, with few exceptions. The infant mortality rate varied in 1983 from a high of 205 per thousand in Afghanistan to a low of 7 per thousand in Sweden and Japan.

Current statistical information about infant mortality in cities is rather limited. In the major cities, it varied in the 1970s from less than 10 per thousand in Goteborg, Stockholm, Tokyo, Nagoya, and Kyoto to more than 70 per thousand in Bogota, Cairo, Istanbul, São Paulo, Rio de Janeiro, and Ahmedabad. The infant mortality rate is typically lower in the major cities than in the country as a whole, except for a few cities like Cairo and for some areas in periods of famine and heavy influxes of refugees to cities.

Infant mortality rates thus closely parallel economic development indicators such as gross national product (GNP) per capita. The latter ranged in 1982 from $80 for Chad to $17,000 for Switzerland. Considering statistical data for the period 1970-82, we find a high correlation for 150 countries between these two variables. One might infer that the high correlation between infant mortality and GNP per capita established at the level of nations holds for giant cities as well, when ranked in terms of per capita income measures.

Differential urban infant mortality rates, of course, are a consequence of many other social-environmental factors such as availability of minimal standard housing (in severe shortage in most Third World cities), nutrition intake, health and medical services, education levels, sanitation, and water quality. We expect that rank-orders of cities on infant mortality and each of these other social indicators would also closely correspond.

## Crime

Considering the enormous gap between the standards of living in the giant cities of advanced and developing countries, an uninformed observer might postulate a higher rate of crime in the poorer countries. In reality, we find the opposite. We need triple comparisons here: first, between American and European cities; second, between European cities and Asian and African cities; third, between Latin American cities and other Third World cities.

Certain facts stand out. Few cities can match American cities in murder. The chances of being murdered in an American city are ten, even 20 times greater than in most European cities. . . . Robbery on a vast scale seems [much more frequent] in American and Canadian cities. The chances of being robbed in Boston are ten times greater than in Liverpool; eight times greater than in Rome. . . . American cities may have the worst incidence of rape of any cities in the world (Fried, 1981, chap. 1).

Ranking cities by the number of homicides, robberies, assaults, or rapes per 100,000 inhabitants shows clearly the higher rates of violent crime in American cities as compared to European, Asian, and African giant cities considered together. This does not appear to be a statistical artifact, since the recording of murder, robbery, or rape is apparently done as consistently in most European cities as in American ones. Statistics on crime in Asia and Africa are, of course, less comprehensive than in the West, but, on the whole, it seems that the combined crime rate is lower in most of these cities than in most European cities. Burglary rates, for instance, are lower in Madras, Calcutta, Bombay, Ahmedabad, Delhi, and Hong Kong than in Paris, Stockholm, Liverpool, Hamburg, Frankfurt, or Vienna. For instance, Ahmed M. Khalifa and Mohamed M. Mohieddin note in their chapter on Cairo that in spite of all the minuses Cairo gets when considering the necessities of a sound and comfortable urban life, the city enjoys a definite plus when it comes to social conduct, security, and street safety. It would be possible to quote many testimonies to this effect from personal experience of many in African and Asian cities.

Latin American urban crime rates are also much higher than those of Asian and African cities. Yet, poverty and deprivation are probably not higher in Rio, Brazilia, Porto Alegre, Recife, Belo Horizonte, Caracas, Maracaibo, Lima, and Quito, than in many poor Asian and African cities. Why then are crime rates several times greater in the Latin American giant cities? Is it related to religion, since it apparently is not related to poverty? Are Islam, Buddhism, and Hinduism stronger than Catholicism in preventing social deviation? What other cultural factors come into play?

The researcher who explains this major difference in crime rates between Latin America's giant cities and those in Asia and Africa will be making an important contribution. It is unlikely that this task can be done using sophisticated statistics because too many cultural and psychological variables are likely involved. The higher rate of urbanization of Latin America is not an explanation since the contrast appears between giant cities. It is true that Latin American giant cities are typically much larger than African ones, but not than Asian ones. A working hypothesis may be that crowding prevents crime. Shanghai, possibly the most crowded city in the world, has a relatively low rate of crime, as do Cairo, Calcutta, Jakarta, Bombay, Hong Kong, Singapore, and most of the Chinese cities.

Crime is rarely committed in front of observers. In a crowded city like Cairo, it is difficult to escape the attention of neighbors and passers-by. In addition, in Asian or African giant cities, cultural traditions favor intervention in case of incident. The knowledge that neighbors and passers-by are not afraid to become involved tends to prevent crime. Consider this testimony from Cairo:

Certainly the most obvious example of citizen involvement is the famous Egyptian *dowsha*, ferocious verbal battles that break out daily on Cairo streets as a result of minor traffic accidents or disputed taxi fares. Occasionally punches are thrown, but policemen are frequently nowhere to be seen. This is by design, according to police officials. The presence of a policeman, they say, would only inflame the situation and, more importantly, passers-by can always be counted on to step in and separate the enraged combatants (Harrison, 1978).

Such behavior can be seen in many crowded Third World cities except in Latin America.

Nor is there a relationship between the crime rate and the number of policemen for 100,000 residents. There are eight times more policemen in Manila than in Djarka—two cities similar from many points of view; there are six times more policemen in Washington than in Barcelona. "What correlation there is tends to be negative: the more policemen there are, the worse crime rates tend to be. The size of the police probably has less impact on crime rates than crime rates have on the size of the police force" (Fried, 1981, chap. 1).

There is for each geographical grouping of cities a great internal diversity. The chance of being murdered in Detroit is several times greater than in Providence, in Palermo than in Milan, in Ahmedabad than in Bombay. There are gaps between officially reported crime and actual crime levels, and these gaps vary from country to country, and from city to city within the same country. In spite of all these statistical difficulties, the basic facts appear clearly, particularly the contrasts between continents. "While robbery is often considered an economic crime, it does not correlate with poverty. . . . On a world basis, in fact, there seems to be more robbery in the richer rather than in the poorer cities" (Fried, 1981, chap. 1).

The explanation of these paradoxical differences would be an ambitious task that cannot be undertaken here, but one point has to be underlined: criminality might be related to the rising expectations of poor people who see substantial wealth nearby. The gap between many very rich and many very poor people in the same metropolitan area is probably larger in most Latin American giant cities than in Asian and African ones. And, from this relative deprivation point of view, cities in the United States would also rank high where everyone can mention a pair of economically disparate districts, like Watts and Beverly Hills in Los Angeles or the Gold Coast and Cabrini Green in Chicago.

## Urban Ataxia

Curious students frequently ask: "Which is the worst giant city in the world?" We are unable to reply to such an intriguing question. What should be the criteria? Based on what indicators? How much quantified analysis is needed? How should we combine the quantified indicators into a meaningful concept?

To this last question, the answer could be *pathology*, or better still, *ataxia*. According to Henry Teune, "pathologies occur if growth is faster than the growth of the boundaries of the niche. The niche is a container and a container has some upper limit. . . . The concept of growth is necessarily tied to some notion of equilibrium with maximum and minimum flows and by implication some idea of optimum based on such principles as the largest size with the least damage to the holding power of the niche" (see vol. 1, chap. 13).

In a pathological situation, negative effects outweigh positive effects. Urban pathology can take the form of traffic congestion, air pollution, water pollution, crowding, and so on, compensated for by certain advantages: more opportunities for jobs, doctors, schools, cultural attractions, or leisure activities. Rapid growth in some fields unaccompanied by growth in other fields creates disequilibrium. Demographic growth without economic growth or without parallel construction of houses, roads, hospitals, and so on engenders an unbalanced development. Urban pathology is not primarily related to city size—as Henry Teune goes on to point out—but to the degree of harmony between its various parts. A city can be giant and have few pathological functions; it can be relatively small and seriously pathological: Barcelona is a giant city, but apparently in good equilibrium. Nouakchott is relatively small, but in a pathological condition, according to Ignacy Sachs' diagnosis (in vol. 1).

We might borrow from biology the term *ataxia* which designates the pathological noncoordination of body movements. Very rapid growth of a city belonging to a relatively poor country is usually characterized by such noncoordination of activities in various sectors. Urban sociologists, behaving like medical doctors, may diagnose the sickness of ataxia for many giant cities in the Third World, and even in the first and second worlds.

Is Calcutta the world's giant city in the worst condition of ataxia? Images of famines, riots, streams of refugees, and crushing poverty are everywhere associated with this city. These images are reflected in the austere United Nations report, *Population Growth and Policies in Mega-Cities: Calcutta* (1986: 1):

Calcutta has been frequently cited by journalists throughout the world as an example of urban pathology, or of what the future may hold for other cities in developing countries that fail to control rapid urban growth. Numerous articles have discussed the scores of thousands of pavement dwellers on the Calcutta streets, the visible evidence of widespread unemployment and absolute poverty, and the severe congestion and infrastructure deficits . . .

More than two thirds of the population have monthly incomes of less than the equivalent of $35 and at least 1.5 million persons are unemployed. More than one third of the city's inhabitants live in 3,000 unregistered slums or squatter settlements. The city's water supply system and sewerage network constructed by the colonial Government in the nineteenth century are inadequate and obsolescent, and severe drainage problems make much of the city impassable during the monsoon season. Because of its limited road space (the last major road inside the city was built half a century ago) and the large number of slow-moving vehicles, traffic problems are among the worst in India. All these problems are made more acute by the city's very high population densities (more than 1,235 persons per hectare in certain central wards) and by its very low income levels, which make cost recovery very difficult.

### Other testimonies also can be taken into consideration:

Turning back the pages of history we find many British rulers and their associates expressing their loathing for Calcutta. Robert Clive had called it "the most wicked place in the Universe." Rudyard Kipling, noted for his pro-imperialist sympathies, bestowed on it the title of "the City of Dreadful Night." Winston Churchill expresses the same loathing for Calcutta when he said that he was glad to have seen it simply for the reason that "it will be unnecessary for me ever to see it again." In more recent times some people ... have called it "the nightmare city." ... A team sent by the World Bank in the spring of 1960 was shocked by what it found in Calcutta and recorded this shock in its report. Some seven years later, a group of nine Anglo-American town planners reported: "A city in a state of crisis. We have not seen human degradation on a comparable scale in any other city in the world" (Narain and Ganguly, 1986:2).

Are conditions of life more pathological in Shanghai than in Calcutta? This is not the verdict Rhoads Murphey formulates in his chapter dedicated to this city. Nevertheless, he does write that, "Shanghai has created the country's and one of the world's worst pollution. . . . Clear days in Shanghai have become rare, and the incidence of cancer has shot up. . . . The air looks, smells and feels as bad as the worst situations elsewhere." He quotes an expert who estimates that Shanghai puts out about 1.6 hundred tons of fly ash and 1.5 hundred tons of sulphur dioxide per month for each square kilometer of the urban area. Murphey continues:

The Huangpu may now be the most polluted stream in the world. . . . The entire river area, and the lower reaches of the Woosung River smell terrible, worst during dry or low water periods, in winter, but bad at all times. . . . The Huangpu is Shanghai's main source of water, but it has become essentially a chemical cocktail composed of raw sewage. . . . Sewage is an even more serious related problem, because the urban area has almost no vertical relief, most of it being only a few feet above mean high tide in the Huangpu.

One would conclude from reading Murphey's chapter that he does believe Shanghai is probably the most polluted giant city in the world.

An article in a weekly British magazine confirms, in different language, Murphey's depressing analysis:

Shanghai is packed as tightly as a matchbox. Up to 425,000 people squeeze into a square mile of Shanghai's inner city. . . . Most homes have little space for anything but beds, with three generations often sleeping in a single room. What housing there is in Shanghai lacks the basic amenities taken for granted in other major cities of the world. . . . Too many people fight for too few goods in too little space, creating a nasty blend of pollution, poverty, over-crowding, shortages and corruption. . . . There is no word for "privacy" in Chinese, and any Shanghai neighborhood shows why. The housing crisis packs the community tightly. . . . The average Shanghaiese lives without toilet or bathing facilities in a room about the size of a double bed. He nestles with his wife in park bushes on summer nights for lack of privacy at home, bringing along his marriage license to show inquiring police. He breathes air that is polluted 10 times worse than American standards allow. . . . He drinks water that is mainly chlorinated sewage and eats vegetables laced with industrial toxins, factors contributing to stomach cancer. . . . He fights with three others for every available space on the bus. . . . Raw sewage leaks into the streets of Shanghai's older sections, overflowing septic tanks. . . . Housewives walk a block or more to draw water from public standpipes, lugging heavy buckets, or balancing them on bamboo poles. They cook on primitive coalstoves in communal kitchens or outdoor corridors clouded in coal dust. . . . As a center for huge oil refineries and chemical and metallurgical plants, it also is home of the "Yellow Dragons," local slang for the sulphurous clouds that pour out of the factory chimneys (Weisskopf, 1965:16).

These two long quotations suggest that Shanghai may be one of the most pathological cities of the world, at least in terms of pollution and over-crowding.

Observers of Lagos may be tempted to advocate that this city has the most extreme case of ataxia and has become a metropolis out of control. In their chapter on Lagos, Michael L. McNulty and Isaac Ayinde Adalemo write, "Lagos teems with inadequate services, uncollected garbage, unmoving traffic, inefficient institutions, and unbridled corruption in the public and private sector." Power failures have become chronic, public transport has been inundated, port facilities have been stretched to their limits, and the city's government has threatened to break down amidst charges of corruption, mismanagement, and financial imcompetence.

Some would place Mexico City at the top of their list of worst cities where chronic congestion and highly polluted air makes this metropolis infamous. With over 3 million cars, 7,000 diesel buses, and 130,000 factories, total air emissions in Mexico City nearly doubled between 1972 and 1983. Pollution levels are six times higher than acceptable human habitat limits. Serious air pollution in Mexico City is estimated to be responsible for approximately 90 percent of respiratory illnesses afflicting its residents (Friedrich, 1984).

If we asked a panel of urban experts to rank cities in order of degree of ataxia, they would spend a lot of time on theoretical and methodological problems, and considering cities like Manila, Jakarta, Teheran, Kinshasa, Rangoon, Addis Ababa, or Ho Chi Minh City would conclude that a lot of empirical work should be done before trying to rank giant cities. They would, nevertheless, sign a manifesto called "A Litany of Urban Ills": "Almost any account of Third World urbanization of cities reads like a litany of seemingly intractable problems. What is more, by interchanging a few names and adjusting some figures slightly, the litany is depressingly similar throughout much of Asia, Africa, and Latin America" (McNulty, 1976:215).

There are similarities, but also differences, among giant cities. To compare means to check differences behind analogies. This is our intention with these two volumes on *The Metropolis Era*.

Comparing giant cities is a promising and exciting field of research. But we should not forget that each major city—being generated by a complex interaction of history, geography, economics, climate, ecology, culture, and politics—is, in a sense, unique. Rome has been unique three times: as the center of the Roman empire, as the nucleus of the Catholic church, as the capital of the Italian Republic. Byzantium-Constantinople-Istanbul is unique. Leningrad is unique. Beijing is unique. Vienna is unique. Amsterdam is unique, and so are hundreds of other major cities around the world. Each of the ten giant cities assessed in the chapters that follow has its own personality. But, knowledge advances by comparing even what is unique.

# Bibliography

Beavon, Keith S. O., and Chris M. Rogerson. 1986. "The Changing Role of Women in the Urban Informal Sector of Johannesburg." In *Urbanisation in the Developing World*, edited by David Drakakis-Smith. London: Croom Helm.

Berry, Brian J. L., and John D. Kasarda. 1977. *Contemporary Urban Ecology*. New York: Macmillan.

Castells, Manuel. 1977. *The Urban Question: A Marxist Approach*. Translated by Alan Sheridan. Cambridge, MA: MIT Press.

Dogan, Mattei, and Dominique Pelassy. 1984. *How to Compare Nations: Strategies in Comparative Politics*. Chatham, NJ: Chatham House.

Fried, Robert C. 1981. "World Handbook of Cities: Comparative Indicators of the Quality of City Life." Department of Political Science, University of California at Los Angeles.

Friedrich, Otto. 1984. "A Proud Capital's Distress." *Time* (6 Aug.): 26-35.

Golden, Hilda Hertz. 1981. *Urbanization and Cities: Historical and Comparative Perspectives on Our Urbanizing World*. Lexington, MA: Heath.

Hackenberg, Robert A. 1980. "New Patterns of Urbanization in Southeast Asia: An Assessment." *Population and Development Review* 6:391-419.

Hall, Peter. 1984. *The World Cities*. 3d ed. London: Weidenfeld & Nicolson.

Harrison, Nathaniel. 1978. "Cairo's Choked Streets: Maddening . . . but Safe." *Christian Science Monitor* (6 Aug.):1+.

Illich, Ivan. 1973. "Energie, vitesse et justice." *Le Monde* (5-7 June).

Lengellier, Jean Pierre. 1978. "Lagos." In *Cites giantes*, edited by Jean Planchais. Paris: Fayard.

McNulty, Michael L. 1976. "West African Urbanization." In *Urbanization and Counter-Urbanization*, edited by Brian J. L. Berry. Urban Affairs Annual Reviews, vol. 11. Newbury Park, CA: Sage.

Narain, Igbal, and Bangendu Ganguly. 1986. "Pathology of Large Cities: Calcutta." Paper presented at a meeting of the Research Committee on Social Ecology, XI World Congress of Sociology, 18-22 August, Delhi.

Orselli, Jean. 1975. *Transports individuels et collectifs en région Parisienne*. Paris: Berger-Levrault.

Portes, Alejandro, and John Walton. 1981. *Labor, Class and the International System*. New York: Academic Press.

Preston, Samuel H. 1979. "Urban Growth in Developing Countries—A Demographic Appraisal." *Population and Development Review* 5:195-215.

Prevelakis, Georges. 1984. "Athens." In XXV Congres International de Geographie, *Cahiers du Centre de Recherche sur Paris et l'Ile de France*, 122-24. Paris.

Richardson, Harry W. 1980. "Polarization Reversal in Developing Countries." *Papers of the Regional Science Association* 45:67-85.

Sauvy, Alfred. 1968. *Les quatre roues de la fortune, essai sur l'automobile*. Paris: Flammarion.

Teilhet-Waldorf, Saral, and William H. Waldorf. 1983. "Earnings of Self-Employed in an Informal Sector: A Case Study of Bangkok." *Economic Development and Cultural Change* 31:587-607.

Tokyo Metropolitan Government. 1985. *Statistics of World Large Cities*. Tokyo.

United Nations. Department of International and Social Affairs. 1985. *Estimates and Projections of Urban, Rural and City Populations, 1950-2025: The 1982 Assessment.* (ST/ESA/SER.R/58). New York.

———. 1986. *Population Growth and Policies in Mega-Cities: Calcutta.* Population Policy Paper no. 1. (ST/ESA/SER.R/61). New York.

Wallerstein, Immanuel. 1974. *The Modern World-System: Capitalist Agriculture and the Origins of the European World-Economy in the Sixteenth Century.* New York: Academic Press.

Weisskopf, Michael. 1965. "Private Squalor and Public Lives." *Guardian Weekly* (3 Mar.):16.

Wolman, Abel. 1965. "The Metabolism of Cities: Water and Air." *Scientific American* 213 (9):178-88+.

# 1

# New York City

George Sternlieb and James W. Hughes

New York City has always been an amalgam of the rich and the poor, straddling an uneasy middle class, separate and distinct from each of them. Stresses between the several groups were never trivial and have frequently bubbled to the surface; as witness stands a long history of urban riots and labor warfare. But within this context, a logic and economic rationale tended to provide common denominators.

An evolving postindustrial world has shattered that balance. New York is coping with the necessity of putting together a new Realpolitik consonant with the limitations and new potentials of its economy. The process is hindered by a failure to comprehend the irrevocable changes in demographics and economic functions which must be accepted for a successful shaping of the future. The city as we have known it and the forms of social and economic organization which have characterized it are simply irrevocable. From the viewpoint of the poor, the city of goods production, or large-scale, relatively unskilled employment, has become the city of redistribution—of transfer payments and welfare. For the elite there is a city—far from new, but increasingly vigorous—of information processing and economic facilitation, of consumption rather than production. Lost between these two poles and fast disappearing are the middle groups. They find both the life-styles and economic opportunities of suburbia (and increasingly exurbia) affordable and much more fulfilling than city life. Pressures of population growth have subsided in New York, unlike the situation in Third World cities. In contrast with older industrial hubs of the West, a successful new economy is emerging. New York's real crisis is not embodied in care of the poor, street cleaning, or even crime rates, potent though these factors may be. It is rather the sociopolitical difficulties involved in shifts of function. Can the city adapt to its postindustrial future? Or will it be submerged by pressures of the past?

There is an enormous difference in goal structures between the groups—the poor and the elite—who increasingly dominate the urban political arena, perhaps most easily characterized in terms of attitudes toward housing and municipal expenditures.

The poor seek inexpensive shelter, looking toward the city to provide housing directly or manage the private markets to assure low rents and minimal increments in shelter costs. The extreme difference with the elite is captured by the stress on housing as shelter. To the elite caught up in postshelter society, housing increasingly is viewed as a form of investment. To them, the goal is capital enhancement through increase in housing value. They have the resources for housing purchase either in fee simple, or condominium/cooperative forms of ownership. They have the tax status that practically demands housing ownership. Thus to the affluent, housing as shelter from the elements is far less important than housing as a tax shelter. While the poor are victimized by housing cost/price increases, for the affluent, high rents justify the investment.

Municipal politics increasingly has as one of its irritants the rival demands of the two groups: one seeking neighborhood improvement, enhanced values, and ownership—the other fearful that any change will increase their rents, that changes for the benefit of the affluent might endanger their own facilities. While the situation sketched here is far from perfectly defined—as witness the love affair of affluent renters with rent control—the schism is becoming increasingly evident.

Variants in housing policy goals are paralleled by attitudes toward the municipal budget. To the poor, municipal expenditures on health, education, and welfare are among the few rays of light in an otherwise dismal environment. To the new urban elite groups, these are extraneous expenditures. In their place are rather environment-enhancing—and real estate price-increasing—investments in parks, recreational facilities, libraries, and the like. The yin-yang of this conflict dominates budget meetings in every major city in America.

These conflicts are merely two facets of a struggle increasingly securing definition: the rival roles of the city as the port of last resort for those who have fallen off or were never able to secure passage on the economic train, versus the city as an entity whose care and enhancement, by definition, require a restructuring to recapture the affluent. The common interest is often obscured. Social expenditures increasingly depend on locally raised revenues. This situation requires the provision of a variety of incentives for land-use activities which will generate more revenue than costs. However, the process conflicts with the provision of amenities and necessities for the poor and, perhaps even more strikingly, with their sense of priorities. Fear leads to immobility, which in turn may endanger the city's future.

# International and
# National Economic Framework

Cities in the United States, particularly within the aging industrial heartland, have been struggling to adapt to a changing worldwide economic environment. The great metropolitan manufacturing areas are lagging, their long-term regenerative capacity severely tested. Before delving into the immediacies of New York City and its response to the new imperatives, a broader economic perspective is required. This section briefly highlights the changing employment patterns and industrial structures of the United States and five industrialized nations, and provides a preliminary overview of New York City's employment matrix in the context of that of the United States.

## Employment and Industrial Structure: International

Broad economic cycles have always set a challenge for national economies and their cities, particularly in the post-1973 period, when the energy revolution brought sharp changes to the industrialized world. A decade of economic shocks hastened the process of economic evolution; nowhere is this more evident than in the United States.

The gross parameters of civilian employment in the United States compared with some of the major developed nations provide insight into the phenomena at work. The absolute and relative growth of U.S. employment from 1970 to 1982 has little parallel in the four major Western European countries, nor for that matter in Japan. Total employment increase in the United States was very comparable to the *total* employment base of France or Italy, and barely lagged the *total* number of jobs extant in Great Britain or Germany. While the U.S. job increase was 26.5 percent, the pattern was negative for the four European countries combined, with a net reduction of 0.4 percent. Germany and Great Britain each lost more than one million jobs over the 12-year period, with total civilian employment declining by 3.9 percent and 5.6 percent, respectively. Japan, in contrast, represents a far more salubrious picture. Absolute Japanese employment growth approached six million jobs, or 11.6 percent. This rate of increase, however, is less than 40 percent of the U.S. equivalent; in absolute job growth it represents less than 30 percent of the gain of the United States.

While use of 1982—the recession trough—as a terminal year in this data sweep may introduce some distortions, the American economy has been a job-generating machine. Thus, U.S. cities are not constrained by total national job-creation stagnation; it is the changing mix of employment—and its ultimate geographic locations—which provide the challenge.

Comparisons with the same set of major industrial nations underscore the disparate sectoral growth patterns. The basic picture is relatively static in the goods-producing sectors for *all* countries. The United States from 1970 to 1982 experienced minor growth; the several European countries collectively were stagnant at best, while Japan's gain in goods-producing employment, though slightly larger than that of the United States, was still modest. (Note that Japan includes utilities in its definition, unlike the United States.)

It is not surprising that the service-producing sector demonstrated absolute growth in all the nations. What is startling is the unique level and scale of expansion in the U.S. Thus while the pattern of services enlargement is international, the United States leads the procession with an absolute increase of 20 million jobs in this sector alone, nearly twice the total in France, Germany, Italy, Great Britain, and Japan combined. And, as the United States faces a future in which labor-intensive, goods-producing employment increasingly is seen as the province of newly industrializing nations, it is the service sector which must compensate. To the degree that this statement has validity, the United States has progressed much further (has less adjustment to make) than its peers.

It is of interest to note the pattern in agriculture. The United States, with the exception of Great Britain, has the smallest proportion of its labor force involved in this sector. And the absolute level is declining. Again, with the exception of Great Britain, the problems of agriculture displacement, as opportunities in this domain shrink, may be much more substantial for the other major industrial nations. But in the U.S. a potential source of new urban recruits has been virtually depleted. The great era of the rise of the city as an industrial mecca, strengthened by agricultural displacees, is at an end in the more mature industrial nations, unlike the situation in the Third World.

## The National Transformation and New York

Table 1.1 provides a preliminary overview of New York City's changes within this broader transformation and emphasizes the changes that have taken place. Despite substantial national total growth, New York's employment has experienced long-term contraction. As will be noted later, decline actually halted in 1976, with modest growth occurring subsequent to that point. Nonetheless, the 1983 employment level was still 10 percent below that of 1969. In the aggregate, the city did not reflect the overall national job surge.

New York City did experience expansion in two of the fastest growing national industrial sectors—finance, insurance, and real estate, and services—and remained virtually stable in the broader service-producing sector. But the city faltered markedly in goods producing, losing almost half (47.7 percent) its manufacturing employment. Within the national and international economic transformation, New York City has been hit by harsh dislocations to its economic base. While analyses presented subsequently will show that the city has adapted much more positively in a shorter-term time frame, clearly mass employment growth has been captured by other geographic regions as illustrated by population shifts, the subject of the following section.

## TABLE 1.1
### Nonfarm Payroll Employment—United States and New York City, 1969-1983 (in thousands)

| | United States Total | | | | New York City | | | |
|---|---|---|---|---|---|---|---|---|
| | | | Change: 1969-83 | | | | Change: 1969-83 | |
| *Industry* | *1969* | *1983* | *N* | *%* | *1969* | *1983* | *N* | *%* |
| Total | 70,384 | 89,978 | 19,594 | 27.8 | 3,797.7 | 3,344.2 | −453.5 | −11.9 |
| Goods Producing | 24,361 | 23,646 | −715 | −2.9 | 932.3 | 521.0 | −411.3 | −44.1 |
| Mining | 619 | 1,021 | 402 | 64.9 | 2.0 | 1.7 | −0.3 | −15.0 |
| Construction | 3,575 | 3,947 | 372 | 10.4 | 104.5 | 87.2 | −17.3 | −16.6 |
| Manufacturing | 20,167 | 18,678 | −1,489 | −7.4 | 825.8 | 432.1 | −393.7 | −47.7 |
| Service Producing | 46,023 | 66,332 | 20,309 | 44.1 | 2,865.4 | 2,823.2 | −42.2 | −1.5 |
| Transportation and public utilities | 4,442 | 4,941 | 499 | 11.2 | 323.9 | 236.0 | −87.9 | −27.1 |
| Wholesale and retail trade | 14,705 | 20,513 | 5,808 | 39.5 | 749.1 | 608.1 | −141.0 | −18.8 |
| Finance, insurance and real estate | 3,512 | 5,454 | 1,942 | 55.3 | 465.6 | 493.8 | 28.2 | 6.1 |
| Services | 11,169 | 19,680 | 8,511 | 76.2 | 779.8 | 966.8 | 187.0 | 24.0 |
| Government | 12,195 | 15,744 | 3,549 | 29.1 | 547.0 | 518.5 | −28.5 | −5.2 |

SOURCE: U.S. Department of Labor (1969; 1983a).

## Population: The Broader Context

The changing population and demographics of New York City do not exist in a vacuum; they tend to reflect broader processes operating nationally, and internationally as well. Before examining the dominant United States phenomena, it is useful to view total population growth within the context of selected Organization for Economic Cooperation and Development (OECD) nations.

### Total Population: United States
### Versus OECD Nations

In the absence of national population growth, city shrinkage would not be an unexpected occurrence. While the *rate* of population growth in the United States has decelerated in the past 30 years, it still remains at the leading growth edge of the major industrialized nations. With the exception of Canada, only Japan exceeded the United States increase of 12.3 percent from 1969 to 1980 among major OECD nations. Population levels of Germany, Switzerland, the United Kingdom, and Belgium were essentially stable, with growth increments hovering around 2 percent.

The growth rate of the United States was more than 50 percent greater than that of OECD Europe, and more than 150 percent greater than that of the European Economic Community (EEC). Thus population expansion in the United States is such that the pool of *potential* city residents could stabilize despite redistributional forces; but this has not been the case. Regional and metropolitan shifts have more than offset overall growth, and have worked to the detriment of the older industrial cities as well as New York.

### Regional Population Shifts

Within the overall context of population growth in the United States, there has been a shift in broad regional settlement patterns. Differential regional population growth is one dimension of the forces having an impact on New York City.

Longer-Term Patterns

As a backdrop to present concerns, it is useful to examine the patterns of regional change for the two decades prior to 1970, the baseline of the present transformation on a regional and divisional base. Between 1950 and 1960, the United States as a whole experienced a growth rate of 18.5 percent and an absolute increase of 28 million people. The growth was relatively evenly shared (in total numbers) among major regional clusters (between 7.2 million and 7.9 million) with the exception of the Northeast, which expanded by only

5.2 million people. Areas trailing the national growth *rate* most severely were the industrialized Northeast (13.2 percent) and agricultural states of the West North Central (9.5 percent) and East South Central (5.0 percent) divisions. In contrast, the West was the fastest-growing area of the nation, increasing in population by 38.9 percent.

Between 1960 and 1970, regional disparities began to sharpen, with the Northeast and North Central regions lagging in absolute growth and percentage change. In the face of a shrinking national growth increment (24 million people), the South's net gain in population (7.9 million) was greater than that of the previous decade. It was the only region to experience an increasing level of absolute growth, despite continued dissolution of its labor-intensive farming.

Recent Acceleration

Trends evident in the two decades prior to 1970 foreshadowed the general pattern of events to take place in the 1970s, but not their scale and magnitude. A gradual evolution rapidly accelerated. The 1970 to 1980 growth rates of the South (20.0 percent) and West (23.9 percent) were more than a hundred times greater than that of the Northeast (0.2 percent) and over five times greater than the North Central region (4.0 percent). Lagging most severely were the highly industrialized states of the Middle Atlantic and East North Central divisions, the historic manufacturing belt of America stretching from New York to Chicago.

Improved relative performance of the West South Central and East South Central divisions gives some indication that the phenomenon of rural agricultural displacement has ended; the latter is no longer available to bolster the sagging populations of northern industrial cities. The postindustrial era marches on the heels of agricultural displacement in the United States. Results are mirrored in central city decline, completely unlike the evolution currently taking place in much of the Third World. Former farm states have "unloaded" their redundant agricultural populations, setting the stage for improved growth performance.

The energy and natural-resources crises of the 1970s improved the economic status of states which export these vital commodities. Not only is this situation reflected in the southern divisions cited above, but also in the Mountain states—whose growth rate (37.1 percent) was the highest in the nation. For the first time, Mountain growth eclipsed that of the Pacific division (19.8 percent) and the oil- and natural gas-rich territories of the West South Central division.

Thus, rise of the Sunbelt and stagnation of the Northeast and North Central states represent significant changes in America's population disposition. The data emphasize one of the hazards of forecasting: that of being right in direction but wrong in time and scale. While earlier trends made apparent the shifts in regional growth (at least in hindsight), the future actually arrived much faster than was anticipated. The bulk of the national population growth from 1970 to 1980 was the province of the South (54.0 percent) and West (36.9 percent), while the Middle Atlantic region, which encompasses New York, showed a net resident loss of 1.8 percent.

So New York City is set in a regional population context of slow growth; at the same time, another basic force influences the city—metropolitan stabilization.

## Metropolitan-Nonmetropolitan Shifts

For the past half century, major growth nodes of American society were its metropolitan centers. Large metropolitan agglomerations (as defined in 1981) were the nation's dominant growth locations during the 1960-70 era. Their average annual population increase (1.7 percent) exceeded that of the United States (1.3 percent), and far outdistanced nonmetropolitan territories (–0.2 percent). Yet even this convention was challenged in the 1970s.

Despite the energy crisis of the decade, from 1970 to 1980, large metropolitan areas were transformed into settings of slow growth, with an average annual rate of population increase of 0.6 percent. At the same time, smaller metropolitan areas experienced growth rates of 1.4 percent. Even the latter were challenged by the revolutionary surge of nonmetropolitan growth, in which population increases also averaged 1.4 percent annually. Some of this shift undoubtedly signifies the drive toward exurbia and can be attributed to the lagging pace of metropolitan definition by the U.S. Census Bureau. However, the latter may be only a secondary explanation for the phenomenon.

The large metropolitan complexes of the Northeast dominate the post-1970 experience of metropolitan decline with a loss of 4.4 percent. Also exhibiting similar symptoms, although not yet shrinking, are their equivalents in the North Central states. In sharp contrast is the status of large metropolitan areas in the South and West, with a sustained pattern of growth through the 1970s.

Major focal points of growth in the Northeast and North Central states are nonmetropolitan areas, which exceeded even the positive growth performances of smaller metropolitan areas. In the South and West, the largest increments of population growth on an absolute basis accrued to small metropolitan settings. Linkage of regional and metropolitan growth patterns become apparent when it is realized that the Northeast has nearly three-fifths of its total population concentrated in four large metropolitan areas. The virtual national belt in large metropolitan growth therefore had a much greater effect in the Northeast than the South, where only one-fifth of the population resides in large metropolitan settings. Similarly, revitalization of the nonmetropolitan sector has negligible positive effects in the Northeast, where only one-seventh of the population is nonmetropolitan; in contrast, in the South, one-third of the population is nonmetropolitan.

The South and West have the advantage of maturing late, while older areas of the country are frozen in the patterns of yesteryear. Avoiding the question of which shift—regional or metropolitan—is the primary causal factor, we can be certain that their connection has been a major force and has had major impact on New York City. Post-1980 data are still fragmentary. There is some indication of recentralization—but the demographic scene is still predominated by the previous decade.

New York in Context

When data for the 20 largest metropolitan areas (as defined in 1981) are further dissected, the losses appear alarming. Seven of ten major metropolises in the Northeast and North Central regions experienced population losses from 1970 to 1980 (table 1.2). Yet in the South, only Washington, DC, and Baltimore reflect slow growth. More than compensating for the latter's performance has been the phenomenal growth of Houston, Miami, Atlanta, and Dallas, indexed by decade growth rates exceeding 25 percent.

As expected, the West exhibits a pattern similar to the South's. Indeed, the San Diego Standard Metropolitan Statistical Area (SMSA) experienced a growth rate of 37.1 percent, partially reflecting its attractiveness as a magnet for military retirees.

Rates of population decline of the older regions' metropolises are far from trivial. New York, for example, lost 5.4 percent of its 1970 population, with similar rates afflicting the Pittsburgh (–5.7 percent) and Cleveland (–5.5 percent) metropolitan areas. Even the nondeclining metropolitan areas of the Northwest and North Central states (excluding Minneapolis-St. Paul) are growing at a rate far slower than all the southern and western metropolises.

Historically, the sheer growth in population, particularly through migration, generated much of the social and economic stress of older urban centers. This pattern has changed very markedly, providing hope and new challenge. Problems of coping with increased housing demand, of overcrowded schools and overstressed physical facilities may be somewhat alleviated by the new conditions of population stability and decline; but in their place is the question of fiscal balance—of the economic wherewithal within older metropolitan settings to service the remaining population and to support overly massive infrastructure. These questions take on even more significance as equally important patterns of change surface inside metropolitan areas.

## Intrametropolitan Shifts

The history of America's major cities is one of meeting and surmounting problems of growth and change. But for the first time, the nation's central cities in total are losing population, and doing so quite markedly, while their corresponding suburban rings expand considerably. Even the latter positive pattern does not strictly hold for the nation's large, aging metropolitan areas in the Northeast and North Central regions. The large central cities of the "Frostbelt," while afflicted with substantial population losses, often fail to

TABLE 1.2

Population for the 20 Largest Metropolitan Agglomerations—
United States, 1960-1980 (in thousands)

| Region and Area | Population | | | Change 1960-70 | | Change 1970-80 | |
|---|---|---|---|---|---|---|---|
| | 1960 | 1970 | 1980 | N | % | N | % |
| Northeast Region | | | | | | | |
| New York SCSA | 15,405 | 17,035 | 16,120 | 1,630 | 10.6 | −915 | −5.4 |
| Philadelphia SCSA | 5,024 | 5,628 | 5,549 | 604 | 12.0 | −79 | −1.4 |
| Boston SCSA | 3,193 | 3,526 | 3,448 | 333 | 10.4 | −78 | −2.2 |
| Pittsburgh SMSA | 2,405 | 2,401 | 2,264 | −4 | −0.2 | −137 | −5.7 |
| North-Central Region | | | | | | | |
| Chicago SCSA | 6,794 | 7,726 | 7,868 | 932 | 13.7 | 142 | 1.8 |
| Detroit SCSA | 4,122 | 4,669 | 4,618 | 547 | 13.3 | −51 | −1.1 |
| Cleveland SCSA | 2,732 | 3,000 | 2,834 | 268 | 9.8 | −166 | −5.5 |
| St. Louis SMSA | 2,144 | 2,411 | 2,355 | 267 | 12.5 | −56 | −2.3 |
| Minneapolis-St. Paul SMSA | 1,598 | 1,965 | 2,114 | 367 | 23.0 | 149 | 7.6 |
| Cincinnati SCSA | 1,468 | 1,613 | 1,660 | 145 | 9.9 | 47 | 2.9 |
| South Region | | | | | | | |
| Washington, D.C. SMSA | 2,097 | 2,910 | 3,060 | 813 | 38.8 | 150 | 5.2 |
| Dallas-Ft. Worth SMSA | 1,738 | 2,378 | 2,975 | 640 | 36.8 | 597 | 25.1 |
| Houston SCSA | 1,571 | 2,169 | 3,101 | 598 | 38.1 | 932 | 43.0 |
| Miami SCSA | 1,269 | 1,888 | 2,640 | 619 | 48.8 | 752 | 39.8 |
| Baltimore SMSA | 1,804 | 2,071 | 2,174 | 267 | 14.8 | 103 | 5.0 |
| Atlanta SMSA | 1,169 | 1,596 | 2,030 | 427 | 36.5 | 434 | 27.2 |
| West Region | | | | | | | |
| Los Angeles SCSA | 7,752 | 9,981 | 11,496 | 2,229 | 28.8 | 1,515 | 15.2 |
| San Francisco SCSA | 3,492 | 4,631 | 5,182 | 1,139 | 32.6 | 551 | 11.9 |
| Seattle, SCSA | 1,429 | 1,837 | 2,092 | 408 | 28.6 | 255 | 13.9 |
| San Diego SMSA | 1,033 | 1,358 | 1,862 | 325 | 31.5 | 504 | 37.1 |

SOURCES: U.S. Bureau of the Census (1981a, b).
NOTE: Standard consolidated statistical areas (SCSAs) and standard metropolitan statistical areas (SMSAs) defined by Office of Management and Budget as of 30 June 1981.

account for all the decline of their metropolitan areas (table 1.3). For example, between 1970 and 1980, New York City and Pittsburgh lost 825 thousand and 96 thousand people, respectively. However, the New York metropolitan area lost 915 thousand people while the Pittsburgh SMSA lost 137 thousand, totals in excess of their central city losses. Thus, suburban decline is beginning to appear in select metropolitan settings (although the presence of smaller, aging subcities within this context plays a significant role).

The actual magnitude of some central city declines is remarkable when viewed over a longer time (table 1.3). From 1950 to 1980, St. Louis lost 47 percent of its population, Buffalo 38 percent, and Pittsburgh and Cleveland 37 percent—with the largest proportion of declines from 1970-80.

Cities in the United States, particularly those born in the industrial era, are experiencing the painful adjustments of shrinkage. New York City's population losses are significant in this context, but its economic fortunes have taken

TABLE 1.3
Population Change, Selected Cities—United States, 1950-1980
(in thousands)

| City | Population | | | Change 1950-80 | | Change 1970-80 | |
|---|---|---|---|---|---|---|---|
| | 1950 | 1970 | 1980 | N | % | N | % |
| Boston | 801 | 641 | 563 | −238 | −29.8 | −78 | −12.2 |
| Buffalo | 580 | 463 | 358 | −222 | −38.3 | −105 | −12.7 |
| Chicago | 3,621 | 3,369 | 3,005 | −616 | −17.0 | −364 | −10.8 |
| Cincinnati | 504 | 454 | 385 | −119 | −23.5 | −68 | −15.0 |
| Cleveland | 915 | 751 | 574 | −341 | −37.3 | −177 | −23.6 |
| Detroit | 1,850 | 1,514 | 1,203 | −646 | −34.9 | −311 | −20.5 |
| Minneapolis | 522 | 434 | 372 | −151 | −28.9 | −63 | −14.6 |
| New York | 7,892 | 7,896 | 7,071 | −821 | −10.4 | −825 | −10.4 |
| Newark | 439 | 382 | 329 | −110 | −25.0 | −53 | −13.8 |
| Philadelphia | 2,077 | 1,950 | 1,688 | −383 | −18.5 | −262 | −13.4 |
| Pittsburgh | 677 | 520 | 424 | −253 | −37.4 | −96 | −18.5 |
| St. Louis | 857 | 622 | 453 | −404 | −47.1 | −169 | −27.2 |

SOURCES: U.S. Census (1950, 1970, 1980); U.S. Bureau of the Census (1957, 1981b).

a different course. The "new" economy now emerging is analyzed in the following section.

# New York City's New Economy

New York epitomizes the changing economic functions of the central city in the United States. Absolute employment declines in goods producing and declining relative shares in service employment have characterized virtually all cities. But New York, from an economic point of view, may be the most successful adaptor to the new American imperatives. Changes in its form and function are briefly outlined below.

## The Economic Turnaround

One of the best measures of economic vitality is changes in payroll employ-ment. The first half of the decade of the 1970s was marked by precipitous employment losses, with annual declines averaging close to 100 thousand jobs between 1969 and 1976 (figure 1.1). But just as the city was an advanced indicator of the adjustment pattern of the national economy in terms of decline, so was its position in the expansion which started in 1976. Though interrupted by the harsh recession of 1981-82, the resurgence has yielded growth increments averaging 30 to 40 thousand jobs per year between 1976 and 1983 (see figure 1.2).

The sheer vitality of the New York metropolitan area is emphasized when we compare it with Los Angeles—typically viewed as the premiere "new"

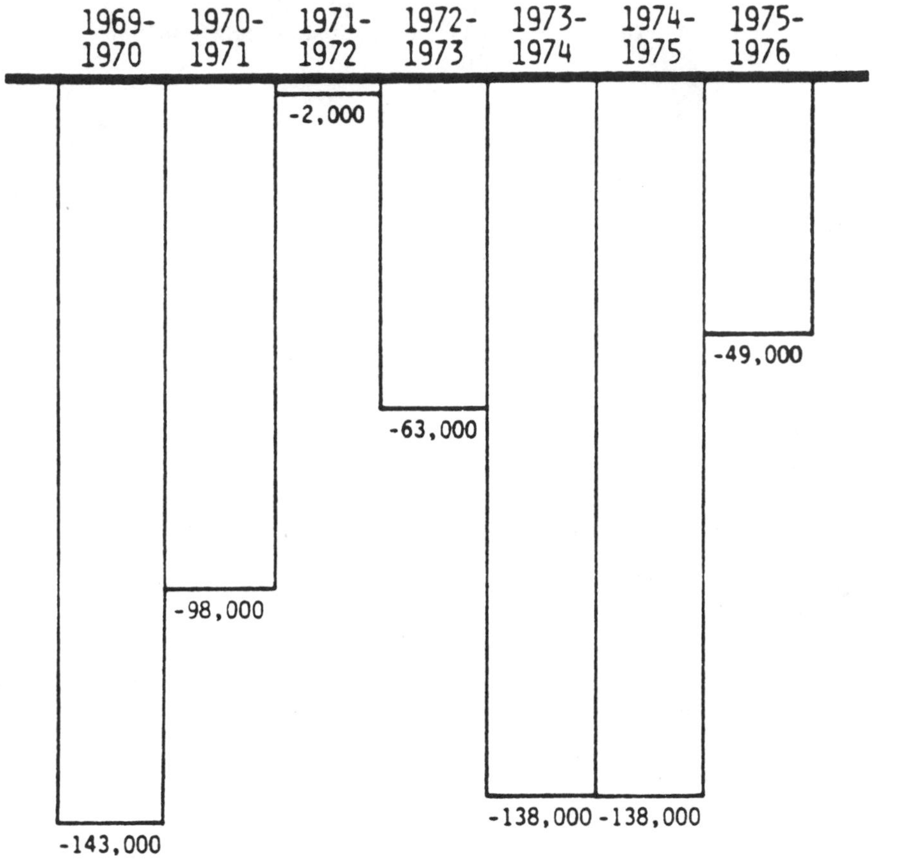

Figure 1.1: Changes in Payroll Employment—New York City, December 1969-76
SOURCE: New York Regional Office, Bureau of Labor Statistics, U.S. Department of Labor (1984).

configuration—where, from December 1979 to 1983, there was a loss of 55 thousand jobs. The case is made even stronger when we turn to a model of the old American industrial hegemony, the Chicago-Gary metropolitan area, which in the same period lost 263 thousand jobs. In contrast, the New York-Northeastern New Jersey metropolitan area secured nearly a quarter of a million new jobs. Indeed, New York had the highest absolute employment growth of all U.S. metropolitan areas with over a million jobs between 1982 and 1983 (table 1.4). A city that had lost 631 thousand jobs from 1969 to 1976 (figure 1.1) suddenly became the focal point of America's greatest metropolitan growth. While Houston was actually losing employment between 1982 and 1983, and the torchbearers of the Southwest, such as Dallas-Fort Worth, were recording relatively modest gains, the dominance of the New York metropolitan area was restored.

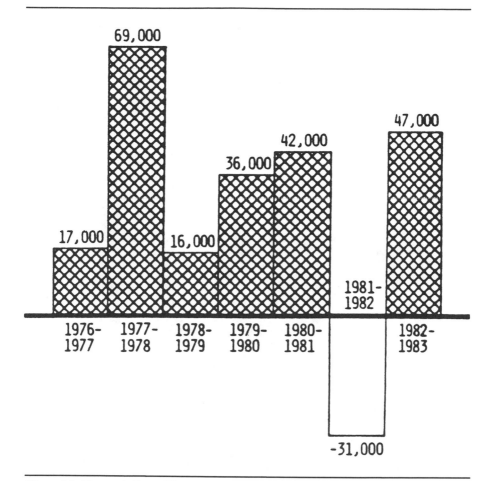

**Figure 1.2: Changes in Payroll Employment—New York City, December 1976-83**

SOURCE: New York Regional Office, Bureau of Labor Statistics, U.S. Department of Labor (1984).

## The Emerging Economic Structure

But this growth was achieved only through radical shifts in employment structure—shifts which in turn have rendered obsolete portions of the city's population. America's older cities, unlike many of their European equivalents, have been much more the creation of business activity, particularly manufacturing, rather than administrative or governmental function.

The synergism between flows of immigrants—both domestically, off the farm, and from abroad—and the manufacturing opportunities available in the earlier industrial city created the massive "smokestack" concentrations which for so very long dominated the image of urban America. These concentrations were predicated on agglomerations of productive power, massed population, and industrial technology. New York shared in the sheer

TABLE 1.4

Payroll Employment Change for Metropolitan Areas
with Over One Million Jobs—United States,
December 1979-1983 (in thousands)

| Area | Employment December 1983 | Change in Number | | |
|------|--------------------------|---------|---------|---------|
| | | 1979-82 | 1982-83 | 1979-83 |
| Atlanta | 1,093.9 | 61.7 | 65.4 | 127.1 |
| Dallas-Ft. Worth | 1,621.7 | 113.8 | 73.9 | 187.7 |
| Houston | 1,512.6 | 120.5 | −27.0 | 93.5 |
| Washington, DC | 1,675.1 | 37.2 | 52.3 | 89.5 |
| District of Columbia | 598.5 | −20.3 | 5.3 | −15.0 |
| Boston | 1,543.2 | 23.0 | 38.8 | 61.8 |
| New York-NE New Jersey | 6,988.9 | 101.7 | 142.7 | 244.4 |
| New York City | 3,408.0 | 46.5 | 46.9 | 93.4 |
| Rest of area | 3,580.9 | 55.2 | 95.8 | 151.0 |
| Nassau-Suffolk | 1,002.3 | 45.7 | 33.8 | 79.5 |
| San Francisco-Oakland | 1,585.0 | −5.1 | 19.0 | 13.9 |
| Philadelphia | 1,967.8 | −24.1 | 40.2 | 16.1 |
| City of Philadelphia | 759.5 | −42.0 | 2.6 | −39.4 |
| Minneapolis-St. Paul | 1,106.6 | −43.3 | 43.4 | 0.1 |
| Los Angeles-Long Beach | 3,619.5 | −140.4 | 85.1 | −55.3 |
| Chicago-Gary SCA | 3,250.6 | −238.9 | −23.6 | −262.5 |
| Chicago | 3,037.3 | −228.0 | 3.2 | −224.8 |
| Detroit | 1,581.6 | −252.3 | 47.3 | −205.0 |

SOURCE: New York Regional Office, Bureau of Labor Statistics, U.S. Department of Labor (1984).

energy generated by the new industrial era in combination with the essential raw material of cheap labor.

But that era is rapidly contracting in America, and in no place more dramatically than in New York City. Manufacturing employment in New York in 1950 was well in excess of one million (figure 1.3). By the mid-1980s, barely 400 thousand manufacturing jobs remained. Sharp declines are evident across virtually the entire manufacturing profile, and these continue to the present. Only publishing, of the 20 major industrial employment industries, has held its own.

The chief loser has been nondurable goods. The city's dominance in apparel manufacturing initially shifted to cheaper labor markets within overnight trucking distance, then moved to the deep South, and finally dispersed in a worldwide stream. Korea, Taiwan, Brazil, Hong Kong, and other non-U.S. locations now dominate this sector. Local manufacturers— and with them local manufacturing jobs for the relatively unskilled—have been and are departing. In their place is a new economy.

## Rise of the Service Economy

As indicated earlier, 1976 was New York City's pivotal point, the year in which the rise of new activities more than overcame loss of the old. Figure 1.4

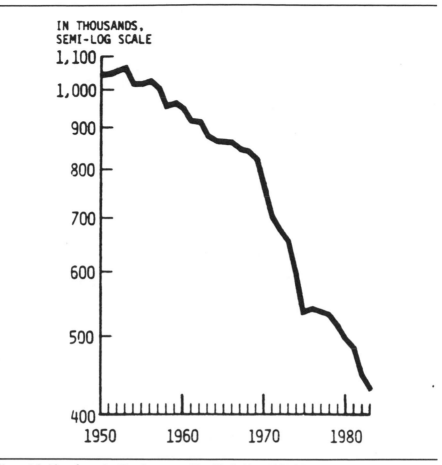

IN THOUSANDS,
SEMI-LOG SCALE

Figure 1.3: Manufacturing Employment—New York City, 1950-83

SOURCE: New York Regional Office, Bureau of Labor Statistics, U.S. Department of Labor (1984).

summarizes the shifts from December 1976 through 1983. Loss of more than 100 thousand manufacturing jobs together with shrinkage in wholesale and retail trade as well as transportation and public utilities (both perhaps victims in whole or in part of the declining population base) was more than made up by an increment in service employment in excess of 200 thousand.

The new leitmotiv of the city is its increased role as a world headquarters, symbolized by a growth of 85 thousand jobs in finance, insurance, and real estate. America's involvement in international trade was expanding—and New York City was increasingly its center.

Figure 1.5 graphically portrays the abrupt nature of the changes in New York City's economic raison d'être. By 1980, employment in finance and business services together passed that of manufacturing within the city. The subsequent years indicate an enormous broadening of the gap. A whole new economy became increasingly ill-matched to the resident masses of the city.

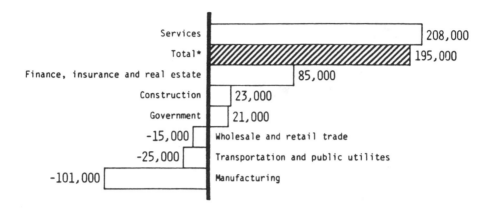

*Includes mining not separately presented.

Figure 1.4: Changes in Payroll Employment—New York City, December 1976-83

SOURCE: New York Regional Office, Bureau of Labor Statistics, U.S. Department of Labor (1984).

New York City's new profile is one in which private sector employment provides nearly six times as many jobs as does government, and nearly six of seven private sector jobs are service producing. Indeed, all goods-producing employment now has been reduced to the scale of government employment.

If we view data on New York City's share of metropolitan area employment, the new areal job calculus becomes evident. New York City accounts for 49 percent of all private sector employment, but only 37 percent of goods-producing jobs. Its leadership in finance, insurance, and real estate is evident, with 71 percent of the regional jobs located within the core city. In securities, the proportion is 92 percent.

It is significant that of all the service-producing sectors, New York's share is greatest in social services. The clash of the social services imperatives (and needs of the resident population) versus the rise of the new postindustrial opportunities will be detailed later.

The changing role of cities versus their suburban areas, even in municipalities as vigorous as New York, is evident. More than two-thirds of total job growth in the metropolitan area from 1977 to 1983 took place *outside* the city. Only in finance has New York continued its potent dominance. Once again, this situation was only equalled by the provision of social services which enjoyed a job growth of over 51 thousand in the metropolitan area—New York secured 40 thousand of the increment.

While the city and metropolitan area lost jobs in goods producing, New York City's losses were in excess of 83 thousand—in the rest of the area, barely 13 thousand. Thus while New York had 37 percent of the goods-producing jobs in 1983, it accounted for 86 percent of the job loss in that sector in the preceding six years (1977-83). Goods producing has been related to suburbia, exurbia, other regions of the U.S., and most significantly to other nations. Yet,

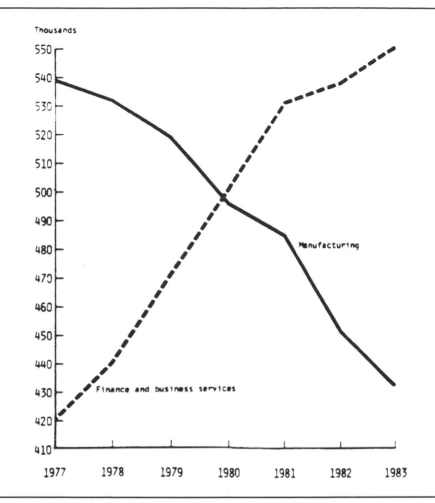

Figure 1.5: **Payroll Employment in Manufacturing and Finance and Business Services—New York City, 1977-83**

SOURCE: New York Regional Office, Bureau of Labor Statistics, U.S. Department of Labor (1984).

there is substantial indication of a successful transition, at least from a macroperspective, of New York City's economy from one dominated by industrial opportunity to the new postindustrial/information-based era.

The city's job base is towering, but highly specialized—and little in the dynamics of change would seem to alter this pattern as we look into the future. This is the case even when employment in high-technology industries, often viewed as having a potential for reversing the outflow of goods-producing activities, is analyzed. Regardless of which definition of high-tech is employed, there is little indication of factors which will radically add to employment within the city. And in the more dynamic service-information economy, high-technology gains have been relatively small. As we look to the future, technological shift may even have a negative impact.

## *Future Economic Opportunities and Threats*

Radical as have been the changes in New York City's job base, they may not yet reflect the impact of future technology. There is a progression as a function of technological competence and information technology which suggests that certain common elements of back office work may be shifted, relatively easily, from high cost core areas. Back office operations in New York City's large commercial banks—Chemical, Chase, Manufacturers Hanover—have long been located in Long Island, while Citicorp has back office facilities in South Dakota and Florida as well as in New York City and Long Island. As banks, brokerage firms, and insurance companies become national operations, location of their back office functions may increasingly be dispersed across the country rather than being based in New York City or the surrounding region.

The equivalent of the sewing machine operator or leather goods finisher of yesterday's labor market in New York City may well be the basic clerk, stenographer, or keypunch operator of today. The skills and orientations of the latter differ from the former—but even they may be disappearing as well. An analogy to the evolution of New York City's apparel industry is foreseeable within the information-processing sector where the shift of back office activities to areas of cheaper rents and better labor force characteristics— places less subject to the disruptions of the central city in general—is reminiscent of the apparel industry of a generation ago. The front office may remain, but the bulk of support activities, as a function of improved communications and transportation, can be situated elsewhere.

The skill requirements for these activities become even more rarified than presently is the case. The mismatch between present needs and present city residents is evident; in the future the incongruity may prove even more substantial. The suburbs have college-educated women seeking part- and full-time employment and at a great distance from the threat of unionization as well. Cities by definition impose a very high level of support costs as a function of their relative complexity and infrastructure, to say nothing of their age and social support problems. Much of this can be avoided by the flight to suburbia and less-developed areas.

In the midst of its incredibly successful turnaround—at a time when the New York City realty market may be the most expensive in the entire world and when the vision of a world city has replaced that of regional equivalents— the potential for economic thinning of the service sector should not be overlooked. The city will have to compete even harder to stay in place, much less expand. This requires a government with the credibility to turn to a poor and needy electorate and ask for programs and indeed even subsidies with which to attract and retain the skilled and rich. This in turn involves a level of legitimacy which seems to be evading present political leadership. If New York City cannot maintain its "export" jobs, the negative multiplier potential is enormously evident.

Within this context, the distance between *the city of tomorrow* and *residents of yesterday* requires exploration. The postindustrial economy has bypassed the residuals of an industrial population sector. A dispersed information-based economy will further widen the gap in the future. An *urban economy* can shift much more rapidly than the capacities of an *urban population*; this growing incongruity is described in the following section.

## The Demographic Dilemma

The American city of the postindustrial era is very different in capacity and need from its predecessor. The march off the land into the urban centers of the last century provided the key raw material for urban industrial development. Cities had little in the way of public services but they did have jobs. In many cases these were brutal, poor-paying occupations. The very desperation of the agrarian displacees, however, caused them to view these in a positive light. The U.S. city not only had a unique scale of labor force concentration, but had very cheap labor as well. Cities grew in potency and industrial hegemony in concert with population growth.

In the United States, and we would suggest possibly in other developed economies, these reinforcing patterns have now aborted. New York City, as perhaps the most visible focal point of postindustrial activity, yields insight into the dynamics at work.

### The Long-Term Pattern of Population

For data on the U.S. population, the New York metropolitan region, and New York City from 1820 through 1980, see table 1.5. The broad sweep of time indicates the city's early dominance, with percentage increments in population exceeding that of the nation until the mid-1880s. While city and nation were roughly matched through the 1940s, the decade of World War II is clearly the turnover point. The rest of the New York metropolitan region which had achieved parity in growth with the core at the turn of the century suddenly became the expansion area, sweeping past the city. The latter, for the first time in its history, lost population on a decennial base in the decade of the 1950s. But even New York's suburbs began to show their age in the 1970s, with a population loss of 3 percent. In contrast, while growth slowed nationally in the 1970s, the population of the United States still expanded by 11 percent, but New York City's declined by an equal percentage. Even its hinterland, the metropolitan region outside the city, began to suffer from the same erosion. The era of dominance of the giant industrial conurbations of the industrial era was coming to an end.

### Racial and Ethnic Shifts

New York City from its very inception has been a melting pot of all nations. Much of U.S. history has been unique in terms of the ease of inmigration.

(While this process was sharply contracted in the early 1920s, it should be noted that current net legal inmigration nationally is still on the order of a half million persons per year.)

Every ethnic group coming into New York has been viewed as a threat and danger to the status quo, with persistent questions raised on their assimilative capacity as well as doubts on their contribution to the economy as a whole. More affluent immigrants could afford the trek to the farm lands. Those who had less in the way of resource tended to terminate their journeys, not uncommonly nearly penniless, in the central city. And of all central cities, New York was the potent in this regard as the chief port of entry.

A major function of the American industrial city of yore was to take unskilled immigrants, harness them to industry, and provide an infrastructure adequate in its scale to sustain the first generation and tutor its successors. The definition of success was for the immigrant to look on new "greenhorns" with the same disdain as had been exhibited to his own generation, an enormously potent process. The graduation ceremony typically was a shift, either in the first or second generation, to a suburban residential location, making way for new groups. Ethnic groups varied in the pace of adaptation and certainly even the most successful left their less fortunate brethren behind. But the demographic/economic "fit" was there.

The issue at hand is whether current inheritors of the city will find it an equally positive springboard. The answer to this issue is not at hand. Certainly, however, its successful resolution is dogged not only by the changing role of the central city—the new job structure earlier described—but also by race.

From 1920 through the last decade, the most important segment of migration was the internal flow of population, evidenced as we view the proportion of blacks within New York City (table 1.6). As late as the turn of the century blacks represented less than 2 percent of the population. The shift off the land, particularly marked in the 1920s and 1930s, combined with substantial birthrates, as well as war-time-generated job opportunities, produced a massive swelling. By 1950 nearly 10 percent of the city's population was black; currently it is estimated to be over 25 percent.

One metaphysical issue of urban analysis is the question of whether, had there not been a racial shift, America's cities would have been very different in their residential roles—or indeed their economic functions as well?

There is still the question of whether the decline of whites by one-quarter—more than 1.2 million people in the decade of the 1970s—was related to racial/ethnic pressures within the city, regardless of the underlying motivations. Tables 1.7 and 1.8 indicate the massive shifts still taking place in the city's ethnic composition. New York will shortly become majority-minority. Hispanics are rapidly challenging blacks in absolute number and have passed them in growth. While other ethnic groups, particularly a very substantial inmigration of Asians, are expanding, their numbers are still relatively small, but their upward mobility is already evident.

TABLE 1.5
Population and Change from Preceding Census—United States,
New York Metropolitan Region,[a] and New York City,
1820-1980

| | Population (in thousands) | | | Percentage Change from Preceding Census | | | |
|---|---|---|---|---|---|---|---|
| Census Year | U.S. | Total NYMR | NYC | Rest of NYMR | U.S. | Total NYMR | NYC | Rest of NYMR |
| 1820 | 9,638 | 364 | 152 | 212 | | | | |
| 1830 | 12,866 | 484 | 242 | 242 | 33 | 33 | 59 | 14 |
| 1840 | 17,069 | 681 | 391 | 290 | 33 | 41 | 62 | 20 |
| 1850 | 23,192 | 1,072 | 696 | 376 | 36 | 57 | 78 | 30 |
| 1860 | 31,443 | 1,755 | 1,175 | 580 | 36 | 64 | 69 | 54 |
| 1870 | 39,818 | 2,274 | 1,478 | 796 | 27 | 30 | 26 | 37 |
| 1880 | 50,156 | 2,907 | 1,912 | 995 | 26 | 28 | 29 | 25 |
| 1890 | 62,948 | 3,807 | 2,507 | 1,300 | 26 | 31 | 31 | 31 |
| 1900 | 75,995 | 5,191 | 3,437 | 1,754 | 21 | 36 | 37 | 35 |
| 1910 | 91,972 | 7,208 | 4,767 | 2,441 | 21 | 39 | 39 | 39 |
| 1920 | 105,711 | 8,660 | 5,620 | 3,040 | 15 | 20 | 18 | 25 |
| 1930 | 122,775 | 11,021 | 6,930 | 4,091 | 16 | 27 | 23 | 35 |
| 1940 | 131,669 | 11,839 | 7,455 | 4,384 | 7 | 7 | 8 | 7 |
| 1950 | 150,697 | 13,154 | 7,892 | 5,262 | 14 | 11 | 6 | 20 |
| 1960 | 178,464 | 15,125 | 7,782 | 7,343 | 18 | 15 | −1 | 40 |
| 1970 | 203,212 | 16,694 | 7,895 | 8,799 | 13 | 10 | 1 | 20 |
| 1980 | 226,505 | 15,571 | 7,015 | 8,557 | 11 | −7 | −11 | −3 |

SOURCE: Tobier (1982). Reprinted with permission.
a. The New York Metropolitan Region (NYMR) includes the five boroughs of New York City; five New York counties—Nassau, Putnam, Rockland, Suffolk, and Westchester; and nine New Jersey counties—Bergen, Essex, Hudson, Middlesex, Monmouth, Morris, Passaic, Somerset, and Union.

Other newcomers to the city, lacking skills commensurate with its new economy, find it increasingly inhospitable. In a city whose affluence is marked by most tourists, roughly a quarter of the population is below the poverty level and more than one in seven is on welfare.

## Geographic Changes Within the City

None of the subregions of the city, divided into five boroughs, is unmarked by these shifts. The traditional blue-collar enclaves of the Bronx and Brooklyn have seen the greatest population losses, with the Bronx losing fully one in five of its 1970 population in the subsequent ten-year period, and nearly half of its whites. Even Manhattan, despite its economic resurgence in the later 1970s, still lost 7.3 percent of its population—10.9 percent of its whites, and a striking 18.6 percent of its blacks. While the newer borough of Queens and the relatively small enclave of Richmond were more stable, they too reflect the ethnic turbulence within the city.

TABLE 1.6
Total and Black Population—New York City, 1900-1980
(in thousands)

| Year | Total Population | Black | |
|------|------------------|-------|------|
| | | N | % |
| 1980 | 7072 | 1788 | 25 |
| 1970 | 7895 | 1668 | 21 |
| 1960 | 7782 | 1085 | 14 |
| 1950 | 7892 | 749 | 9 |
| 1940 | 7455 | 458 | 6 |
| 1930 | 6930 | 328 | 5 |
| 1920 | 5620 | 152 | 3 |
| 1910 | 4767 | 92 | 2 |
| 1900 | 3437 | 61 | 2 |

SOURCE: New York City Department of City Planning.

## Population, Households, and Housing

Despite the absolute declines in population, New York City still suffers from a severe housing shortage. Corporate executives complain bitterly of rents in new structures in excess of $2,000 per month for an 800-square foot non-super-luxury accommodation. Condominium and cooperative sales within Manhattan, again outside of the super-luxury category, are averaging nearly $300 a square foot. On the other side of the spectrum is an abandonment rate of 20 thousand units a year and shattered hulks occupied only by squatters. The peculiar calculus of declining population and sustained housing pressures is at work. The inconsistency was made up by a shrinkage in household size, which declined to a 1980 level of only 2.4 persons. The number of households was barely altered; fewer bodies were occupying as much or more space. Manhattan was at the forefront of this phenomenon with a 5.6 percent increase in housing units, while its population declined by 7.3 percent. And yet the pressures increased as average household size in 1980 achieved a remarkable level of under two persons. The number of nonminority group children in the city has declined markedly, with public primary and secondary schools now in total having roughly only a quarter of their attendees white (non-Hispanics).

## The City of the Poor Masquerading as the City of the Rich

Over the past 25 years, New York has changed from having a median family income above the national average to an average family income well below it; this trend holds for all racial groups. For the whole city in 1982, the median income was 82 percent of the nation's median. In 1959, it was 11 percent above the national median.

TABLE 1.7

Changes in Racial and Ethnic Composition—New York City,
1978-1984 (in thousands)

| Race | Year | | | Percentage Change | | |
|------|------|------|------|------|------|------|
| | *1978* | *1981* | *1984* | *1978-81* | *1981-84* | *1978-84* |
| Total | 6948 | 6872 | 6950 | −1.1 | +1.1 | 0 |
| White | 3805 | 3642 | 3478 | −4.3 | −4.5 | −8.6 |
| Black | 1547 | 1527 | 1684 | −1.3 | +10.3 | +8.8 |
| Puerto Rican | 804 | 765 | 774 | −4.9 | +1.2 | −3.8 |
| Other | 336 | 430 | 634 | +27.8 | +47.7 | +88.7 |
| Not reported | 445 | 509 | 380 | NA | NA | NA |

SOURCE: Stegman (1985). Reprinted with permission of the publisher.
NOTE: Data in this table are from the New York City Housing and Vacancy Survey conducted every three years. Data in Table 1.6 are from decennial censuses. There are several minor deviations in results.

TABLE 1.8

Racial and Ethnic Composition—New York City, 1978-1984
(in percentages)

| | *1978* | *1981* | *1984* |
|------|------|------|------|
| White | 58.5 | 57.2 | 52.9 |
| Black | 23.8 | 24.0 | 25.6 |
| Puerto Rican | 12.4 | 12.0 | 11.8 |
| Other | 5.2 | 6.8 | 9.7 |

SOURCE: Stegman (1985). Reprinted with permission of the publisher.
NOTE: See note, Table 1.7. Each year = 100 percent.

The problem is most evident among the approximately 70 percent of all New York City households who are renters. As such they dominate not only housing politics of the city as evidenced by strong rent control ordinances, but also nonresidential development as well. The latter typically runs head-on into the property interests which have grown around tenancy. Alternate use of buildings, demolition, any form of making way for the new postindustrial elite are very difficult to achieve within the city. The risk of development's aborting is substantial enough to require very large gains to offset its hazards.

Some indication of the incongruity between the mass of New York City residents and its newly defined economic functions is indicated by the key parameters of the rental market in contrast to homeowners. See table 1.9 for median household income of renters and owners in current dollars by race and ethnic origin for the city. Even New York's white renters had income levels only half those of its more affluent owner households; minority levels were far worse.

The pattern of decline is all too clear. From 1969 to 1980 there was a reduction in constant dollar income of some 28.3 percent. Even in the years of relative prosperity from 1980 to 1983, as the city added substantially to its employment rolls, the decline among renters was still 4.4 percent. The loss has been most striking in the city's key demographic growth sectors, blacks and

Hispanics, with the latter group showing an income shrinkage of nearly 40 percent in the 11 years from 1969 through 1980. While the minority of the city's households who are owners become more affluent—regardless of race— only white renters are ahead of inflation.

The erosion of real incomes for the bulk of New York's households now is a predominantly outer-borough phenomenon. In the most current years, Manhattan has moved to near stability. The Borough of Manhattan with its postindustrial economy is flourishing. But despite the existence of a few enclaves in Brooklyn and Queens which are part of its economy, the other boroughs, and with them the bulk of the municipal population, show little evidence of the economic good fortune which the core has enjoyed.

In the face of the politics of shrinking real dollar incomes, the city has been driven to harden its housing policies, yielding a near immobilization. The poor and middle-income groups cannot hazard displacement, or rent increases commensurate with the market. But the effort to control rents has been a losing struggle. The shrinkage in incomes has more than overcome the safeguards against rent increments.

Median gross rent-to-income ratios are substantially above those of the nation (table 1.10). From 18.9 percent in 1950, they reached a high of 29.3 percent in 1984. An increasing proportion of New Yorkers simply cannot afford their rents. The long-held tradition of a quarter of one's income going for shelter clearly has long been bypassed. As of 1984 nearly a quarter of the city's renter households were forced to make shelter payments in excess of half their incomes. While some of this may be mitigated by welfare payments, the pressures are evident. The relatively small though glittering locomotive of Manhattan's banking and production service economy must pull an increasingly lengthy train of people who have little relationship to it.

In turn staffing of the new towers is increasingly becoming the role of suburbanites—nonvoters in the city. Three-quarters of the job growth from 1977 to 1981 went to noncity residents. The lack of fit, of congruity between the urban economic growth sectors and its resident population expands. The political results of this mismatch are felt presently and may be even more compelling in the future.

## Problems and the Future

New York City has renewed its magnetism to the young elite of the country. The most prestigious law firms vie for new graduates at salary levels of $50,000 to start—an income triple that of the median household. At the other end of the spectrum, there is clear evidence of new immigrant groups moving into the basic retailing/service sector. Largely isolated between them are groups qualifying neither in skills nor in family structure/orientation to either of these growth nodes. Table 1.11 indicates the scale of the problem; it compares labor force participation—the proportion of people aged 16-64 either employed or seeking jobs—in 1983 for the United States as a whole and the City of New York.

TABLE 1.9

Median Household Income by Race and Tenure—New York City,
1980 and 1983

| | Current Dollars | | | Constant 1967 Dollars | | |
|---|---|---|---|---|---|---|
| | | | Percentage Change | | | Percentage Change |
| | 1980 | 1983 | 1980-83 | 1980 | 1983 | 1980-83 |
| All households | $12,687 | $16,166 | +27.4 | $5,349 | $5,602 | +4.7 |
| White | 16,023 | 20,255 | +26.4 | 6,755 | 7,018 | +3.9 |
| Black | 10,613 | 13,035 | +22.8 | 4,474 | 4,517 | +1.0 |
| Puerto Rican | 7,171 | 8,596 | +19.9 | 3,023 | 2,979 | −1.5 |
| Other | 12,300 | 13,928 | +13.2 | 5,185 | 4,826 | −6.9 |
| Consumer Price Index | 237.2 | 288.6 | +21.7 | | | |
| All renters | $11,001 | $12,797 | +16.3 | $4,638 | $4,434 | −4.4 |
| White | 13,474 | 17,326 | +28.6 | 5,680 | 6,003 | +5.7 |
| Black | 9,142 | 10,957 | +19.9 | 3,854 | 3,797 | −1.5 |
| Puerto Rican | 6,460 | 7,698 | +19.2 | 2,723 | 2,667 | −2.1 |
| Other | 10,903 | 12,129 | +11.2 | 4,597 | 4,203 | −8.6 |
| All owners | $20,333 | $25,183 | +23.9 | $8,572 | $8,726 | +1.8 |
| White | 21,142 | 26,298 | +24.4 | 8,913 | 9,112 | +2.2 |
| Black | 17,517 | 23,225 | +32.6 | 7,385 | 8,047 | +9.0 |
| Puerto Rican | 16,708 | 21,308 | +27.5 | 7,044 | 7,383 | +4.8 |
| Other | 21,907 | 26,941 | +23.0 | 9,236 | 9,335 | +1.1 |

SOURCE: Stegman (1985: 122). Reprinted with permission of the publisher.

In general, labor force participation is far lower in the central city than in the nation, without regard to race. When we focus on the latter, however, a number of incongruities are evident. Black men have the lowest labor force participation, at 65.8 percent, of the three male groups shown. While it is somewhat higher for black women than for New York's white women, their employment is still substantially under the national equivalent.

The situation is even more strikingly incongruous as we turn to Hispanics. Much of New York's population under that rubric is from Puerto Rico; much of the national Hispanic incidence outside the Northeast region is of Mexican origin. The latter seem to partake much more vigorously in the labor market than do the former.

Even more troubling are data on youth from 16 to 19. New York City simply is not a hospitable place for adolescent entrants into the labor force. Nationally, for example, over half of whites 16 to 19 are in the labor force—the equivalent in New York City is less than a third. The national incidence rate for black teens is 36.4 percent; Hispanics fall between whites and blacks. While data are lacking for black youth in the city, data for the greater New York labor market area indicate an involvement of only a fourth, barely two-thirds the national figure for blacks, and well under the halfway mark for whites.

TABLE 1.10
Rent-Income Ratios—New York City, 1950-1984

| | Median Gross Rent-Income Ratio, All Renter Households (percentage) | | | | | |
|---|---|---|---|---|---|---|
| | 1950 | 1965 | 1970 | 1975 | 1981 | 1984 |
| | 18.9 | 20.4 | 20.0 | 24.7 | 28.0 | 29.3 |

| Gross Rent-Income Ratio | Percentage of Households with Rent-Income Ratios over Specified Levels | | | | | |
|---|---|---|---|---|---|---|
| | 1950 | 1965 | 1970 | 1975 | 1981 | 1984 |
| 25% or more | NA | 34.9 | 35.5 | 49.1 | 56.6 | 60.2 |
| 30% or more | 25.0 | 25.3 | NA | 38.1 | 45.6 | 48.2 |
| 35% or more | NA | 18.7 | 23.2 | 30.8 | 37.1 | 39.5 |
| 40% or more | NA | 14.5 | NA | 25.2 | 30.5 | 32.9 |
| 50% or more | NA | NA | NA | NA | NA | 23.2 |

SOURCE: Stegman (1985). Reprinted with permission of the publisher.

TABLE 1.11
Employment Patterns—United States and New York City, 1983
(in percentages)

| Racial Groups | Labor Force Participation | | Unemployment Rates | | Employment Rates | |
|---|---|---|---|---|---|---|
| | United States | New York | United States | New York | United States | New York |
| White | | | | | | |
| Men | 77.1 | 68.5 | 8.8 | 8.4 | 70.4 | 62.7 |
| Women | 52.7 | 40.7 | 7.9 | 7.5 | 48.5 | 37.7 |
| Both sexes, age 16-19 | 56.9 | 31.1 | 19.3 | 26.0 | 45.9 | 23.0 |
| Black | | | | | | |
| Men | 70.6 | 65.8 | 20.3 | 16.9 | 56.3 | 54.7 |
| Women | 54.2 | 45.9 | 18.6 | 10.8 | 44.1 | 40.9 |
| Both sexes, age 16-19 | 36.4 | 25.2[a] | 48.5 | 50.2[a] | 18.7 | 12.5[a] |
| Hispanic | | | | | | |
| Men | 80.8 | 68.8 | 13.5 | 12.5 | 69.9 | 60.2 |
| Women | 48.5 | 32.2 | 13.8 | 12.4 | 41.8 | 28.2 |
| Both sexes, age 16-19 | 45.3 | NA | 28.4 | NA | 32.5 | NA |

SOURCE: U.S. Department of Labor, Bureau of Labor Statistics (1983a).
a. New York Labor Market Area; central city data not available. All other New York statistics for central city.

Incongruity in unemployment rates is even more striking. The base, 1983, was a very mixed year, with the recession bottoming out in the spring on the national level, and substantial employment growth, particularly at the end of the year. New York's economy as a whole fared far better than that of the nation, yet unemployment for black youths was an appalling 50.2 percent.

When we focus on the city, the hostility of the job market for black men versus black women is evident: the former had an unemployment level double that of white men; black and white women were only 3.3 points apart, clearly not at the same levels of incongruity. Again, Hispanic men within the city were nearly precisely intermediate.

While there are no separate unemployment data for Hispanic adolescents either for the city or U.S. labor market as a whole, data on blacks indicates an unemployment level of fully one-half. The absorptive capacity of New York's economy for this group, damned by race, age, and sadly enough, by educational attainment, is at a very low point indeed.

The sum of these data indices is shown in the right-hand side of table 1.11, which gives employment rates. Far more New Yorkers are, for any of a variety of reasons, simply not on the employment rolls when compared with the nation, regardless of race. But nationally—and in New York, very particularly—the proportion of black men who are employed is at a very low point indeed—with barely half of those residing in New York employed on average during 1983. Hispanic men within New York did better, at 60.2 percent, versus whites, at 62.7 percent.

Employment levels for women are somewhat different, with black women actually leading in terms of the proportion employed, about two-fifths; whites over a third; Hispanic women, over a fourth. Much of this latter incongruity, however, seems to indicate the better "fit" of women to the postindustrial economy particularly for blacks, and cultural preferences among New York's largely Puerto Rican Hispanic population.

While data are lacking for New York City's minority group adolescent employment, the product of low labor force participation rates and unemployment data for the region suggest a relatively trivial proportion finding their way into entry-level employment. Comparisons over time are not available because of changes in definition, but a glance at data available over the last decade indicates a substantial subsequent degeneration of minority employment activity.

The trauma of youth in a city which no longer is hospitable to them can only be hinted at. A dreadful parameter is suggested by homicides which accounted for nearly half of all deaths in 1981 for males aged 15-24.

There is some measure of resemblance to a series of geological strata as we view the world's cities, each representing a different time stream. The metropolises of high birthrates, of agricultural displacement, are currently beset by problems of growth, of provision of infrastructure—sewerage, water, basic shelter—in a fashion reminiscent of urbanization in the United States of the past century. The issues of fit at that time—of matching socioeconomic demographic characteristics to job base—were cruel and painful. But by and large, the process worked remarkably effectively. Past experience provides little insight, however, as we turn to the premier postindustrial city, New York.

As we have suggested before, many of the earlier urban focal points of industry have seen an erosion of their economic base and catastrophic population shrinkage. The latter's scale, though monumental, has still not been able to yield a reasonable relationship between residual jobs and residual populations. Thus, as we view a Newark, a Cleveland, a St. Louis, we see cities that have shrunk in head counts within the last score of years by anywhere from a third to a half, and there is every danger of continued erosion in jobs and population.

New York City's situation is far more salubrious. There has been a remarkable thickening of economic functions, nearly adequate in size to replenish its peak job base. The population shrinkage to date is on the order of one in eight. While this mitigates some of the problems of the more purely industrial complexes, it leads to yet other issues of accommodation. These, we have suggested, are focused on the increasing mismatch between requirements of the new growth sectors and New York's resident population. How do we reconfigure a city, much of whose voting bloc seemingly is uninvolved in its own employment base? Many of the voters are poor, yet the city's future demands not only a national but, increasingly, an international elite.

To the poor, support services are essential, particularly welfare and other forms of social services. The bulk of New York City's voter population is still involved in a shelter society: basic accommodations and costs are key priorities. Increasingly, however, the city's economic base depends on a postshelter society. Proximity to the fun and games of the core, direct linkage to work place, and a capacity to avoid the increasingly frictional and dangerous public transit facilities are their key priorities. The affluent are concerned about taxes; not merely the provision of literacy for their children, but facilities competitive with those of the best of the nation. As the city's political wars grow heated, the one major common denominator is fear of crime. But even in this arena there are severe differences in approach to its constraint.

Unlike many other cities in the United States—and we would suggest elsewhere as well—New York City has the economic vitality with which to reshape itself. The problem is the development of political consensus. The issues of reconciling not merely class elements, but those of race and ethnicity as well, are keys to its future development.

# Bibliography

Auletta, Ken. 1979. *The Streets Were Paved with Gold.* New York: Random House.

Brecher, Charles, and Raymond D. Horton, eds. 1983. *Setting Municipal Priorities, 1984.* New York: New York University Press.

Gallagher, David. 1984. *Manhattan Loft Space and Manufacturing Jobs: The Outlook.* New York: New York Interface Development Project.

Harrigan, John J. 1973. *Political Change in the Metropolis.* 3d ed. Boston: Little, Brown.

New York City. Human Resources Administration. 1985. *Dependency: An Economic and Social Data Report for New York City: 1975-1984.* New York.

New York City Council on Economic Education. 1982. *1983-1984 Fact Book: Tables and Charts on the New York Metropolitan Region.* New York.

———. 1984. *Challenges of the Changing Economy of New York City: New York City's Changing Role in the Regional Economy.* New York.

Richardson, Harry. 1985. *Economic Prospects for the Northeast.* Philadelphia: Temple University Press.

Starr, Roger, 1985. *The Rise and Fall of New York City.* New York: Basic Books.

Stegman, Michael A., with Doug Hillstrom. 1985. *Housing in New York: Study of a City, 1984.* Report prepared for the City of New York, Department of Housing Preservation and Development, Office of Rent and Housing Maintenance, Division of Rent Policy and Regulation. New Brunswick, NJ: Center for Urban Policy Research, Rutgers University.

Sternlieb, George, and James W. Hughes. 1975. *Post-Industrial America: Metropolitan Decline and Inter-Regional Job Shifts.* New Brunswick, NJ: Center for Urban Policy Research, Rutgers University.

———. 1983. "The Three Faces of the New York Region: Housing and the New Economy." *Regional Economic Digest* 1 (Spring):1+.

———. 1984. *Income and Jobs USA: Diagnosing the Reality.* New Brunswick, NJ: Center for Urban Policy Research, Rutgers University.

Sternlieb, George, James W. Hughes, and Connie O. Hughes. 1982. *Demographic Trends and Economic Reality: Planning and Markets in the '80s.* New Brunswick, NJ: Center for Urban Policy Research, Rutgers University.

Tobier, Emanuel. 1982. "Population." In *Setting Municipal Priorities 1982*, edited by Charles Brecher and Raymond D. Norton. New York: Russell Sage Foundation.

U.S. Bureau of the Census. 1957. *County and City Data Book, 1956. A Statistical Abstract Supplement.* Washington, DC: U.S. Government Printing Office.

———. 1981a. "Major Metropolitan Complexes Show Slower Overall Population Growth during 1970s, 1980 Census Analysis Reveals." *United States Department of Commerce News* (8 Apr.).

———. 1981b. *1980 Census of Population. Standard Metropolitan Statistical Areas and Standard Consolidated Statistical Areas: 1980.* PC80-S1-5. Washington, DC.

———. 1981c. "Three Cities of 100,000 or More at Least Double Population between 1970 and 1980, Census Bureau Reports." *United States Department of Commerce News* (3 June).

U.S. Department of Labor. Bureau of Labor Statistics. 1969. *Employment and Earnings* (Jan.-Dec.).

———. 1983a. *Employment and Earnings* (Jan.-Dec.).

———. 1983b. *Geographic Profile of Employment and Unemployment.* Washington, DC.

Viger, Lawrence, 1984. *Commuting in the New York City Metropolitan Area, 1970 and 1980.* BLMI Report no. 9. New York: Department of Labor, State of New York.

# 2

# Los Angeles

## Ivan Light

In 1980 the urbanized area centering on the City of Los Angeles spread outward to include portions of five adjacent counties. In addition to the Los Angeles-Long Beach Standard Metropolitan area (SMSA), this politically fragmented area included 131 incorporated cities of which 50 had 1980 populations in excess of 50 thousand and 14 in excess of 100 thousand. This urbanized area lay within a 96.5 kilometer radius of downtown Los Angeles, but included only 12,320 square kilometers of land area because of the Pacific Ocean on the West (SPNB Research Department, 1984:6). Within these 12,320 square kilometers of urbanized area, 11,332,400 persons resided in 1980.

Population growth in the Los Angeles five-county area has always depended heavily on internal and international migration. In 1980, 19 percent of the Los Angeles urbanized area's population was foreign born compared to 6 percent of the United States population. Additionally, only 42 percent of area residents were born in California. Only about one-fifth of area residents were born in the Los Angeles urbanized area, a proportion no larger now than a generation ago (McWilliams, 1946:165). Two growth waves built the population of the Los Angeles urbanized area. Between 1929 and 1936, 100 percent of population growth resulted from migration and none from natural increase. Thereafter, natural increase became an increasing component of area growth, reaching approximate parity with migration in 1964. Between 1964 and 1972 the share of natural increase in the area's growth consistently

AUTHOR'S NOTE: I am grateful to Karen Lam, Rebecca Morales, Keiko Nakajima, and Larry Tawa for their assistance.

increased while absolute population growth decreased. In 1972 this trend reached its nadir. Population growth in the Los Angeles urbanized area stopped (Pastier, 1970). However, after 1972 growth resumed, largely as a result of migration.

Of the largest migration, a significant proportion occurred because of the resumption of international immigration to the United States in the wake of liberalized immigration laws adopted by Congress in 1965 and fully effective in 1968 (Keely and Elwell, 1981). These laws facilitated immigration from Latin America and Asia. Thereafter, Los Angeles became a favored destination for immigrant Mexicans, Filipinos, Chinese, and Koreans. Additionally, refugees from Vietnam and Iran turned Los Angeles into the world's major center of these nationals outside of their own countries. In the decade of the 1970s, Los Angeles also received a net immigration of between 400 thousand and 1.1 million illegal immigrants, mostly Mexicans but also including Central Americans (Soja et al., 1983:219). In the decade 1970-80, the population of the five-county Los Angeles region increased by 1,525,031. Since, in this same period, the number of Hispanics and other persons—the chief immigrant populations—increased by 1,672,199, immigrants accounted for more than the entire population increase of the decade.

Aggregate growth statistics do not expose the intriguing patterns of population growth within this vast urbanized area. In general, growth of Los Angeles in this 20-year period followed the "hole in the doughnut" syndrome generally identified with older cities of the North and Central regions of the United States (Light, 1983:421-29). To examine this patterning in Los Angeles, one needs to distinguish the region's older core from its newer rings (see table 2.1). Here the awkwardly shaped City of Los Angeles represents the core, its encapsulating Los Angeles County the inner ring, and the four adjacent counties the outer ring. As table 2.1 indicates, between 1960 and 1980 growth of the City of Los Angeles was relatively least, county growth next, and growth of the outer ring strongest. In 1960, only 22 percent of the area's population resided in the outer ring; by 1980, 35 percent did so.

In terms of ethnic composition, two population movements predominated (SCAG, 1984b). First, the absolute number and percentage of non-Hispanic whites declined in the city and county. Thus, in the City of Los Angeles the percentage of whites declined 27.5 percent from 1960 to 1980, an absolute decline of 465,130 whites in the 20-year period. In the County of Los Angeles, whites declined by 29.6 percent, an absolute loss of 1,079,634. Since absolute growth of the white population in the four adjacent counties did not exceed 1,021,636, the region experienced a small loss of white population. Naturally, all the net emigration from Los Angeles City and County did not wind up in the region's outer ring. Some whites left the region altogether. Nonetheless, it is also apparent that in this 20-year period, the white population significantly relocated toward the region's perimeter.

Second, growth of Hispanic and Asian populations more than compensated the egress of whites from the city and county. The City of Los Angeles experienced a net gain of 657,197 Hispanic and other (mostly Asian and

TABLE 2.1
Population Changes—Southern California, 1960-1980
(in thousands)

| Location | 1960 | 1970 | 1980 | Index |
|---|---|---|---|---|
| Los Angeles | | | | |
| Population | 2,479 | 2,816 | 2,967 | 120 |
| % White | 71.9 | 59.0 | 44.4 | 73 |
| % Black | 13.8 | 17.7 | 21.5 | 185 |
| % Hispanic | 10.7 | 18.3 | 27.5 | 308 |
| % Other | 3.6 | 5.0 | 6.6 | 220 |
| Los Angeles County | | | | |
| Population | 6,042 | 7,036 | 7,478 | 124 |
| % White | 79.0 | 65.9 | 49.4 | 77 |
| % Black | 8.8 | 12.0 | 16.2 | 228 |
| % Hispanic | 9.9 | 18.1 | 28.8 | 360 |
| % Other | 2.3 | 3.9 | 5.5 | 296 |
| Adjacent Counties: | | | | |
| Orange, Ventura, San Bernardino, Riverside | | | | |
| Population | 1,713 | 2,936 | 4,020 | 235 |
| % White | 96.9 | 78.4 | 66.7 | 162 |
| % Black | 2.1 | 2.3 | 2.9 | 324 |
| % Hispanic | NA | 17.2 | 17.3 | |
| % Other | 1.0 | 2.1 | 13.1 | 3074 |
| 5-County Total | 7,755 | 9,973 | 11,498 | 148 |

SOURCE:  Los Angeles, City of (1977: 2); U.S. Census (1960, 1970, 1980).
NOTE:  Index = 1980/1960 x 100.

Pacific Islander) persons in the period 1960-80, and a net loss of 465,130 whites. In the County of Los Angeles, net egress of whites was 1,079,634 and net influx of Hispanics and others was 1,827,607. Since immigration produced most of the gain in Hispanics and others, population stability in the City and County of Los Angeles really depended on immigration without which population declines would have occurred.

An influx of Asians caused the expansion of "other" population in the outer ring, reflecting three sources. First, the socioeconomic mobility of native-born Asians permitted a higher proportion to reside in suburbs. Second, immigrant Koreans and Vietnamese settled in Orange County where the Vietnamese in particular obtained production jobs in electronics manufacturing. Third, the economic status of immigrant Asians was distinctly above that of the Los Angeles region's common citizen. Therefore, many immigrants moved directly to affluent suburbs, "leapfrogging" the low-income whites, blacks, and Hispanics who piled up in the region's core (Light, 1983:322-23, 352).

Growth of the black population in this period was chiefly the result of natural increase rather than migration. Black population expanded faster in the outer ring than in the County of Los Angeles, and faster in the county than in the city. Suburbanization of blacks reflected the increased proportion of middle-class households among them, a social mobility noted by theorists of the "new orthodoxy" on black mobility (Oliver and Glick, 1982). However, suburbanization of blacks "proceeded faster in Los Angeles than in any other metropolitan area" in the United States during the 1970s (Rabinovitz, 1977:67).

## Economic Basis of the Los Angeles Urbanized Area

The Los Angeles urbanized area entered the 1960s with an economic base specialized around aircraft production, automobile and related assembly, tourism, and film production and distribution. In the next 20 years this relatively specialized base broadened into what Soja and colleagues (1983:211) labelled a "diversified and decentralized industrial/financial metropolis." In general, economic growth between 1960 and 1980 depended on industrial diversification rather than augmentation of the region's share of a national industry. Thus, in 1980 "over half" of the United States's motion picture and television employment centered on Los Angeles (SPNB, 1979:1). Although still the cinematic capital of the United States and the world, Los Angeles had actually lost relative ground in this industry since, according to McWilliams (1946:340), Hollywood had produced 90 percent of the United States's motion pictures in 1945.

Similarly, although manufacturing was still the largest industry in 1980, its predominance peaked in 1959 when 38.1 percent of the five-county labor force worked in this industry (table 2.2). By 1980 only 30 percent of the region's labor force still worked in manufacturing. Pacing the relative decline of manufacturing were declines in other blue-collar industries, notably construction, transportation, public utilities, and mining. Conversely, white-collar and technical industries increased their share of the region's labor force in the 20-year period. Services, finance, insurance, real estate, and retail trade all grew faster than the five-county average (table 2.2). Of these industries, services registered the greatest absolute increase, adding 750,000 workers. These new service workers represented 40 percent of employment growth from 1960 to 1980, but only 24.8 percent of total employment in 1980. Conversely, the region's industrial mainstay, manufacturing, only added 21 percent of new employment, but still constituted 30 percent of total employment in 1980.

Industrial growth paralleled residential growth in the core and rings. (See table 2.3.) Between 1959 and 1980 for Los Angeles, Los Angeles County, and the outer ring, employment growth was greatest on the periphery and weakest in the core. Total employment increased 137 percent in the city, 174 percent in the county, and 469 percent in the outer ring. This concentric pattern persisted in every industry. However, absolute employment growth was 1,129,478 in Los Angeles County (excluding the city) and only 991,538 in the four adjacent counties. Although city data are incomplete, the same rank order of relative

TABLE 2.2
Employment by Industry—Five-County Los Angeles Region,
1959 and 1980 (in thousands)

| Industry | 1959 | | 1980 | | Index |
|---|---|---|---|---|---|
| | N | % | N | % | |
| Agriculture | 6 | 0.3 | 23 | 0.5 | 415 |
| Mining | 15 | 0.7 | 17 | 0.4 | 116 |
| Contract construction | 141 | 6.8 | 232 | 5.2 | 164 |
| Manufacturing | 794 | 38.1 | 1287 | 30.0 | 162 |
| Transport and public utilities | 134 | 6.4 | 252 | 5.7 | 189 |
| Wholesale trade | 152 | 7.3 | 317 | 7.1 | 209 |
| Retail trade | 370 | 17.8 | 824 | 18.5 | 223 |
| Finance, insurance, and real estate | 118 | 5.6 | 348 | 7.8 | 297 |
| Services | 354 | 17.0 | 1104 | 24.8 | 311 |
| Total employment | 2098 | 100 | 4445 | 100 | 211 |

SOURCE: U.S. Bureau of the Census (1962, 1983).
NOTE: Index = 1980/1959 x 100.

growth recurred there in manufacturing, wholesale trade, retail trade, and services as well (table 2.3). That is, the index of job growth in Los Angeles County exceeded the city index, and the index in the four adjacent counties exceeded that of Los Angeles County. Job growth proceeded faster in the outer than the inner ring, and faster there than in the core.

In view of numerous industrial plant closings in the core (Soja et al., 1983:217), the case of manufacturing, the region's industrial mainstay, is especially interesting. In the City of Los Angeles, growth of employment in manufacturing industries represented 13 percent of total employment growth. In the outer ring, on the other hand, manufacturing employment growth amounted to 25 percent of total employment growth. Thus, although manufacturing was declining in importance in the five-county region, it was declining more rapidly in the core than on the periphery (table 2.3).

Despite its industrial decline, the Los Angeles five-county region generated many new jobs. Comparing growth of population and employment in the core and rings, one finds that employment grew relatively faster than population in every sector, but employment's margin of superiority was greater on the periphery (table 2.3). In both the City and County of Los Angeles, population growth and job growth were in approximate balance, but on the periphery population growth was more than twice employment growth. Since the average worker supported 2.6 persons, the ratio of increase in population to increase in employment in the outer ring fell slightly short of the regional mean. In the city and county, on the other hand, the ratio of population increase to employment increase was less than half the regional mean, suggesting that the core generated more jobs for nonresidents than did the outer ring. Therefore, periphery residents more frequently required interzonal commuting to work places than did core residents. In 1960 and 1980 periphery residents more frequently commuted to their jobs by car.

TABLE 2.3
Growth of Employment by Industry—Los Angeles Area,
1959-1980 (in thousands)

| Industry | Los Angeles | | Los Angeles County | | 4 Adjacent Counties[a] | |
|---|---|---|---|---|---|---|
| | Jobs | Index | Jobs | Index | Jobs | Index |
| Agriculture | | | 7 | 263 | 11 | 802 |
| Mining | | | 0 | 98 | 3 | 146 |
| Contract construction | | | 28 | 125 | 63 | 306 |
| Manufacturing | 49 | 118 | 246 | 134 | 247 | 428 |
| Apparel | | | 41 | 189 | 5 | 406 |
| Transport and public utilities | | | 73 | 162 | 46 | 374 |
| Wholesale trade | 1[b] | 101 | 109 | 180 | 57 | 468 |
| Retail trade | 33[b] | 122 | 227 | 174 | 227 | 450 |
| Finance, insurance, and real estate | | | 148 | 240 | 82 | 790 |
| Services | 75[b] | 176 | 504 | 262 | 245 | 662 |
| Total employed | 381 | 137 | 1511 | 174 | 992 | 469 |
| Population | 488 | 120 | 1435 | 124 | 2307 | 235 |

SOURCE: U.S. Bureau of the Census (1962, 1983).
a. Orange, Riverside, San Bernardino, and Ventura.
b. 1958-77.

## Commerce in the Core

Although high-technology firms paced the economic growth of the periphery, within the stagnating urban core, Los Angeles' open shop "cheap labor" tradition gave rise to a primitive, labor-intensive economy (McWilliams, 1946:277). On its industrial side, this restructured economy emphasized low-wage, nonunion firms based on ready availability of distressed female and immigrant workers (Soja et al., 1983:211). The prototypical such industry was garment manufacturing. Investigators for California's garment industry strike force (Subber and Tchackalian, 1976:2) found 2,210 garment firms in Los Angeles County; only 10 percent were "subject to union-management agreement." Apparel manufacturing contributed 40,609 new jobs to the economy of Los Angeles County (table 2.3)—17 percent of all new employment in manufacturing. Because garment factories were small and hard to enumerate, these statistics probably underestimate true employment in this industry. Garment factories employed between 5 and 100 seamstresses under conditions that rampantly violated sanitary and wage provisions of the labor code (Bernstein, 1980; Merina, 1979).

The women worked long hours in unhealthful lofts and basements; received no vacation, pension, or health benefits; and earned less than the statutory minimum wage. Shunned by native-born workers, these low-wage production jobs were nonetheless attractive to Mexican women who had earned less in their homeland. Additionally, an estimated 65 percent of employees in garment sweatshops were undocumented (Subber and Tchackalian,

1976:2) and in no position to complain to police about long hours, low wages, or substandard working conditions.

Although most workers were Mexican women, Chinese from Hong Kong, and Taiwan, Koreans, and Latin Americans were their principal employers. Some of the entrepreneurial Asian and Latin immigrants had owned businesses in their countries of origin. Most had not. Among Koreans, the most entrepreneurial of the new immigrants, about one-fifth of business owners indicated they had owned a business in South Korea.

In general, immigrant entrepreneurs were well-educated former white-collar workers, especially professionals. In 1980, 66.5 percent of Americans age 25 or older were high school graduates and 16.2 percent were college graduates. However, among entrepreneurial Vietnamese, Israelis, Chinese, Indians, and Koreans, the education level was much higher. For example, Koreans reported 77.8 percent were high school graduates and 34.2 percent were college graduates; reports for Hong Kong Chinese were 80.3 and 42.7 percent, respectively (U.S. Bureau of the Census, 1984). Among nonentrepreneurial Mexican immigrants, on the other hand, only 21.3 percent were high school graduates and 3 percent were college graduates.

The labor-intensive economy of the region's core revived small retail and service business. Retail and service establishments and employees increased more rapidly on the periphery than in the core (table 2.4). In view of more rapid population growth on the periphery, this discrepancy was unsurprising. However, mean employment size of retail establishments also grew less rapidly in the core than on the periphery. The same phenomenon recurred more strikingly in service industries where mean employment per firm *declined* in the core "but continued" to increase on the periphery.

The revival of small commercial firms in the core reflected two processes. First, an influx of affluent but underemployed Asian immigrants increased the supply of entrepreneurs in the labor force, thus swelling the number of "mom and pop" stores. Second, the twin problems of crime and poverty discouraged chain stores from locating in the urban core. Mainstream corporations preferred to invest in suburban shopping malls. Therefore, in Los Angeles, as in other major American cities, small business flourished in slow-growth areas abandoned by big business. Although immigrant entrepreneurship caused wheels of commerce to grind where they would otherwise have been still, thus stimulating employment and earnings in central Los Angeles, foreign-born small business owners developed into a "middleman minority" buffering the whites above and blacks and Mexicans below (Bonacich and Modell, 1980:chap. 2). As elsewhere, middleman minorities in Los Angeles made easy targets for the socioeconomic frustrations of the disadvantaged people they serviced.

## Industrial Technology

According to the Security Pacific National Bank Research Department (1984:4), the Los Angeles region had the "greatest concentration of mathema-

TABLE 2.4
Retail and Service Establishments—Los Angeles Area,
1967-1977 (in thousands)

| Location | Retail Trade | | | Selected Services | | |
|---|---|---|---|---|---|---|
| | 1967 | 1977 | Index | 1967 | 1977 | Index |
| Los Angeles | | | | | | |
| Establishments | 25 | 25 | 102 | 26 | 41 | 159 |
| Employees | 157 | 183 | 117 | 134 | 174 | 129 |
| Employees/establishment | 6.4 | 7.3 | 114 | 5.2 | 4.3 | 83 |
| Los Angeles County | | | | | | |
| Establishments | 57 | 59 | 103 | 53 | 84 | 158 |
| Employees | 368 | 452 | 123 | 234 | 358 | 153 |
| Employees/establishment | 6.4 | 7.7 | 120 | 4.4 | 4.3 | 98 |
| 4 Adjacent Counties[a] | | | | | | |
| Establishments | 24 | 30 | 123 | 16 | 35 | 218 |
| Employees | 126 | 232 | 185 | 47 | 115 | 243 |
| Employees/establishment | 5.2 | 7.8 | 150 | 2.9 | 3.2 | 110 |

SOURCE: U.S. Bureau of the Census (1972, 1977, 1983).
NOTE: Index = 1977/1967 x 100.
a. See Table 2.3.

ticians, scientists, engineers, skilled technicians, and high technology industry in the United States." Soja, Morales, and Wolff (1983:212) also identified the Los Angeles region's science and technology sector as "the largest such employment concentration in the U.S., perhaps in the world." Admittedly, this concentration lacked the intellectual luster of San Francisco and Boston which housed more of the nation's research scientists (Dye, 1984). However, in terms of applied science and technology, the Los Angeles region contained two of the nation's four leading centers: Orange County and Los Angeles County. The Los Angeles concentration was larger than that of either Boston or San Francisco in 1980, and the Orange County concentration, although smallest of the four, was growing faster than any of the other three.

In the early 1980s, electronics manufacturing and research were rapidly expanding industries. Examining the growth of seven technology-related industrial sectors between 1972 and 1979, Soja and colleagues (1983:212) reported 50 percent growth and the addition of 110 thousand jobs. By 1979, these seven industries contributed 332,450 jobs to the region's economy: 72 percent in Los Angeles County, 24 percent in Orange County, the balance elsewhere in the periphery. Leading this industrial growth were electronics, especially components and accessories, aircraft and aircraft parts, and communications equipment. The aerospace/electronics industry alone accounted for more than one-fourth of the Los Angeles region's value added by manufacturing and a third of its total manufacturing employment (SPNB, Research Department, 1981:7).

High-technology employment in Orange County centered around the immense land development operated by the Irvine Company, a private real estate business (O'Dell, 1984). In 1984, Orange County housed 1,432 computer companies and provided 98,800 high-technology jobs, employing 11 percent of the county's nonagricultural labor force (Applegate, 1984). Orange County also housed 552 electronics companies, 56 percent of the region's total (Applegate, 1984:6). Under the blanket of a public/private development plan masterminded by the Irvine Company, Orange County expanded its educational facilities to provide the skilled labor and research personnel needed to staff a growing high-technology sector. Unlike Los Angeles County, chiefly dependent on aerospace contracts, Orange County stressed medical technology—optical, dental, and surgical equipment, pharmaceuticals, hospital supplies, and technical instrumentation. Additionally, Orange County's "bio-med" industry contained several firms owning valuable patents in genetic engineering technology.

Unlike other urbanized regions, Los Angeles' manufacturing sector was oriented toward aerospace/electronics rather than consumer products (SPNB, Research Department, 1981:7). Directly or indirectly, all the aerospace industries and a substantial proportion of Los Angeles' electronics industries depended on military contracts, a dependence labor union officials blamed for the hawkish political views of their rank and file (*Los Angeles Times*, 1984). Soja, Morales, and Wolff (1983:214) concluded that the Los Angeles region, with 3 percent of the U.S. population, received about 10 percent of all U.S. defense contracts in the late 1970s. Among defense contractors operating in Los Angeles were Hughes Aircraft, Northrup, McDonnell Douglas, Canon Electric, Lockheed Aircraft, Control Data, and TRW. These and other region contractors have designed and assembled many weapons of nuclear war including the B-1 bomber, MX and Trident missiles. The region's military industries also produce support aircraft, warships, and tactical missiles.

Because of dependence on Department of Defense contracts, employment in the aerospace/electronics/technology industries has been volatile, and inversely related to prospects for détente with the Soviet Union (Chrysler and Doan, 1985). Between 1967 and 1976 aerospace/electronics employment in the Los Angeles region endured stagnation and even decline in the wake of popular antimilitarism and the Nixon Administration's policy of détente. In 1976, unemployment among engineers was high, and many looked for nonmilitary employment. However, after 1977 military spending resurged and produced annual employment gains of 25 thousand in aerospace firms (SPNB, Research Department, 1981:7). Rearmament initiatives of the first Reagan Administration (1980-84) accelerated the boom in defense industries, thus converting a surplus of engineers into an acute shortage which encouraged immigration of scientists, engineers, and technicians from foreign countries.

## Transportation Issues and Planning

The Los Angeles urbanized region entered the 1960s with a transportation system already heavily committed to private automobiles (Findley, 1958:106). Three decades of federally supported highway construction had permitted creation of the world's most extensive system of motor highways (Light, 1983:214-15; Nelson, 1983:275-86). Well adapted to the region's mild climate, the freeway system was popular with people who had, in any event, little choice but to use it in view of the decrepit state of the Pacific Electric Line, the area's once peerless street car line. When the Pacific Electric Line ran its last car in 1963, only railroad sentimentalists mourned its passage. In practical terms, residents and planners alike agreed that the private automobile offered comfort, convenience, and speed unobtainable on any public transportation system.

Two decades of lopsided dependence on private automobiles shook but failed to dislodge this faith. (See table 2.5.) Between 1961 and 1979, the number of motor vehicles in the five-county region grew from 4,398,652 to 8,881,230. In the same 18-year period, highway mileage expanded from 40,112 to 44,322.

Because vehicles increased more than did road surface, traffic congestion increased. Aggravating this deterioration, the percentage of workers driving to work increased in the center and rings (table 2.6). Traffic congestion expanded the duration of rush hours, increased idling time on freeways, lengthened the time required for journeys to work, and increased the money cost of commuting. Nonetheless, commuters preferred cars to buses. In 1978, the California Department of Highways declared that Los Angeles freeways had reached "saturation" at rush hours. The agency predicted a "freeway traffic crisis" unless relief measures were introduced (Herbert, 1978), but the public refused to accept designation of even one traffic lane in five for exclusive use of carpool vehicles. Five years later, Los Angeles residents still bewailed traffic congestion (Waters, 1984), but public officials had introduced no measures to remedy the problem whose dimensions, it must be added, fell short of the chaos predicted earlier.

Traffic congestion had different effects in the core and the ring. Los Angeles County entered the 1960s with an appreciable congestion problem (Herbert, 1972) which slowly worsened in the next 20 years. In contrast, the urbanized ring entered the 1960s with no traffic congestion and had only begun, by the end of the 1970s, to approach the congestion level attained in the core 20 years earlier. In Los Angeles County motor vehicles per mile of road grew by 69 vehicles from 1960/61 to 1978/79 (table 2.5), a 138 percent growth in vehicular density on county roads. On the other hand, in the four adjacent counties, vehicles per mile of road increased by 94. Although this increase was 301 percent of the 1961 number, more than twice the rate of increase in the core, absolute vehicular density on the roads of the ring was still only 56 percent of that in the core as late as 1979. Central city business interests believed that traffic congestion discouraged new business enterprises from opening in central Los Angeles, thus reducing economic growth in the core.

Los Angeles's automobile-dependent transportation system developed with the fervent support of national business interests which profited from road construction, suburban housing, land development, petroleum, tires, and automobile sales. In 1930, General Motors (GM) Corporation, Firestone Tire Company, Standard Oil of California, and allied interests formed the

TABLE 2.5
Motor Vehicles Registered and Miles of Roads—
Los Angeles Five-County Region, 1961-79

| Location | 1961 | 1967 | 1979 | Index |
|---|---|---|---|---|
| Los Angeles County | | | | |
| Vehicles registered | 3,404,147 | 4,268,450 | 5,256,029 | 140 |
| Miles of roads[a] | 18,669 | 18,365 | 20,944 | 112 |
| Vehicles/miles of road | 182 | 232 | 251 | 138 |
| Residents/vehicle | 1.8 | 1.6 | 1.4 | 78 |
| 4 Adjacent Counties | | | | |
| Vehicles registered | 994,505 | 1,666,210 | 3,625,201 | 328 |
| Miles of roads[a] | 21,443 | 17,903 | 23,378 | 109 |
| Vehicles/miles of road | 46 | 93 | 140 | 301 |
| Residents/vehicle | 1.7 | 1.8 | 1.2 | 71 |

SOURCE: County Supervisor's Association of California, 1963, 1969, 1981.
a. 1960, 1967, 1978.

TABLE 2.6
Private Vehicles Used in Journey to Work—Los Angeles Area,
1960 and 1980 (in thousands)

| | 1960 | | 1980 | |
|---|---|---|---|---|
| Location | Vehicles | Percentage of Workers | Vehicles | Percentage of Workers |
| Los Angeles | 701 | 71 | 1095 | 80 |
| Los Angeles County | 1795 | 77 | 2884 | 86 |
| 4 adjacent counties | 479 | 80 | 1640 | 91 |

SOURCE: U.S. Bureau of the Census (1963, 1983).

National City Lines (NCL), a holding company, to acquire shares in municipal trolley lines in American cities. NCL compelled electric trolley managements to switch to diesel-powered buses, thus enriching the conspirators who built the frames and motors, manufactured the tires, and provided the fuel for diesel buses. In 1949 a federal jury in Chicago convicted Firestone, GM, and Standard Oil of California of criminal conspiracy to acquire, motorize, and resell electric streetcar lines. By then, the conspirators had already replaced electric streetcar systems with GM buses in 45 cities including New York, Philadelphia, Baltimore, St. Louis, Oakland, Salt Lake City, and Los Angeles (Light, 1983:210; Snell, 1974).

Destruction of public transportation in Los Angeles illustrates the conspirators' modus operandi. In 1938, General Motors and Standard Oil organized Pacific City Lines, an NCL affiliate, to acquire, motorize, and resell electric railroads in California. Pacific City Lines acquired control of Los Angeles' Pacific Electric Railway, then the world's largest trolley system. Pacific Electric served 56 cities and eight million passengers yearly in the Los Angeles region. Pacific City Lines replaced trolley service with diesel buses and pulled up the railway tracks. Loss of tracked right-of-ways compelled buses to compete on unequal terms with automobiles so buses proved less effective in traffic than did trolleys (Snell, 1974:38). Therefore, dieselization reduced the attractiveness of mass transit, and accelerated a spiral of destruction which ultimately destroyed the trolley system, reducing the people of Los Angeles to dependence on private automobiles and freeways.

After the destruction of Pacific Electric, automotive interests long opposed local efforts to revive a mass transit system in supplementation of the private automobile. The last antitransit intervention of the "highway lobby" occurred in 1970. The lobby consisted of automobile and truck manufacturing firms, construction employers and labor unions, oil companies, the Teamsters' Union, officials of federal and state highway departments, the American Automobile Association, and banks. These interests benefited financially from federal and state highway construction (Light, 1983:211). In 1970 the highway lobby raised $333,445 to defeat a California ballot initiative (Proposition 18) which would have permitted diversion of state gasoline tax revenues from highway construction and maintenance to rapid transit and air pollution reduction (Rosenblatt, 1970). Proponents of diversion had raised only $22,721 so, in the electoral campaign preceding the voting, the highway lobby had hegemonic access to advertising and promotional media. This advantage defeated the initiative with 54.1 percent of the voters opposing and 45.9 percent supporting the transit measure.

However, in 1974, just four years after the anti-Proposition 18 initiative, Los Angeles business interests supported a milder transit measure (Proposition 5), and oil companies did not oppose it (Whitt, 1982:132, 160). California voters approved Proposition 5. However, in Los Angeles County, voters turned down Proposition A, a rail and bus measure, *even though* the same business interests that had supported Proposition 5 on the California ballot had also raised $563,000 in support of County Proposition A (Whitt, 1982:T2). The outcome suggests that Los Angeles voters were less sympathetic to mass transit than California voters and that even support of local business interests could not overcome voter resistance.

According to Whitt (1982:198, 209) Los Angeles's "financial and administrative capitalists" decided that automobile dependency was undermining the center to the benefit of the periphery. Accordingly, Whitt (1982:131, 197) maintains, they abandoned their historic prohighway political position in 1974—but failed to bring the voters into compliance with their protransit position. Whitt's view lacks historical depth as Findley (1958:69) earlier showed that central city business interests already supported mass transit in the 1920s because of the harmful effects of automobile transport on central city commerce and property values.

In fact, quite early in the 1920s, Los Angeles business spokesmen advocated subway construction "as a means of halting the decentralization of business in Los Angeles" (Findley, 1958:86). Their plan failed, but the same alignment surfaced in the campaign for Metro Rail, a $3.3 billion subway proposal under consideration in 1983-84. The proposed subway would run 18.6 miles from downtown Los Angeles, along the Wilshire Boulevard corridor, a high-rise office sector, toward the San Fernando Valley, a residential suburb populated by office workers. Supporters claimed the subway would reduce traffic congestion, fuel consumption, and air pollution; control parking costs; serve presently neglected populations of aged, disabled, and young people; and help reshape the Los Angeles region into a more cohesive form (Trombley, 1984a). Opponents found little in the experience of other cities' recently constructed subway systems to warrant confidence that the predicted benefits would materialize on completion. Opponents also questioned the subway's cost, ridership projections, route, effect on neighborhoods, and, in view of the terrain's earthquake faults, even its engineering feasibility (Trombley, 1984a, b, c, e).

However, these complex arguments concerned only the public interest. Central city businesses frankly supported Metro Rail on grounds of economic self-interest (Trombley, 1984f). One context for candor was the controversy over reductions in congestion to be expected following Metro Rail's completion. A research report had concluded that, without relief, Los Angeles County's automobile-dependent transportation system would experience "increasing difficulty" in accommodating commuter traffic on the freeways, thus encouraging business tenants to relocate outside Los Angeles County (Trombley, 1984b). Some Metro Rail supporters privately agreed that the subway would, in fact, do little to relieve road congestion. But, they favored it anyway on the grounds that a subway line would help sustain the downtown high-rise building boom in Los Angeles and would "encourage economic development at points along the [subway] route" (Trombley, 1984f). " 'We're simply trying to protect the investment that's here' said Christopher Stewart, President of the Central City Association, an organization of 250 downtown companies" (Trombley, 1984a:6). The *Los Angeles Times* agreed with Stewart's assessment:

Whatever else the subway may or may not do, it will enable many more workers and shoppers to enter and leave the downtown business district, so the building boom of recent years can continue.

Hence, [Metro Rail] is considered indispensable by the corporate executives who influence many of the important planning decisions in Los Angeles (Trombley, 1984e:6).

Central city interests supporting Metro Rail "spread largesse" among a dozen U.S. representatives and senators whose support was critical in obtaining promises of $2 billion in federal subsidies for the Los Angeles subway (Trombley, 1984d). Congressional beneficiaries of this largesse threw their influence behind the proposal—even at the cost of public incompatibility with earlier statements of political philosophy (Dreier, 1984). However, as elsewhere in the United States (Yago, 1983:586), the Los Angeles subway proposal encountered political opposition among local taxpayers who would be required to shoulder 14 percent of the cost from sales tax increases. In a general context of taxpayer resistance to any and all government cost increases, the political future of Metro Rail was intensely problematic in 1984.

Mirroring this uncertainty was the failure of Proposition A to obtain voter approval in Orange County in June 1984 (Bunting, 1984). The Orange County Transportation Commission (1984) had requested a 1 percent sales tax increase to support its 15-year "transportation investment plan." Claiming that highway construction had not kept pace with population and business growth in Orange County, the transportation commission (1984:1) called for "improvements to our existing freeways, construction of new highways, improvements to arterials . . ., construction of a rapid transit system, improvements to the bus system, and a greater emphasis on ridesharing." The transportation commission promised that improvements would promote reduced highway congestion, fuel consumption, vehicle usage, and maintenance costs; and improved transit access to employment centers, business climate, and air quality. On the other hand, if the tax increase were not approved, the transportation commission warned, the county's streets and highways would "turn into virtual parking lots by the year 2000." The commission cautioned, "For every ten cars waiting in front of you today, there will be fifteen in the future." Orange County's largest landowner, the Irvine Corporation donated $100,000 toward passage of Proposition A, the single largest political contribution in the history of Orange County (Bunting, 1984). Nonetheless, in June 1984, Orange County voters refused to authorize a transportation-related tax increase. In 1985 a survey of 2,241 randomly selected adult respondents in Los Angeles County disclosed that more than 66 percent favored "construction of a mass transit system" but only 40 percent supported higher taxes to pay for it (Roderick, 1985a).

## Los Angeles in the National and World Economies

In 1981, the Los Angeles urbanized region contained the largest port complex on the Pacific Coast and the third largest in the United States. In dollar volume of international trade, the Los Angeles area was exceeded only by New York City (SPNB Research Department, 1984:4). Because Los Angeles is the second largest city (ranking behind Tokyo) of the Pacific rim trading sphere, and a U.S. gateway for trade with the Far East, economic development of the Pacific Rim especially benefited Los Angeles. In 1976, two-thirds of Los Angeles's trade was with Asia. Japan, Hong Kong, South Korea, and Taiwan together accounted for nearly half the trade that passed through Los Angeles (SPNB Research Department, 1977:20). Combined export and imports passing through the Los Angeles Customs District increased from a valuation of $6.2 billion in 1972 to $38.3 billion in 1982, a 611 percent increase. This increase considerably exceeded the inflation of the dollar in the United States as well as the increases recorded in New York and Chicago in this decade (SPNB Research Department, 1984:9).

Southern California's exports passed through the ports of Los Angeles and San Diego, registering exports of $16.2 billion in 1980. Of total exports, 93 percent passed through Los Angeles, the remainder through San Diego (SPNB Research Department, 1977:20). In 1980, transportation equipment was Southern California's leading export in terms of dollar volume as well as percentage growth; aircraft and parts accounted for 86 percent of that export. Military aircraft represented about half the equipment exported. The second largest export was machinery and mechanical equipment: computers, office machines, and internal combustion engines represented 48 percent of exports in this category. Electrical machinery and equipment was the third-ranking export. Electronic tubes and transistors were the outstanding products, accounting for 47 percent of exports in this category. Cotton, grains, fruits, and nuts were important agricultural exports of the San Joaquin Valley.

Consumer products represented the largest category of imports entering the Los Angeles Customs District. In 1980 Southern California customs districts (Los Angeles and San Diego) registered $21 billion of imported goods entering the United States (SPNB, 1981b). Motor vehicles, consumer electronics, and wearing apparel alone constituted two-thirds of consumer imports in 1976 (SPNB Research Department, 1977:21). Crude oil was the next most important import at about 20 percent of total 1976 imports. Prior to the 1973 oil embargo, crude oil exports had represented only about 5 percent of regional imports. After the embargo, oil imports drastically increased, exacerbating the region's negative balance of trade. By 1980 petroleum imports were increasing much more slowly than they had in the mid-1970s, thus easing but not eliminating the negative balance of trade through the Los Angeles Customs District.

Waterborne commerce made a significant contribution to the economy of the Los Angeles five-county region; one study concluded that it "directly generated " 121,476 jobs in the region (SPNB Research Department, 1977:22). If so, these jobs would represent about 5 percent of all job growth in the region from 1960-80. Of course, only some of these jobs depended on production or consumption in Southern California. Most jobs developed because of Los Angeles's role as entrepôt of international commerce. In 1976, about 7 percent of total U.S. foreign trade passed through Los Angeles though only 3.5 percent of the U.S. population resided in the region. On this rough basis, it appears that about half the jobs attributable to import-export commerce arose from local consumption or production, the other half from transshipment of goods produced and consumed outside the region.

Nonetheless, the region's imports and exports importantly reflected specific industrial and infrastructural characteristics of the Los Angeles region itself. On the export side, Los Angeles shipped abroad some of the military and civilian aircraft, electronic equipment, computers, machinery, and movies it produced. On the import side, Los Angeles became an important customer for foreign petroleum and fuel-efficient foreign automobiles. In 1980 foreign automobiles represented 51 percent of those sold in Los Angeles—more than twice the U.S. average; 37 percent were made in

Japan. Similarly, the region's big imports of petroleum ($2.7 billion in 1980) reflected the high rate of oil consumption in this automobile-dependent but petroleum-producing region. In fact, the principal response to higher petroleum prices during the 1970s was a massive consumer switch to fuel-efficient foreign cars. This response let area residents conserve gasoline without reducing their driving or centralizing their scattered residences in relation to their work places. Coupled with opening of the Alaska oil production fields, petroleum conservation accounted for the big reduction of petroleum imports achieved during the late 1970s (SPNB, 1981b:2).

## Political and Cultural Role of Los Angeles in the World System

Los Angeles's cultural and political importance probably exceed its economic importance on the world stage. First, Los Angeles in 1980 remained the world capital of popular culture. As the home of the late Walt Disney and headquarters of Walt Disney Enterprises, Los Angeles deluged the world with the animated cartoon characters Disney invented. Fully using the latest communications technology of this century, Walt Disney achieved an astounding penetration of world cultures. As a result, many more people in the world can identify Mickey Mouse than can identify Hamlet. Additionally, through the Hollywood film industry, Los Angeles has broadcast its version of reality all over the globe with immense consequences. Throughout the world virtually everyone has viewed Hollywood western, adventure, police, horror, or romantic films. In tiny villages throughout the Third World, semiliterate peasants learn about the outside world from Hollywood movies. Indeed, to a substantial extent, the revolution of rising expectations arose precisely from the collision of Hollywood fantasies with the harsh realities of Third World life.

Second, Los Angeles became politically important to the world in 1980 because President Ronald Reagan is its political product. As a result, the policies, personnel, intellectual outlook, and origins of Reagan's administration reflect the metropolitan region from which he emerged and without which his administration would not have been possible. Before his election as U.S. president in 1980, Reagan passed a political apprenticeship of 16 years, half as governor of California. In his 16-year apprenticeship, Reagan depended for support on Southern California business circles, absorbing and projecting the ideologies and economic needs current in them. Although one cannot reduce the Reagan administration's policies of arms race and Cold War to Reagan's political apprenticeship, like all political leaders, Reagan emerged from a social context. This context was a metropolitan region heavily dependent on weapons manufacture and the Cold War ideology that legitimized that manufacture.

Because Hollywood was located in their region, Southern California business magnates had convenient access to right-wing movie stars like Reagan. In 1966, a group of conservative Los Angeles tycoons decided Reagan was the man to communicate their political message to the masses (Cannon, 1982:103; Dugger, 1983:26). Reagan's utility to them depended on his earlier movie career, a distinctly Southern California industry. In his Hollywood career, Reagan had produced a series of B-movies. Fortunately for him, these movies had awarded Reagan the kind of public persona he could exploit in his political career. Unlike Lon Chaney, Lou Costello, or James Cagney, Ronald Reagan portrayed tough but kind, honest, uplifting, and patriotic characters, roles which readily supported his later political career. When his movie career faltered, Reagan went to work for General Electric Corporation. As host for eight years of the "General Electric Theater," originating in Hollywood, Reagan studied and delivered to millions of American homes the political and social outlook of a conservative employer and arms contractor (Cannon, 1982:93-94). In his own estimation, the General Electric years gave "the Great Communicator" the public-speaking experience he later used so effectively in his televised debates with Jimmy Carter.

Reagan's ties to Southern California business were obvious in his appointments of Californians Caspar Weinberger and George Shultz to his cabinet. Both appointees had been senior officers of the Bechtel Corporation, a Los Angeles-based construction firm specializing in nuclear power plants (Dugger, 1983:144). Even in respect to their intellectual origins, Reagan's controversial economic policies connected to Los Angeles. The "supply side" economic theory on which Reagan based his 1982 tax cut legislation emerged from an obscure economist employed at the University of Southern California, the largest private university in Los Angeles.

## The Office Construction Boom

Prior to 1960, downtown Los Angeles contained few office buildings (Nelson, 1983:194). Their tenants were local corporations whose operational facilities were located elsewhere in the region (Haas and Heskin, 1981:547). Indeed, until 1959, city ordinances prohibited buildings over 13 stories in the interest of earthquake safety. Around Los Angeles's central business district, small for a city of its size, there spread 70 square miles of low-density, heavily blighted inner city (Parson, 1982:399). The future of this area was uncertain. One possibility was to expand public housing to improve the accommodation of the one-third of the Los Angeles population who resided in slums. As Parson (1982:399-405) has shown, this possibility had political momentum in the wake of the Housing Act of 1949. However, business and real estate interests in Los Angeles launched a massive political counterattack in the 1950s.

This counterattack diverted redevelopment programs from public housing toward urban renewal. Instead of building public housing, municipal redevelopment agencies began to acquire sites, raze existing structures, and sell the cleared property to private developers. For the most part, developers erected facilities to serve business and upper-income consumers on sites that had previously housed low-income residents. For example, in 1961, the Community Redevelopment Agency of Los Angeles (CRALA) began to acquire land on Bunker Hill in central Los Angeles. The CRALA cleared Bunker Hill of 7,310 low-income residences and an undetermined number of small businesses, making space for an opera house, corporate headquarters, and luxury housing (Parson, 1982:403). The Bunker Hill project established the character of urban development in the next quarter century. By 1984, 60 municipalities in Los Angeles county had used redevelopment authorities to clear 136 square miles of land at a cost of $3 billion in tax funds. On this redeveloped land were erected 31 shopping malls and centers, at least 20 high-rise office buildings, 20 apartment complexes (of which two-thirds were for high-income households), 17 industrial parks, 8 hotels, 7 banks, 6 city halls, 2 hospitals, a brewery, a golf course, a gambling hall, and "a smattering" of parks, museums, and community centers (Clifford, 1984). Instances of corruption surfaced.[1]

In 1984, the City of Los Angeles announced an agreement between the Community Redevelopment Agency and Maguire Thomas Partners, a real estate developer. This agreement authorized construction of three new office buildings with 3.2 million square feet of office space (Kaplan, 1984). The Library Square Project boasted one cylindrical tower 70 stories high, and a second 65 stories high. The taller structure will be 8 stories higher than the 62-story First Interstate Bank Building, previously the largest in Los Angeles. Its shorter twin will be three stories higher. Summing up 19 years of redevelopment, the *Los Angeles Times* observed that when redevelopment began in 1961 downtown land sold for "less than $10 a square foot." Redevelopment had changed that shabby picture. "By 1980, the [downtown] landscape was bristling with fancy hotels and shimmering office towers. Thanks to redevelopment, the downtown real estate market was reborn" (Clifford, 1984:15).

Because of its favorable location, Los Angeles was well positioned to benefit from increased trade across the Pacific Ocean. However, location was not the only nor even the principal commercial advantage of Los Angeles. After all, San Francisco, San Diego, and Seattle had equally attractive locations, but from 1960-80 these cities did not attract foreign and out-of-state capital to the same extent as did Los Angeles. Los Angeles's distinguishing attraction was its probusiness redevelopment strategy which made office space available cheaply. Public subsidy encouraged out-of-state domestic corporations and foreign corporations to locate administrative facilities and headquarters in Los Angeles. Beginning in the early 1960s, the influx of corporate offices supported an office construction boom that, after a contraction in 1974-79, continued until the mid-1980s when rumors of oversupply again surfaced. "Between 1960 and 1970, Los Angeles moved from ninth to fifth on the list of corporate headquarter cities" (Haas and Heskin, 1981:547).

In the course of this building boom, the region's once paltry skyline developed five imposing clusters of high-rise office buildings: downtown Los Angeles, Century City, Beverly Hills, the airport area, and Westwood. Over 30 million square feet of office space was constructed between 1972 and 1982 (Soja et al., 1983:224). Some of this office construction arose from relocation of major American corporations, notably Atlantic Richfield, which moved its corporate headquarters to Los Angeles. By 1983 downtown Los Angeles alone contained the headquarters of 13 major corporations with another 36 elsewhere in the region (Soja et al., 1983:225). On the other hand, all this growth did not prevent the decline of the central Los Angeles business district relative to peripheral locations in the region. Between 1960 and 1983, the number of the United States's 500 biggest corporations with headquarters in Los Angeles remained unchanged at 11. In the four adjacent counties, the number of *Fortune* 500 (*Fortune*, 1961; 1983) firms grew from three to seven so the region's entire gain in headquarters of the biggest domestic firms occured on the periphery.

Foreign capital also discovered Los Angeles (Faris, 1977). An appreciable share of the office construction boom occurred because foreign capital decided that Los Angeles offered an outstanding location for financial and control facilities (Haas and Heskin, 1981:548). International banks and accounting firms boosted the demand for office space in Los Angeles high-rises. Additionally, foreign corporations purchased real estate in downtown Los Angeles, thus acquiring equity shares of the building boom (Turpin, 1979). Soja, Morales, and Wolff (1983:225) estimate that foreign corporations owned 21 of the 75 most valuable parcels of real property on the expanding edge of downtown Los Angeles.

## Air Pollution

Air pollution became a serious concern in Los Angeles County during World War II when clouds of eye-stinging smog became visible in hot summer months (Zeman, 1944). Causes of smog were incompletely understood at the time, but early analysis showed its behavior and composition were different from the air pollution which afflicted European cities and U.S. cities in the eastern region. In 1947, Los Angeles County authorized the formation of an Air Pollution Control District (LAAPCD) responsible for controlling smoke and sulphur dioxide and investigating the chemical properties of Los Angeles's unique air pollution. In the early 1950s, LAAPCD researchers discovered that hydrocarbons and oxides of nitrogen react photochemically in the presence of sunlight to form ozone. Since the rate of conversion depends on the intensity of sunlight, Los Angeles suffers its most serious ozone problem during the summer. In addition, researchers found that certain hydrocarbons (olefins and aromatics) together with oxides of nitrogen were primarily responsible for the eye irritation, pulmonary impairment, decay of rubber, and vegetation damage caused by Los Angeles smog (Nelson, 1983:66-69; SCAQMD, 1982:8-9).

Because of its volume and toxicity, ozone was the South Coast Air Basin's worst atmospheric problem (SCAQMD, 1982:3). Los Angeles had the highest ozone concentrations in the United States. Additionally, ozone concentrations in the basin rose high enough to damage plants and materials (especially rubber and paint) frequently, injure the health of sensitive people often, and harm healthy individuals occasionally. However, ozone was not the only toxic substance in Los Angeles smog. In addition to ozone, a colorless but odorous gas, Los Angeles smog also contained carbon monoxide, sulfur dioxide, nitrogen dioxide, lead, sulfate, nitrate, hydrogen sulfide, vinyl chloride, and suspended particulates—a mixture of natural and synthetic materials (SCAQMD, 1983a:6-23). Whether in gross or small particulates, this pollutant is responsible for most pollution-induced reduction of visibility, but it accounts for only a fraction of the total health and property damage attributable to smog, most of which is invisible (SCAQMD, 1982:21).

The regional successor to the LAAPCD, the South Coast Air Quality Management District, distinguished stationary and mobile sources of polluting chemicals. Stationary sources include petroleum refining and marketing, chemical and metallurgical industries, painting, and miscellaneous sources such as forest fires. Mobile sources include automobiles, trucks, motorcycles, aircraft, ships, and portable utility equipment. (See table 2.7.) During 1979, an average of about 11.3 thousand tons per day of these major pollutants was deposited in the air basin. Carbon monoxide accounted for about 68 percent of the total, followed by organic gases (14 percent), oxides of nitrogen (11 percent), total suspended particulates (5 percent), and oxides of sulfur (2 percent). Mobile sources were higher in carbon monoxide and oxides of nitrogen emissions. Stationary sources were higher in sulfur dioxide emissions. Both sources emitted approximately equal volumes of reactive hydrocarbons. Since reactive hydrocarbons and nitrogen oxides are precursors of ozone, both sources shared responsibility for this toxic pollutant, but mobile sources produced more ozone than did stationary sources (SCAQMD, 1982:10).

By inhibiting normal dispersion of atmospheric pollutants, the topography and climate of the Los Angeles region complicated the region's air pollution problem (Keith, 1980:1, 69). Bounded on the west by the Pacific Ocean, on the north and east by the San Gabriel, San Bernardino, and the San Jacinto mountains, the South Coast Air Basin is especially susceptible to the formation of atmospheric inversion layers (Nelson, 1983:36-38). An inversion layer is a meteorological condition in which a layer of warm air lies on top of a layer of cooler air. Inversion layers reverse the normal inverse relationship between temperature and altitude. Inversions promote smog formation because they hinder vertical mixing in the atmosphere, thus inhibiting the dispersion of pollutant emissions (SCAQMD, 1982:12). When inversion conditions exist, the layer of warm air acts like a huge plastic sheet draped over the urbanized area: trapped under the sheet, pollutants do not disperse, and continued emissions intensify the density of pollution. Inversion layers form naturally on two of three days in the South Coast Air Basin. Although inversion layers dissipate more rapidly in the summer than in the winter,

TABLE 2.7
Emission of Major Chemical Pollutants in Annual Average
Tons per Day—Los Angeles Air Basin, 1979

| Pollutant | Stationary Source | | Mobile Source | | Total Tons/Day |
|---|---|---|---|---|---|
| | Tons/Day | Percentage of Total | Tons/Day | Percentage of Total | |
| Carbon monoxide | 590 | 8 | 7060 | 92 | 7650 |
| Reactive organic gases[a] | 680 | 44 | 850 | 56 | 1530 |
| Nitrogen oxides | 410 | 33 | 830 | 67 | 1240 |
| Sulfur oxides | 200 | 74 | 70 | 26 | 270 |
| Total suspended particulate | 520 | 85 | 90 | 15 | 610 |

SOURCE: South Coast Air Quality Management District (1982: Table 1).
a. All organic gases except methane, methylene chloride, methyl chloroform, and a number of freon-type gases.

photochemical smog is actually thicker in the summer because the intense summer sunshine provides the energy to convert primary chemicals into ozone (SCAQMD, 1982:13).

The seriousness of any local air pollution problem depends jointly on emissions and meteorological conditions. With its automobile-dependent transportation system, bright sunshine, and frequent inversion layers, the Los Angeles five-county region coupled high emissions with unfavorable meteorological conditions, a formula for air pollution. Federal research on all U.S. counties in 1980 revealed that Los Angeles had "the worst ozone problem in the U.S., one of the worst nitrogen dioxide problems, and a significant carbon monoxide problem" (SCAQMD, 1982:28). In terms of number of days exceeding the federal ozone standard, Riverside, San Bernardino, and Los Angeles counties placed first, second, and third among counties in the United States with 132, 130, and 129 days exceedence, respectively. Strikingly, the fourth ranking county was Fairfield in Connecticut in which 36 days excedence occurred in 1980 (SCAQMD, 1982:75). In respect to annual mean of atmospheric nitrogen dioxide, Los Angeles and Orange counties placed first and fifth among U.S. counties with .070 and .051 parts per million, respectively. On the other hand, no county in the Los Angeles Air Basin had any days exceedence of the federal standard for sulfur dioxide, a product of stationary polluters. In terms also of total suspended particulate, the highest county in the Los Angles Air Basin failed to place in the top five U.S. counties, and measured effluent was also relatively low.

Although dismal, this record climaxed two decades of control efforts in the course of which Los Angeles developed and enforced antipollution laws much stricter than federal or California laws. Moreover, in this entire period, the problem of emission grew persistently worse as the number of vehicles in the five-county region increased. Between 1970 and 1980, for example, the population of motor vehicles in the Los Angeles urbanized area increased 31 percent, thus counteracting decreases achieved in per capita emissions

(SCAQMD, 1982:10). For this reason, authorities regarded as *successful* measures which achieved per capita reductions in pollutant emissions even when the total volume of pollutants in the atmosphere remained above federal standards. In the case of sulfur dioxide, per capita reductions in emissions did succeed in bringing the basin into compliance with state and federal air quality standards *despite* population growth. Since sulfur dioxides were chiefly emitted by industrial polluters, the stagnation of heavy industry helped regulators to achieve this success. Pollutants from mobile sources declined but remained above federal and state levels. Between 1955 and 1960, the Los Angeles monitoring station recorded an average annual one-hour maximum of 55.2 parts per million of ozone; between 1978 and 1983, the station only recorded 33.1

However, since the ozone level in the Los Angeles Air Basin was still the nation's worst in 1980, the long-term per capita reduction, significant as it was, was not enough to solve the ozone problem. Moreover, as Los Angeles was improving its air quality, outlying counties were experiencing deterioration of theirs, a fact commonly overlooked in official documents. This deterioration occurred because the movement of jobs and population toward the periphery naturally increased vehicular emissions there, thus causing deterioration of air quality. For example, while average annual daily one-hour maxima of ozone were declining in Azusa—a Los Angeles County monitoring station—in peripheral San Bernardino, ozone levels were increasing, reaching the Azusa level in about 1981. Achieving per capita emissions reductions, and even reductions in the total volume of pollutant emission, the Los Angeles basin failed to bring pollution below the threshold at which pollutants damage health, vegetation, and property. Because of growth in the vehicular population, mobile sources of pollution consistently expanded to offset regulatory gains. Stricter controls on mobile polluters were necessary, but politically unpopular among vehicle owners.

Administrative centralization strengthened enforcement of pollution control legislation. With the formation of the South Coast Air Quality Management District (SCAQMD) in 1976, a unitary agency finally emerged to coordinate the six county agencies that had previously conducted individual control efforts (Nelson, 1983:60, 69). This coordination produced "significantly more stringent control of both stationary and vehicular sources of pollution" (SCAQMD, 1983b:3). As a result of enhanced enforcement, real progress began to be made in pollution control, net of population growth. Concentrating on ozone, the region's biggest problem, SCAQMD data (1983:14ff) demonstrate that the basin's ozone level was rising until 1978, one year after SCAQMD began operations, then fell precipitously. Other data document the same improvement in nitrogen dioxide, carbon monoxide, lead, and sulfate.

Taking credit for abrupt improvements relative to federal standards, SCAQMD (1983b:6) called attention to its "active rule development program" which had produced "more than 50 rules and regulations" requiring emissions sources to control pollution. Additionally, SCAQMD implemented a major program requiring use of special nozzles to capture vapors released at gasoline stations when vehicles are fueled.

SCAQMD mounted continuous monitors on industrial smokestacks to measure and record emissions. SCAQMD designed and implemented a permit fee for stationary source emissions based on the amount and health effects of emissions. This revenue measure shifted the monitoring cost from taxpayer to polluter. The SCAQMD (1982:7-9) also developed a strengthened enforcement program involving off-hour surveillance, violation prevention activities, an emergency action schedule, vigorous and consistent prosecution of violators, source testing, and technical assistance to local police and fire departments confronting pollution emergencies. Finally, the SCAQMD sponsored successful state legislation which committed California to a biennial motor vehicle inspection and maintenance program in urban areas falling below air quality minima. Unpopular with motor vehicle owners, this program targeted owners who had unlawfully disconnected catalytic converters required by state law on new vehicles using leaded fuel. Under current legislation, all vehicles sold in California must contain catalytic converters permitting use of unleaded fuel.

## Water Supply

Although California has generous water resources, the state's rivers and deltas lie far from the Los Angeles five-county area, a semidesert environment with average annual rainfall less than 20 inches. Therefore, urbanization of the Los Angeles region recurrently bumped against the limited supply of local water. Each time the region's expanding population approached the limit of what available water could support, the municipal growth bloc confronted the possibility that population growth would end and with it enhancement of property values (Light, 1983:103-7). This possibility encouraged strenuous efforts to maintain entitlement to distant water and financing for lengthy aqueducts. When other water users objected, conflicts over water rights erupted. Success in such conflicts was an indispensable condition of urbanization in Southern California, but each success only postponed a subsequent confrontation with the water shortage that emerged from a new cycle of growth.

The first major success occurred in 1905 and 1907 when, reeling under the impact of an eight-year drought, Los Angeles voters approved successive bond issues that authorized construction of the Los Angeles Aqueduct from the Owens Valley. Opened in 1913, this 338-mile aqueduct supplied enough water to satisfy two million people (Nelson, 1983:77-78). However, in diverting a massive volume of water, the Los Angeles Aqueduct reduced the fertility of the Owens Valley. As the valley population grew, and the area emerged as a major agricultural producer, Owens Valley farmers brought repeated lawsuits against the Los Angeles water company in vain efforts to staunch the diversion of water. To strengthen its legal claim to valley water, the Los Angeles Department of Water and Power bought 450 square miles of the upper Owens Valley, thus becoming the area's principal landowner. On the strength of this ownership, the California Supreme Court upheld Los Angeles's diversion of Owens Valley water (Bigger and Kitchen, 1952). Of water used in the City of Los Angeles, 80 percent still comes from the Owens Valley via the Los Angeles Aqueduct.[2]

Owens Valley water did not permanently slake the thirst of a growing urban region. Following the drought of 1923, Southern California cities turned to the Colorado River for additional water resources to assure their growth. Incorporated in 1928, the Metropolitan Water District of Southern California (MWDSC) came into existence as an umbrella organization of 13 Southern California cities (Nelson, 1983:81-82). Other cities subsequently joined the MWDSC, and in 1979 it had a membership of 126 cities. A member of the MWDSC, Los Angeles retained its title to the Owens Valley water it obtained via the Los Angeles Aqueduct, but shared with the other 125 cities of the region the right to purchase supplementary Colorado River water from MWDSC. In 1931 Southern California voters approved a $220 million bond issue to build a 242-mile aqueduct across the desert from the Colorado River. Depression-era citizens voted for the bonds on the grounds that, even if the water were not immediately needed, constructing the aqueduct would produce 10 thousand jobs. A hundred thousand men applied for work on the project, and 35 thousand found employment on it. Ironically, when water from the Colorado River finally became available in 1941, the MWDSC had to give it away because four years of generous rainfall had completely satisfied local water demand. However, World War II began two decades of rapid population growth, and, by 1949, in the face of a serious water shortage, MWDSC authorities declared a moratorium of water system annexations, depriving desperate San Gabriel Valley municipalities of the right to acquire membership in the only reliable water source, the MWDSC. In 1951 voters approved an expansion of the Colorado River aqueduct to its present capacity, one billion gallons daily. This expansion satisfied regional demand for water in the Los Angeles five-county area as well as in San Diego.

In the meantime, settlement was proceeding in the Southwest, whose agriculture and cities (Albuquerque, Phoenix) also turned to the Colorado River for *their* water. After 20 years of wrangling over water with California, the Central Arizona Project brought suit against the state, alleging that California was exceeding its diversion entitlement under earlier treaties. In 1964 the Supreme Court of the United States ruled in favor of Arizona. The Supreme Court awarded Arizona an entitlement of 2.8 million acre-feet compared to California's 4.4 million acre-feet, and Nevada's 0.3 million. This ruling especially injured the MWDSC because, as junior member of the five California water users, it was compelled to accept the largest reduction in water entitlement, 662,000 acre-feet—enough water to supply the City of Los Angeles. However, *Arizona vs. California* granted the U.S. Secretary of the Interior authority to allocate unused water apportionments to other users, and Arizona did not fully use its apportionment. In 1982, Arizona only used 43 percent of its apportionment. Therefore, California users obtained 1.6 million acre-feet, the unused residuum of Arizona's apportionment, thus cushioning the shock of *Arizona vs. California*. To support agriculture and urbanization, Arizona undertook a huge canal and aqueduct construction project, the Central Arizona Project. Completion of the project in 1985 let Arizona use a greater proportion of its entitlement. However, the MWDSC (1982a:III-1) estimated that Arizona would not use its full apportionment until after 1990. When Arizona fully uses all its water entitlement, the MWDSC must have alternative water sources to make up for the 662,000 acre-feet it will lose.

In 1960 California voters approved bonded indebtedness of $1.75 billion to construct the California Aqueduct. Completed in 1974, this aqueduct extended 473 miles from the Sacramento-San Joaquin Delta southward to Los Angeles. However, the aqueduct was only the initial stage of the State Water Project, ultimately expected to deliver 4.2 million acre-feet of water annually to urban users in Southern California and agricultural users in the San Joaquin Valley (Nelson, 1983:82-83). That amount is enough water to satisfy seven cities the size of Los Angeles. To complete the aqueduct, the State Water Project needed to construct 23 dams and reservoirs, 22 pumping plants, 473 miles of canal, 175 miles of pipelines, and 20 miles of tunnels.[3] Although the aqueduct and storage reservoirs were completed in 1974, completion of the State Water Project still required construction of additional facilities.

Without additional facilities, the State Water Project could deliver only 55 percent of the 4.2 million acre-feet of water it had contracted to provide. Of this water, San Joaquin valley agriculture obtained 34 percent, Southern California 58 percent, and Northern California the balance (SPNB, 1979b). However, public controversies delayed construction of additional facilities.

The most strenuous controversy centered around the peripheral canal. Diverting water from the southern end of the river delta, the California Aqueduct had destabilized the balance of fresh and salt water in the delta, causing salt water intrusions and reversals of normal flows (Malinowski, 1983). These intrusions threatened delta agriculture. To eliminate this problem and permit California to meet its water contracts, the California Department of Water Resources recommended construction of a peripheral canal which, skirting the delta, would divert water from the southern terminus of the Sacramento River, north of the delta. This canal proposal encountered overwhelming political opposition in the delta region and in San Francisco. Northern California water users feared that, with the balance of legislative power in Southern California, the MWDSC would appropriate all the fresh water from the delta to meet Southern California's water needs. Other Northern Californians objected to assuming an additional $66 billion of bonded indebtedness to complete a canal which, even if it did not harm them, would not benefit them either. Northern California's environmentalists opposed the canal because of its possible adverse impact on scenic north coast wild rivers, a recreational resource. Canal opponents also claimed that big business (land developers, agricorporations, and oil companies) would be the main beneficiaries of the canal, with California taxpayers paying the bill.

Finally, opponents denounced the practice of selling water to agribusiness for less than the price to households (SPNB Research Department, 1981:28-30).

In June 1982, California voters rejected the peripheral canal. Rejection of the canal created an awkward situation for the Metropolitan Water District. Confronting a massive 1985 decrease in its Colorado River entitlements, the MWDSC had depended on the State Water Project to make up the deficit and provide additional water to satisfy Southern California's growing population. But without the peripheral canal, the state could not deliver more water to Southern California. As a result, MWDSC's loss of Colorado River entitlements left it facing a reduced supply of water with which to satisfy a growing population. Although conservation and water reclamation offered some opportunity to ease the shock, MWDSC (1982b:I-1) gloomily concluded that "there could be water shortages in the future." To "reduce the severity of future water shortages," MWDSC (1983d:I-3) vowed to "take all steps necessary to insure completion of the State Water Project." In the meantime, the California Department of Water Resources had assembled new plans for enabling the state to fulfill its water delivery obligations (MWDSC, 1983d). The new plans envisioned four "through delta" projects to replace the peripheral canal scheme. However, in 1985 it remained uncertain whether these new plans would satisfy Northern California water users or whether, in the absence of Northern California's agreement, the MWDSC could obtain voter approval of the facilities needed to permit the state to meet its contractual water-delivery obligation. If not, attorneys for the MWDSC were prepared to argue that the governor of California already had the legal authority to order the improvements without voter approval, but this route obviously depended on ultimate success before the United States Supreme Court.

## Housing Supply and Cost

Housing costs have long been higher in Los Angeles than in California, and higher in California than in the rest of the United States. In 1960, cost of an average house in Los Angeles was $5,300 above the average California price, and $8,200 above the United States average (table 2.8). During the 1960s, house prices in Los Angeles County and the four adjoining counties as well as in California caught up with those in the City of Los Angeles, but prices elsewhere in the United States remained about 60 percent of Los Angeles costs. In the mid-1970s house prices surged ahead in the Los Angeles area. By 1980, the average house in the United States cost only 44 percent of the average one in Los Angeles, and the average house in California cost 74 percent. Median incomes were not sufficiently higher in Los Angeles to explain this discrepancy. In 1979, median income was $21,125 in Los Angeles County, $21,537 in California, and $19,917 in the United States. Therefore, in 1980 the average house cost 5.6 times the median income in Los Angeles, 3.9 times in California, and 2.4 times in the United States.

TABLE 2.8
Specified Vacant-for-Sale Value in Unadjusted Dollars—
Los Angeles Area, California, and United States,
1960-1980

| Location | 1960 ($) | 1970 ($) | 1980 ($) | Index |
|---|---|---|---|---|
| Los Angeles | 21,700 | 25,500 | 118,000 | 544 |
| Los Angeles County | 17,700 | 24,900 | 103,900 | 587 |
| 4 adjacent counties | 14,800 | 26,300 | 95,100 | 642 |
| California | 16,400 | 24,400 | 87,500 | 534 |
| United States | 13,500 | 15,400 | 51,600 | 382 |

SOURCE: Based on data from U.S. Census of Housing (1960, 1970, 1980).
NOTE: Index = 1980/1960 x 100.

According to the Security Pacific National Bank Research Department (1981:17), escalating housing costs in Southern California resulted from "intense demand for housing" coupled with restrictions on supply. Intense demand resulted from net immigration, faster than average growth of the population aged 20-35, and decline in average household size. Decline in average household size arose because of an increased proportion of divorced and never-married persons in the population. On the demand side, the bank (1981:17-18) pointed out that the housing stock had increased only 11 percent from 1975-80 while the Southern California population increased 17 percent in the same period. The bank blamed two national recessions, high interest rates, environmental protection measures, and local government growth limitations for the failure of supply to keep pace with demand (see also Lawrence, 1978).

Although these factors plausibly explain the advance of home prices in Los Angeles, they do not provide a satisfactory account of why the prices surged ahead of those in California and the United States. Particularly on the demand side, most of the factors Security Pacific National Bank advanced were as powerful in other metropolitan areas as they were in Los Angeles. Even if interest rates, environmental protection measures, and local growth limitations depressed housing supply, there is no evidence that the latter two factors were appreciably more restrictive in Los Angeles than elsewhere. Certainly, they were not twice as restrictive, but Los Angeles housing costs were more than twice those elsewhere in the United States. If local growth limitations boosted housing prices, they did so in the suburbs where population growth was greatest. Yet, housing prices in the Los Angeles area were highest in the City of Los Angeles where local growth limitations were least influential (table 2.8). Interest rates were approximately the same everywhere in the United States so they cannot explain the more rapid escalation of home prices in Los Angeles than elsewhere.

On the supply side, the Security Pacific National Bank's explanatory factors fare little better. Although Hollywood has always tolerated unconventional life-styles, the rest of Los Angeles was more famous for occult sects and female charismatics than for matrimonial nonconformity (McWilliams, 1946:252-72, 343-47). Hence, a national surge in the popularity of single life-styles cannot explain the surge of housing prices in Los Angeles without evidence that this surge was more powerful there than elsewhere. Similarly, the postwar baby boom was a national phenomenon, not a Los Angeles one. Admittedly, the changing age composition of Los Angeles increased the proportion of persons in home-buying years during the 1970s, but it seems unlikely that this increase was twice as great as that which occurred elsewhere in the United States.

Of the factors Security Pacific National Bank advanced, only net migration offers a plausible basis for explaining the escalation of Los Angeles housing prices in the late 1970s. Proof of this conclusion would require comparing Los Angeles with other U.S. cities experiencing net migration, a project beyond the scope of this chapter. But, there is circumstantial evidence that the character and volume of migration did increase housing costs. First, the acceleration of housing prices coincided with the resumption of migration to Southern California. Second, Los Angeles was the favorite destination of wealthy Asians and Iranians many of whom regarded real property investments in Southern California as a hedge against inflation and political turmoil. Third, the surge of high-technology industries in the mid-1970s encouraged the migration of scientists and engineers who earned high salaries. Saxenian (1983) found that in Silicon Valley (Santa Clara County, CA) housing prices rose rapidly in the 1970s because high-technology industries attracted highly paid workers who outbid locals in the housing market. Since Los Angeles County and Orange County added the equivalent of a Silicon Valley to their labor forces in the 1970s, the addition presumably had the same accelerating effect on housing prices in Los Angeles which it had in Santa Clara County (Soja et al., 1983:214). Fourth, the office-building boom in Los Angeles provided work places for high-income managers and professionals. These workers had a motive to acquire residences near their work places, thus explaining the enhancement of house prices in the City of Los Angeles relative to the ring. Finally, in the 1970s, foreign and multi-national corporations *invested* in Los Angeles properties (Turpin, 1979). These investments increased demand for real property and caused home prices in central Los Angeles to rise correspondingly.

Whatever their cause, accelerating housing prices distressed both the middle and working classes of Los Angeles, albeit in different ways. The working class, renters, and first-time home buyers experienced the accelerating cost of housing as a crisis of affordability (SPNB Research Department, 1977:10). Vacancy rates fell below 2 percent, and rents increased. In central Los Angeles, blue-collar workers faced plant closures, rising unemployment, reductions of real wages, *and* escalating home prices. One result was increased homelessness (Overend, 1983). A suddenly visible class of "bag people" carried their possessions in paper bags or shopping carts, and slept under freeways (Roderick, 1985b). A second result was increased overcrowding as people doubled up to meet huge housing payments. A conventional measure of overcrowding is the proportion of persons residing in housing with more than 1.01 persons per room. An abrupt increase in overcrowding affected the City and County of Los Angeles—especially between 1970 and 1980 (table 2.9). This increase contrasts with continuous reductions in overcrowded housing in the four adjacent counties, California, and the United States.[4]

In principle, homeowners should have been unaffected by the escalation of home prices since, by owning their homes, they had fixed mortgage obligations to meet. In reality, homeowners experienced escalating home prices as a crisis of property taxation. This deflection occurred because California law tied property taxes to the assessed value of homes. As the value of homes rose in response to the influx of Asian investors, engineers, and corporate managers, property taxes rose as well. By 1978-79, property taxes in California were 150 percent of the national norm (SPNB Research Department, 1981). Many homeowners faced loss of their property because of constantly rising property taxes. With the assistance of major banks and real estate interests, an antitax social movement formed among the home-owning majority. The move produced an antitax ballot initiative intended to provide relief to homeowners and business. In June 1978, California voters approved a comprehensive tax reform (Proposition 13) which pegged property taxes to 1 percent of a home's original purchase price. This popular reform rolled back property tax increases that had accrued in the preceding period of home price escalation. Additionally, Proposition 13 guaranteed homeowners against the threat that future increases in assessed valuation of their home could raise their property taxes beyond their ability to pay (Lawrence, 1978).

Of course, Proposition 13 accomplished nothing for renters, 45 percent of the Los Angeles region's population. After the passage of Proposition 13, renters expressed dissatisfaction with a measure that conferred economic benefits on their landlords, but provided them no relief from rent increases. In response to renter discontent, the cities of Santa Monica and Los Angeles both adopted rent control laws which regulated the rate at which landlords could raise rents (Edelman, 1977). By protecting renters against abrupt and destabilizing rent increases, rent control laws extended the political control of housing costs which Proposition 13 had earlier achieved for homeowners. Rent control enraged landlords who failed, however, to dislodge it in five years of trying (Soble, 1984).

## Intergroup Conflicts

A split labor market arises whenever cheap, distressed labor enters metropolitan labor markets (Light, 1983:337). The result is a three-cornered industrial contest in which employers use the cheap labor bloc to restrain wage demands from their current work force. In a context of industrial plant closings, increasing unemployment, and falling real wages, the influx of Mexican and Central American immigrants split the labor market in central

TABLE 2.9
Percentage of Occupied Housing Units with 1.01 or More Persons
per Room—Los Angeles Area, California, and United States,
1960-1980

| Location | 1960 | 1970 | 1980 |
|---|---|---|---|
| Los Angeles | 8.0 | 8.8 | 13.0 |
| Los Angeles County | 8.6 | 8.5 | 11.2 |
| 4 adjacent counties | 10.6 | 7.8 | 6.0 |
| California | 9.5 | 7.9 | 7.4 |
| United States | 11.5 | 8.1 | 4.5 |

SOURCE:  Based on data from U.S. Census of Housing (1960, 1970, 1980).

Los Angeles. Blacks in particular came into conflict with immigrant, low-wage Mexicans over jobs and housing. Overrepresented in unskilled work and low-income housing, blacks complained that undocumented Mexicans were underbidding them at work, outbidding them for housing, and riding free on social welfare programs set in motion by black protests of the 1960s (Oliver and Johnson, 1984:75). Therefore, blacks championed a strict enforcement policy on the Mexican frontier in the interest of excluding undocumented labor competition. Additionally, black political spokespersons favored imposition of civil penalties on employers who knowingly hired undocumented workers.

Oliver and Johnson (1984) performed a secondary analysis of a telephone survey which the Los Angeles Times completed in 1980. This poll (N = 1,295) oversampled blacks and Mexicans in Watts and East Los Angeles, the residential centers of both populations. Black respondents were more likely than nonblacks to perceive Latinos as economic threats, political free riders, and unwilling to learn English. For their part, Latins, especially low-income ones, complained that blacks had too much political power. As Oliver and Johnson (1984:66) noted, this complaint had a factual basis in the quite unequal political representation of the two ethnic minorities. Representing only 8 percent of the California population, blacks held eight seats in the California Legislature; representing 16 percent, Chicanos (Mexican-Americans) had only six seats. In Los Angeles, where Chicanos were more than twice as numerous as blacks, the mayor and three city council members were black, but no Chicanos held elected office.

Asian immigrants also came into conflict with native-born Americans over jobs, housing, and business opportunities. As incidents mounted, the Los Angeles County Commission on Human Relations (1983) convened a special investigation of "rising anti-Asian bigotry." The context for this investigation was the "perception" that hostile bumper stickers, racial insults, murders, assaults, school incidents, discrimination, economic backlash, and hostility to Japanese-American reparation claims were increasingly frequent manifestations of an "anti-Asian climate." Although the commission conceded that these incidents were more frequent in central California, where the Ku Klux Klan was active, it assembled evidence of intergroup tension in Los Angeles too.

Examples of racial tension abounded. In 1981 arsonists destroyed a Chinese newspaper in suburban Monterey Park where a substantial influx of affluent Chinese was under way. In the same year, vandals damaged a Chinese movie theatre in Alhambra, leaving behind notes attributing the damage to the Ku Klux Klan. Residents of Garden Grove, a heavily Asian suburb in Orange County, petitioned their city council to grant no more business licenses to Asians, and to prohibit business signs in foreign languages. Spokespersons complained that Koreans and Southeast Asians were "taking over" Garden Grove. Half the Asian students at California State University, Long Beach, complained that they had been subjected to racial insults, threats, or violence on campus. In 1983, Los Angeles's black community newspaper, *The Sentinel*, published a series of articles on the "problem" of Korean business enterprises in black neighborhoods. The paper accused Koreans of nepotistic hiring, exploitative pricing, and disrespectful treatment of black customers. Finally, "Buy American" campaigns also encouraged anti-Asian incidents since the American public blamed all Asians for Japan's business competition.

## Crime

The Los Angeles public regarded crime as the region's greatest problem. In a 1981 telephone survey of 2,063 people, the *Los Angeles Times* found city residents ranked crime number one among the area's social problems. Respondents also expressed more fear of crime than did residents of other major U.S. cities (Endicott, 1981). Half the Los Angeles respondents reported high fear of crime compared to only 36 percent in other cities. Of those surveyed, 15 percent of Los Angeles respondents indicated they or someone in their household had been a felony victim. Half the respondents in the city and 39 percent in the suburbs reported knowing a felony victim. A majority thought themselves in danger of victimization from some serious crime. Topping the list of worrisome crimes were burglary (44 percent) and rape (40 percent). Fourteen percent of Los Angeles respondents indicated fear of murder.

A 1985 survey of 2,241 randomly selected respondents revealed that the public still regarded crime as Los Angeles's major problem. However, social class and residence area importantly affected public perceptions of the seriousness of crime. People in low-income, central Los Angeles neighborhoods expressed more fear of crime than did people in suburban, upper-income areas (Roderick, 1985a). For example, about half of central city residents reported they felt unsafe walking in their neighborhoods at night; only one-third of suburban respondents reported this problem.

The public's concern about crime had some basis in fact: crime rates were increasing during the 1970s. Although the increases affected the entire metropolitan region, the crime rate and rate of increase were higher in central Los Angeles than on the metropolitan periphery. Rates of serious crimes known to the police almost doubled in the City of Los Angeles between 1970

TABLE 2.10

Serious Crimes Known to Police—Los Angeles Area, 1970-1980

(rates per 100,000 population)

| Location and Percentages | 1970 | 1975 | 1980 |
|---|---|---|---|
| Los Angeles | 5,922 | 8,191 | 10,033 |
| Los Angeles County | NA | 7,209 | 8,298 |
| 4 adjacent counties | NA | 6,759 | 6,946 |
| Los Angeles County as percentage of city | | 88.0 | 82.7 |
| 4 adjacent counties as percentage of city | | 82.5 | 69.2 |

SOURCE: U.S. Bureau of the Census (1972, 1977, 1983).

NOTE: Adajcent counties are Orange, Riverside, San Bernardino, and Ventura.

and 1980 (table 2.10). In Los Angeles County and the four adjacent counties crime rates were also increasing in this period, but the rate of increase was less rapid. As a result, the rate of serious crime in the adjacent counties was only 69.2 percent of the Los Angeles rate in 1980; it had been 82.5 percent in 1975 (table 2.10). Since local differences in reporting or recording might cause spurious interlocal differences in "serious crimes known to the police," it is desirable to compare concrete categories which minimize reporting bias. Isolating two felonies least subject to underreporting, I compare rates of murder and automobile theft in Los Angeles County with rates in two SMSAs located in adjoining counties (table 2.11). Both crimes were much higher in Los Angeles County than in the neighboring SMSAs, thus confirming the greater frequency of felonious crimes in the central city.

The inverse correspondence between crime rate and distance from the central city replicates the relationship long observed in American cities of every size (Light, 1983:376). Disadvantaged and low-income populations invariably produce higher rates of felonious crimes, and, since such people characteristically inhabit central cities, central cities have high crime rates in the United States. Naturally, this consideration does not explain why crime rates increased all over Southern California in the 1970s. On the other hand, the increased socioeconomic segregation that accompanied demographic changes in the Los Angeles region left its center with the higher proportion of disadvantaged, low-income persons in 1980 than it had in 1960. Possibly, crime rates increased faster in the region's center as a reflection of that segregation which deposited a higher proportion of the disadvantaged "at risk" population in the center.

In view of plant closings, increased unemployment, declining real wages, and housing shortage which also had impact on the region's center, increased economic hardship offers a plausible, supporting explanation of increased crime rates in central Los Angeles. But this hypothesis encounters two preliminary objections. First, generations of research have failed to establish an unambiguous causal relationship between unemployment and crime (Light, 1983:377). Second, between 1980 and 1983, homicide rates fell all over

TABLE 2.11
Selected SMSA Crime Statistics—Los Angeles Five-County Region,
1960-1980 (rates per 100,000 population)

| Location | 1960 | 1970 | 1980 |
|---|---|---|---|
| Anaheim-Garden Grove-Santa Ana | | | |
| Total offenses | NA | 3586.5 | 7436.3 |
| Murder | NA | 2.7 | 5.0 |
| Automobile theft | NA | 376.7 | 485.4 |
| Los Angeles-Long Beach | | | |
| Total offenses | 2678.9 | 5036.9 | 8418.7 |
| Murder | 4.4 | 9.4 | 23.3 |
| Automobile theft | 444.3 | 944.6 | 1104.9 |
| Riverside-San Bernardino-Ontario | | | |
| Total offenses | NA | 3877.5 | 8231.9 |
| Murder | NA | 6.2 | 15.5 |
| Automobile theft | NA | 459.6 | 642.5 |

SOURCE:  Data from Federal Bureau of Investigation (1960-66, 1970-72, 1981).

California and the United States (Sanchez, 1984). Although teenaged unemployment declined between 1980 and 1983, the decline in unemployment was
less than the decline in murder rates. Some criminologists claimed that a
reduced proportion of young men in the crime-prone years (16-24) explained
the decline in homicide rates. However, a 1983 study conducted by the
California Bureau of Criminal Statistics found "little relationship between
the size of the high crime-prone youth population and the violent crimes rate"
(Sanchez, 1984). Police and public officials attributed the decline to the
deterrent effect of more vigorous prosecution, harsher sentencing, and more
frequent executions.

## Controlling Urbanization in
## the Los Angeles Region

Growth of the five-county Los Angeles area was, to a substantial extent, an
episode in the realignment of regional urbanization in the United States
toward the South and West and away from the North and East (Light,
1983:107-12). Sources of this realignment were political, economic, and
demographic, but it is unnecessary to analyze them to appreciate the
implications of shifting regional urbanization for Los Angeles. Simply in
view of Los Angeles's geographical location and climate, some urban growth
was unavoidable.

Given unavoidable growth, Los Angeles suffered unavoidable growth
problems, but just which were avoidable and which unavoidable? Growth
problems are difficulties arising from the volume or quality of growth, and
management problems are unnecessary difficulties above and beyond those
attributable to growth. This chapter has identified at least eight problems in

the Los Angeles region: intergroup conflict, abuse of redevelopment authority, economic decline of the central city, traffic congestion, air pollution, unaffordable housing, water shortages, and high crime rates. Population growth was neither a necessary nor a sufficient condition of any problem on this list, but in all the empirical cases population growth did aggravate the social problem.

For example, because of *Arizona vs. California*, Los Angeles would have confronted reduced water entitlement in 1985 even if the region's population failed to grow in the preceding 25 years. Therefore, population growth was not a necessary condition of water shortage. On the other hand, population growth clearly aggravated the water shortage, converting a small problem into a big one. Even in the case of air pollution, clearly linked to the population of motor vehicles, technological improvements were, in principle, capable of reducing vehicle emissions enough to solve the problem of air pollution. Since population growth might have occurred without increased air pollution, growth was neither a sufficient nor a necessary condition of air pollution. In reality, however, population growth did exacerbate the pollution problem which was empirically inseparable from population growth whatever its theoretical independence.

From Los Angeles's point of view, population growth caused three kinds of social problems: those directly responsive to control, those indirectly responsive, and those unresponsive. *Unresponsive problems* were those Los Angeles had no means to eliminate, reduce, or mitigate. None of the problems identified above were unresponsive: Los Angeles could affect all of its growth-related problems. *To affect* means to mitigate, not necessarily to eliminate. *Directly responsive problems* were those Los Angeles had independent means to reduce or mitigate. *Indirectly responsive problems* were those Los Angeles could diminish with the assistance of others, but whose diminution required coordinated, long-run effort. For example, Los Angeles attacked air pollution on its own and in coordination with other regions and political units. On one hand, the Los Angeles region diminished air pollution by consolidating six county agencies into a unitary Southern California Air Quality Management District. This consolidation was a direct control. On the other hand, using its political strength, the SCAQMD induced the California Legislature to require unleaded fuel use by new vehicles sold in the state, thus improving air quality in Los Angeles. This political achievement reflected the ability of Los Angeles to obtain the cooperation of external agencies in its problem-control efforts, an indirect control strategy.

Similarly, the region's inventory of direct and indirect social controls consisted of controls directed at problem causes and manifestations. Los Angeles could respond to growth problems by diminishing growth or mitigating its ill effects. In specific response to problems arising from net migration, the causes of Los Angeles's problems originated beyond the boundaries of the five-county region and were, as a result, subject *only* to indirect controls. For example, some of Los Angeles's population growth originated as illegal migration from Mexico. If controllable at all, the causes of this illegal migration required the joint interventions of the United States and Mexico. Los Angeles could request the United States government to obtain Mexico's cooperation in reducing the volume of illegal migration, an indirect strategy of problem control, but the city could not take direct, independent action to exclude immigration from Mexico. However, the County of Los Angeles did develop a human relations commission to improve communication and understanding between Mexican immigrants and other ethnic blocs. By mitigating intergroup tensions, the commission permitted Los Angeles County to facilitate the adjustment of ethnic groups whose influx it had no power to regulate.

How effective was Los Angeles in using indirect control strategies? In regard to growth-induced problems, the answer is clear: Los Angeles made no effort to preempt growth-induced problems by reducing growth. Indeed, except for the Sierra Club and other environmental organizations, inhibition of population growth was not even a conscious policy alternative inside or outside government. In view of the lack of a national U.S. urban policy, fatalistic acceptance of population growth was admittedly practical, even rational for the Los Angeles region. After all, Los Angeles had no higher authority to approach in a framework of national consensus about how much growth Los Angeles should be expected to accommodate. On the other hand, the pervasive unconsciousness of even the possibility of regulating regional growth prevented political efforts to develop a national urban policy. Therefore Los Angeles made no long-run progress toward improved control of external political and population forces affecting the region's growth.

Rather, the region's authorities devoted their entire effort to the task of managing growth they could not influence, much less control. Growth management took two forms. First, local authorities sought facilities to mitigate problems growth had already induced. For example, MWDSC authorities looked for new water sources to serve additional population. Second, local authorities sought to screen out the costs of growth while admitting growth-induced resources. For example, redevelopment authorities subsidized office space, but demolished low-income housing, thus enhancing tax revenues while eliminating consumers of municipal services.

In either management effort, both direct and indirect control strategies suffered from fragmentation of political authority, a common ailment of metropolitan regions in the United States (Nelson, 1983:chap. 21). In 1980 the U.S. Department of Housing and Urban Development enumerated 232 local governments, 16 other governments, 7 special districts, 8 school districts, and 1 county government within Los Angeles County alone (Light, 1983:425). Yet, this fragmented polity was an economic giant. When compared to the leading nations of the world in 1981, the gross domestic product of the Los Angeles region ranked 15th, just behind the Netherlands but ahead of Australia (SPNB Research Department, 1984:16). However, unlike either the Netherlands or Australia, Los Angeles had no central government.

Political fragmentation precluded indirect control of growth problems. At a minimum, indirect control requires someone to formulate a plan of action, and mobilize support for the plan inside and outside the region. Because of the proliferation of governments and authorities in the Los Angeles five-county region, no agency had the authority or responsibility to formulate and coordinate political strategies of problem control. Therefore, as a general rule, the Los Angeles region was ineffective in formulating long-run plans, mobilizing support for them, and then obtaining the cooperation of external agencies in reducing externally caused problems. The outstanding exception was the Metropolitan Water District of Southern California, an umbrella organization of 126 municipalities. This organization did not attempt to reduce the water shortage by inhibiting population growth in Southern California; rather, it projected growth trends into the future and tried to assemble the means to satisfy the water demand associated with its highest projection. To this end, MWDSC studied the water problem, formulated a long-run state water plan, mobilized political support in Southern California, and came close to obtaining voter approval of the peripheral canal. When California voters rejected the canal, the MWDSC fell back on alternative plans. The lonely efficacy of the MWDSC underscores the general inability of the politically fragmented Los Angeles region to plan, mobilize, and effect political responses to growth problems.

As a result of this failure of regional planning and political mobilization, the burden of managing growth fell on separate municipalities, each fending for itself in a devil-take-the-hindmost scramble. The first strategy of each municipality was to obtain for itself any resources growth provided while screening out growth's problems. A common offense in this regard was the use of redevelopment authority to replace existing housing with office buildings. Since office buildings produced tax revenues while housing produced demands for social services, each municipality sought to rid itself of population, especially low-income population, while attracting office buildings. This strategy subsidized the office space of multinational corporations whose high-income employees bid up the price of housing in the region. Another offense was the common suburban tendency to attract quiet, nonpolluting industries while zoning out low-income residents, apartment housing, welfare institutions, and children. This strategy created islands of tax privilege, a benefit to capital, but it imposed on less wealthy neighbor cities the whole fiscal burden of coping with growth-related problems (Hoch, 1984).

Given incomplete success of the first, the second management strategy was to cope with growth problems the city had been unable to evade. At best, municipalities could defensively mitigate some ill effects of growth without harming neighbors. For example, Santa Monica and Los Angeles made effective use of rent control ordinances to mitigate the affordability crisis in housing. Similarly, the City of Los Angeles empowered a Human Relations Commission to cushion intergroup conflicts; the City of Beverly Hills, a wealthy enclave having no Mexican immigrants, had no problem of intergroup tension. As these examples suggest, successful problem management in municipalities depended on the size of the problem and the resources locally available for their mitigation. Generally speaking, the City of Los Angeles had big problems and scant resources whereas suburban municipalities had small problems and abundant resources (Nelson, 1983:313-15).

In summary, three political features principally inhibited the ability of the Los Angeles region to control growth-related problems. First, no agencies made any effort to reduce growth problems by controlling the causes of growth. This failure deprived the region of a useful, long-run tool of social planning. Second, Los Angeles still suffered from what Crouch and Dinerman (1963:368) long ago identified as "the absence of any unified political leadership." This fragmentation reduced the ability of regional governments to use indirect control strategies to mitigate growth-induced problems. Third, competing municipalities sought to attract new industry, a benefit of growth, while imposing the fiscal and social costs of growth on less wealthy neighbors. Rational from the point of view of the individual municipality, this strategy separated growth-provided resources from growth-induced problems, thus promoting a mismatch of growth resources and growth problems among the region's municipalities. Therefore, the Los Angeles region was more unsuccessful in controlling or managing growth-induced problems than it absolutely needed to have been, and the political fragmentation which frustrated coordination was, in a sense, the region's master problem.

## Notes

1. The California Legislature passed the Community Redevelopment Act (CRA) in 1945; the Los Angeles City Council activated it in 1948. The CRA's function is to study social, physical, and economic conditions in targeted areas, formulate a plan for correcting blighted conditions and execute it (Jerison, 1976:112). Receiving contributions from the U.S. Department of Housing and Urban Development under the Housing and Community Development Act, any CRA has the authority to offer grants and loans to subsidize new construction in blighted areas. In principle, the new construction stimulates higher real estate values and increased property taxes. The increased taxes are supposed to retire the debts (see Light, 1983:255-56, 425-26). Controversy arose because of the tendency of CRAs to subsidize construction of office buildings and luxury hotels in clearing blighted areas and to subsidize development in wealthy suburbs untroubled by blight. Grand jury investigations criticized use of public funds to redevelop unblighted rural land as well as some questionable expenses incurred in the process. Criminal charges have also been lodged against redevelopment officials as, for example, in Compton where two city councilmen were convicted of extortion, fraud, and perjury in connection with land purchase for a redevelopment project (Clifford, 1984:16).

2. However, Inyo County and the City of Los Angeles continue to litigate claims to water use in the Owens Valley (see MWDSC, 1983a).

3. A principal contractor on this project was Bechtel Corporation, two of whose officers, George Shultz and Caspar Weinberger, later became cabinet members during the first Reagan administration (1980-84).

4. Reagan's Undersecretary of Housing and Urban Development claimed overcrowding had increased because Mexicans preferred high-density living (Kurtz, 1984).

# Bibliography

Applegate, Jane W. 1984. "Orange County Sets Its Sights on Becoming the Southern Silicon Valley." *Los Angeles Times* (June 10): V-1+.

Bernstein, Harry, 1980. "U.S. to Join Crackdown on Exploiting of Illegal Aliens in L.A. Garment Industry." *Los Angeles Times* (Mar. 1): II-1.

Bigger, Richard, and James D. Kitchen. 1952. *How the Cities Grew: A Century of Municipal Independence and Expansionism in Metropolitan Los Angeles*. Los Angeles: Haynes Foundation.

Bonacich, Edna, and John Modell. 1980. *The Economic Basis of Ethnic Solidarity: Small Business in the Japanese American Community*. Berkeley: University of California Press.

Bunting, Kenneth F. 1984. "Tax Hike for Transit Sparks Costly Battle." *Los Angeles Times* (May 20): II-1.

Cannon, Lou. 1982. *Reagan*. New York: Putnam.

Chrysler, K. M., and Michael Doan. 1985. "California's Second Look at Its Defense Dollars." *U.S. News and World Report* 98 (25 Feb.): 62-63.

Clifford, Frank. 1984. "Redevelopment's Spread to Richer Areas Stirs Critics." *Los Angeles Times* (Sept. 24): I-1+.

County Supervisor's Association of California. 1963. *California County Fact Book 1963*. Sacramento.

———. 1969. *California County Fact Book 1969*. Sacramento.

———. 1981. California County Fact Book 1980-81. Sacramento.

Crouch, Winston W., and Beatrice Dinerman. 1963. *Southern California Metropolis, A Study in Development of Government for a Metropolitan Area*. Berkeley: University of California Press.

Day, Kathleen, and David Holley. 1984. "Vietnamese Create Their Own Saigon." *Los Angeles Times* (Sept. 30): I-1.

Dreier, David. 1984. "Metro Rail Is Worth the Cost Even to a Fiscal Conservative." *Los Angeles Times* (10 June): IV-5.

Dugger, Ronnie. 1983. *On Reagan: The Man and His Presidency*. New York: McGraw-Hill.

Dye, Lee. 1984. "Looking for Brains? Try S.E. Bay Area." *Los Angeles Times* (29 Oct.): I-3.

Edelman, Edmund D. 1977. "The Crisis in Housing Threatens to Overshadow All Other Issues." *Los Angeles Times* (Dec. 9): II-7.

Endicott, William. 1981. "Public Calls Crime L.A.'s Top Problem." *Los Angeles Times* (1 Feb.): I-1+.

Faris, Gerald. 1977. "Foreign Investments in Local Land Appear on Increase." *Los Angeles Times* (26 Sept.): II-1.

Federal Bureau of Investigation. 1960-66. *Uniform Crime Reports for the United States, 1959-65*. Washington, DC: U.S. Department of Justice. [annually].

———. 1970-72. *Crime in the United States, Uniform Crime Reports for the United States, 1969-71*. Washington, DC: U.S. Department of Justice.

———. 1981. *Crime in the United States, Uniform Crime Reports for the United States, 1980*. Washington, DC: U.S. Department of Justice.

Fielding, Gordon J. 1984. "California Transportation: Inventory and Prospects." *California Management Review* 26: 100-11.

Findley, James Clifford. 1958. "The Economic Boom of the Twenties in Los Angeles." Ph.D. dissertation, the Claremont Graduate School.

*Fortune* 64 (July 1961): 167-86. "The Fortune Directory. The 500 Largest U.S. Industrial Corporations."

*Fortune* 107 (2 May 1983): 226-54. "The 500."

Haas, Gilda, and Allan David Heskin. 1981. "Community Struggles in Los Angeles." *International Journal of Urban and Regional Research* 5: 546-64.

Herbert, Ray. 1972. "L.A. Traffic: Cars Moving at 1965 Pace." *Los Angeles Times* (9 Aug.): II-1.

———. 1978. "Freeway Traffic Crisis Feared." *Los Angeles Times* (13 Feb.): II-1.

Hoch, Charles. 1984. "City Limits: Municipal Boundary Formation and Class Segregation." In *Marxism and the Metropolis: New Perspectives in Urban Political Economy*, edited by William K. Tabb and Larry Sawers, 101-22. 2d ed. New York: Oxford University Press.

Jerison, Irene, ed. 1976. *Los Angeles: The Structure of a City*. Los Angeles: League of Women Voters.

Kaplan, Sam Hall. 1984. "2 New Towers Would Change L.A. Skyline." *Los Angeles Times* (9 Nov.): I-1.

Keely, Charles B., and Patricia J. Elwell. 1981. "International Migration: Canada and the United States." In *Global Trends in Migration: Theory and Research on International Population Movements*, edited by Mary M. Kritz, Charles B. Keely, and Silvano M. Tomasi, 181-207. Staten Island, NY: Center for Migration Studies.

Keith, Ralph W. 1980. *A Climatological/Air Quality Profile. California South Coast Air Basin.* El Monte, CA: South Coast Air Quality Management District.

Kurtz, Howard. 1984. "Official Says Latinos Prefer Overcrowding." *Los Angeles Times* (12 May): I-15.

Lawrence, John F. 1978. "U.S. Housing Buyers Keep Up with Pace." *Los Angeles Times* (28 Sept.): I-1.

League of Women Voters of Orange County [CA]. 1972. *Population Growth in Orange County*. Santa Ana, CA: League of Women Voters of Orange County.

Light, Ivan. 1983. *Cities in World Perspective*. New York: Macmillan.

———. 1984. "Immigrant and Ethnic Enterprise in North America." *Racial and Ethnic Studies* 7: 195-216.

Los Angeles, City of. Department of Community Development, Community Analysis and Planning Division. 1977. *An Ethnic Trend Analysis of Los Angeles County, 1950-1980*. Los Angeles.

Los Angeles, County of. Commission on Human Relations. 1983. "Rising Anti-Asian Bigotry: Manifestations, Sources, Solutions." Transcript of public hearing conducted 9 November.

*Los Angeles Times* (5 Nov. 1984): IV-2. "Arms-Buildup Support Laid to 'Job Blackmail.' "

Malinowski, Jay. 1983. "The Delta: A Study in Transition." *Aqueduct* (Metropolitan Water District of Southern California) 49(1).

McWilliams, Carey. 1946. *Southern California Country, An Island on the Land*. American Folkways. New York: Duell, Sloane & Pierce.

Merina, Victor. 1979. "Pines Assails Garment Industry." *Los Angeles Times* (6 Sept.): II-1.

Metropolitan Water District of Southern California. 1982a. *Future Colorado River Water Supply of the Metropolitan Water District*. Report no. 947. Los Angeles.

———. 1982b. *1982 Population and Water Demand Study*. Report no. 946. Los Angeles.

———. 1983a. "Inyo County, Los Angeles Agree on Owens Plan." *Focus on Water* (Metropolitan Water District of Southern California), no. 9: 4-5.

———. 1983b. "Lively Debate Expected on State Water Proposals Going to Legislature in '84." *Focus on Water* (Metropolitan Water District of Southern California), no. 8: 1-2.

———. 1983c. *The Need for Major Additions to Metropolitan's Distribution System*. Abridged report no. 949. Los Angeles.

———. 1983d. *Water Supply Available to Metropolitan Water District Prior to Year 2000*. Report no. 948. Los Angeles.

Morgan, Judith, and Neil Morgan. 1981. "Orange, a Most California County." *National Geographic* 160: 750-79.

MWDSC. *See* Metropolitan Water District of Southern California.

Nelson, Howard J. 1983. *The Los Angeles Metropolis*. Dubuque, IA: Kendall/Hunt.

O'Dell, John. 1984. "Orange County's Future in Hands of a Few Big Developers." *Los Angeles Times* (29 Jan.): V-1+.

Oliver, Melvin L., and Mark A. Glick. 1982. "An Analysis of the New Orthodoxy on Black Mobility." *Social Problems* 29: 511-23.

Oliver, Melvin L., and James H. Johnson, Jr. 1984. "Inter-Ethnic Conflict in an Urban Ghetto: The Case of Blacks and Latinos in Los Angeles." *Research in Social Movements, Conflict, and Change* 6: 57-94.

Orange County [CA] Transportation Commission. 1984. "15 Year Transportation Investment Plan." Santa Ana, CA: Orange County Transportation Commission.

Overend, William. 1983. "A Time of Crisis for Our Brothers' Keepers." *Los Angeles Times* (May 1): VI-1-3.

Parson, Don. 1982. "The Development of Redevelopment: Public Housing and Urban Renewal in Los Angeles." *International Journal of Urban and Regional Research* 6: 393-413.

Pastier, John. 1970. "Los Angeles Records Slowest Expansion Rate in Its History." *Los Angeles Times* (20 Sept.): F-1.

Rabinovitz, Francine F., with William J. Siembieda. 1977. *Minorities in Suburbs: The Los Angeles Experience.* Lexington, MA: Lexington Books.

Roderick, Kevin. 1985a. "Most in L.A. Are Satisfied Despite the Fear of Crime." *Los Angeles Times* (25 Mar.): I-1+.

―――. 1985b. "New Skid Row Shelter Full: Hundreds Still on Streets." *Los Angeles Times* (26 Jan.): II-1.

Rosenblatt, Robert A. 1970. "How Highway Lobby Ran Over Proposition 18." *Los Angeles Times* (27 Dec.): F-1.

Sanchez, Jesus. 1984. "Experts Puzzled by 3-Year Decline in State Homicides." *Los Angeles Times* (29 Nov.): I-3+.

Saxenian, AnnaLee. 1983. "The Urban Contradictions of Silicon Valley: Regional Growth and the Restructuring of the Semiconductor Industry." *International Journal of Urban and Regional Research* 7: 237-62.

SCAG. See Southern California Association of Governments.

SCAQMD. See South Coast Air Quality Management District.

Schwartz, Joel. 1982. "California's Lifeline. A History of the State Water Project." *Aqueduct* [Metropolitan Water District of Southern California] 48(2): 5-51.

Security Pacific National Bank. 1979a. *Monthly Summary of Business Conditions* 58 (July 31): 1.

―――. 1979b. "Water Supplies for Southern California." *Monthly Summary of Business Conditions* 48 (31 Aug.): 2.

―――. 1981a. "Revenues and Expenditures – State and Local Governments in California." *Monthly Summary of Business Conditions* 50 (31 Mar.): 1-3.

―――. 1981b. "Southern California's International Trade in 1980." *Monthly Summary of Business Conditions* 50 (30 Apr.): 1-2.

Security Pacific National Bank. Research Department. 1977. *Southern California: Economic Trends in the Seventies.* Publication no. 009167. [Los Angeles: The Bank.]

―――. 1981. *Economic Issues in the Eighties: Southern California.* Los Angeles. The Bank.

―――. 1984. *The Sixty Mile Circle.* Los Angeles: Security Pacific Bank.

Snell, Bradford. 1974. *American Ground Transport. A Proposal for Restructuring the Automobile, Truck, Bus, and Rail Industries.* Report prepared for the Subcommittee on Anti-Trust and Monopoly of the Senate Committee on the Judiciary. 93d Cong., 2d Sess. Committee Print.

Sobel, Ronald L. 1984. "Rent Control: A City at War." *Los Angeles Times* (26 Nov.): II-1+.

Soja, Edward, Rebecca Morales, and Goetz Wolff. 1983. "Urban Restructuring: An Analysis of Social and Spatial Change in Los Angeles." *Economic Geography* 59: 195-230.

South Coast Air Quality Management District. 1982. *Air Quality Trends in California's South Coast Air Basin, 1965-1981.* El Monte, CA: South Coast Air Quality Management District.

―――. 1983a. *1982 Summary of Air Quality in California's South Coast Air Basin.* El Monte, CA: South Coast Air Quality Management District.

―――. 1983b. *The South Coast Air Quality Management District. A Progress Report 1977-1983.* El Monte, CA: South Coast Air Quality Management District.

Southern California Association of Governments. 1984a. *Southern California: A Region in Transition.* Vol. 2. *Impacts of Present and Future Immigration.* Los Angeles: Southern California Association of Governments.

―――. 1984b. *Southern California: A Region in Transition.* Vol. 3. *Locational Patterns of Ethnic and Immigrant Groups.* Los Angeles: Southern California Association of Governments.

SPNB. See Security Pacific National Bank.

Subber, William, and Edward Tchakalian. 1976. Memorandum to James J. Quillan, State Labor Commission, San Francisco. (21 July).

Trombley, William. 1984a. "The Bottom Line: Is Metro Rail Worth It?" *Los Angeles Times* (1 July): I-1+.

———. 1984b. "Metro Rail as Catalyst for Growth: Debate Continues." *Los Angeles Times* (4 July): I-1+.

———. 1984c. "Metro Rail's Builders Will Tunnel into the Unknown." *Los Angeles Times* (5 July): I-1+.

———. 1984d. "Private Metro Rail Lobby Spreads Largesse in Capital." *Los Angeles Times* (21 Apr.): I-1+.

———. 1984e. "$3.3-Billion Cost Looms as Metro Rail's Achilles' Heel." *Los Angeles Times* (3 July): I-1+.

———. 1984f. "Traffic, Downtown Growth—The Ways to Metro Rail?" *Los Angeles Times* (2 July): I-1-2+.

Turpin, Dick. 1979. "L.A. a 'Bargain' Counter." *Los Angeles Times* (7 Oct.): IX-1+.

U.S. Bureau of the Census. 1962. *County and City Data Book 1962. A Statistical Abstract Supplement*. Washington, DC: Bureau of the Census, U.S. Department of Commerce.

———. 1963. *Census of Population: 1960*. Vol. 1. *Characteristics of the Population*. Pt. 6. *California*. Washington, DC: Bureau of the Census, U.S. Department of Commerce.

———. 1972. *County and City Data Book 1972. A Statistical Abstract Supplement*. Washington, DC: Social and Economic Statistics Administrator, Bureau of the Census, U.S. Department of Commerce.

———. 1973. *1970 Census of Population*. Vol. 1. *Characteristics of the Population*. Part 6. *California*. Sec. 1. Washington, DC: Social and Economic Statistics Administration, Bureau of the Census, U.S. Department of Commerce.

———. 1977. *County and City Data Book 1977. A Statistical Abstract Supplement*. Washington, DC: Bureau of the Census, U.S. Department of Commerce.

———. 1982. *1980 Census of Population*. Vol. 1. *Characteristics of the Population*. Pt. 6. *California*. Washington, DC: Bureau of the Census, U.S. Department of Commerce.

———. 1983. *County and City Data Book 1983. A Statistical Abstract Supplement*. Washington, DC: Bureau of the Census, U.S. Department of Commerce.

———. 1984. "Socioeconomic Characteristics of U.S. Foreign-Born Population Detailed in Census Bureau Tabulations." *United States Department of Commerce News* (27 Oct.).

Waters, Tim. 1984 "The 'South Bay Curve Curse.'" *Los Angeles Times* (15 Apr.): IX-1.

Whitt, J. Allen. 1982. *Urban Elites and Mass Transportation*. Princeton: Princeton University Press.

Yago, Glenn. 1983. "The Coming Crisis of US Transportation." *International Journal of Urban and Regional Research* 7: 577-601.

Zeman, Ray. 1944. "City 'Smog' Laid to Dozen Causes." *Los Angeles Times* (18 Sept.): I-1.

# 3

# London

## Emrys Jones

London was preeminent among modern cities for more than a hundred years after the first decade of the nineteenth century. It was then that the city reached the magic figure of a million, a population not attained by Paris until mid-century, New York until 1871, Berlin until 1880, and Vienna until 1885. Outside Europe, Tokyo may well have reached a million at about the same time as London. From the second decade of the 20th century, however, London's primacy was challenged by the New York metropolitan region, challenged in turn by Tokyo in the 1950s. Since then Third World cities have entered the lists, and the older cities have declined. Exactly where cities now stand in the league depends largely on where boundaries are drawn.

Figures notwithstanding, London is still in the premier league of world cities, and this chapter considers its continuing role and importance as well as its many problems, using the framework of the great changes of the last two or three decades, and with constant reference to its historic past. The historic core, the City of London, with a continuous history of over a thousand years, not only has to be accommodated as a unique fragment in the administration of the city, but is an invaluable asset to London's modern functions. That London emerged as the greatest European city in the 17th century and became the center of an empire in the 19th has shaped many aspects of its physical growth and environment, functions, and status. In many ways the London of today is different from that of the last century, but today's city has inherited the latter, with its advantages and disadvantages. London is still a national capital; it has switched its financial interests from empire to world; its name still carries a little magic.

Although, like most other giant cities of the world, London has failed to come to terms with the late 20th century, it has at least managed to conserve many of the virtues of past centuries. Among the strands of historical continuity, for example, is that—geographically in the main—it is a world node for communications: Heathrow is one of the world's busiest airports. London has maintained its role of world center for banking and insurance. The city is still a seat of government and coordinator of a commonwealth of nations. London remains a major manufacturing and trade center, though to a much lesser extent than before. And if manufacturing has diminished, London is still enhancing its position as a center for professional services, learning, and medicine, and has maintained an envious position in the arts, particularly music.

In spite of a rapidly changing skyline in the last quarter century, when high-rise office blocks have challenged St. Paul's Cathedral, and despite the towering apartment blocks rising above the inner city residential areas, the predominant impression of London is its human scale—a tradition established by 18th century squares and confirmed by Victorian streets and terraces, and by sprawling interwar suburbs of detached and semidetached houses. The features which captured Rasmussen's imagination when he referred to London as "the unique city" still prevail (Rasmussen, 1967). Preservation of the green squares, the many parks, and the enveloping green belt add to an air of spaciousness. Preservation has underlined historical continuity, and London's wealth of things past has enriched its fabric, making it an unparalleled attraction to tourists and visitors. London is a friendly city with much to offer its own people and those who visit it. We shall see below that its very age and former preeminence pose great difficulties for the city's economic well-being, and London has its share of social and ethnic problems inseparable from the life of a modern cosmopolitan city. But London takes its place in the upper hierarchy in the network of supernational cities as well as being the first city of the United Kingdom. The city faces a period of intense concern even when its government is in balance: London needs to replace much of the old fabric and make good obsolescence; it must ease unemployment problems; it must bring its roads and transport system into a condition to meet the future. A comparison with other great cities may well encourage much of what is being done, set London's problems in a wider perspective, and enable the future to be planned in a better way.

## Physical Extent

While London's attributes are easily described, it is much more difficult to define the city in physical terms. One of the enduring characteristics of London has been the inability of municipal boundaries to keep pace with its growth. This phenomenon has been made almost impossible by the suburban mingling of town and country, and the increasing tendency for many who work in London to live well beyond its confines and still rely on its services in many ways. To some extent the problem of boundary is eased by the presence of the green belt, initiated in 1938 and confirmed by Abercrombie's 1944 Plan for London (1945), as a means of limitation or containment—the modern equivalent of the medieval wall (D. Thomas, 1970). But the corollary was the development of new towns beyond the green belt to take the overspill of London's growth. The inner limit of the green belt very approximately coincides with the limit of Greater London (administered by the Greater London Council), a convenient unit for discussion.

It would be unrealistic not to recognize an outer metropolitan area extending far beyond Greater London with close economic and social ties with the city and from which half a million people commute daily. With a radius of about 65 kilometers from the center, this outer area together with Greater London can be thought of as the metropolitan area, and is so recognized by the census.

Continuing the model of concentric rings, the area within Greater London can be divided into two: an Inner London consisting of 13 of 32 boroughs of the Greater London area, and coinciding approximately with the area of the London County Council which controlled the city from 1888 to 1965. This, too, is Victorian and Edwardian London (pre-1914) with all the characteristics implied by those terms. Outer London comprises the remaining 19 boroughs, a broad belt of suburban housing, the result of unprecedented expansion between the two world wars. Within Inner London we may distinguish a Central London, the heart of London, bounded by the great railway terminals; and within this area again is the City of London, the historic square mile only marginally greater than the area once bounded by the medieval wall.

These concentric zones, representing historic phases in growth, have contrasting physical and socioeconomic characteristics; they are convenient units for discussing most features of London's life and are constantly referred to in the text. Before doing so, some clarification of the government of London is necessary, to understand later the administrative framework of policy-making, control, and financing. Today's government is the result of a 1965 reorganization when the Greater London Council replaced the London County Council with control over an area more consonant with the expanded city (Rhodes, 1970). There seemed, at the time, an overwhelming need for a metropolitan government in which strategic planning of land use, housing, and transport could be coordinated and many of London's problems seen in toto. The boundary chosen seemed to enclose the smallest area which was reasonable to meet this aim. At the same time, 32 boroughs were created as a second tier because many administrative matters could be dealt with more easily at the local level. The borough councils are in fact quite powerful and have a wide range of responsibilities including most aspects of housing, and even education in the outer boroughs. The boroughs serve large populations, 26 of them being over 200 thousand and 5 over 300 thousand. Perhaps the division of responsibilities had within it the weakness now attributed to the Greater London Council, which is responsible for only 16 percent of local government expenditure in London.

The government introduced legislation in 1985 to abolish all metropolitan councils in Britain, putting local authority entirely in the hands of the boroughs but creating a number of semivoluntary bodies to coordinate those functions not confined to individual boroughs. The ultimate effect may well be that power will be even more centralized, as the semivoluntary bodies will be responsible to central government. There is little doubt that most people concerned with the government of London regard this move as a retrograde step (Sharpe, 1984). "The absence of a core authority has no parallel in any other major city in the Western world" (Travers, 1984). Now it may well be that the problems of any city are unique, but as Norton (1984) states, the nature of the problems is not, and the universal acceptance in Europe of a middle tier between local and national authority makes the London solution out of line with the general acceptance that very large metropolitan areas demand a unifying authority which can take an overall view of strategic planning and economic policy. Moreover, such a unified body is the logical link between metropolis and metropolitan region. In spite of this, dissolution was inevitable in 1986 and it remains to be seen how effective the many coordinating bodies will be in safeguarding that interests of the boroughs individually will not be exploited at the expense of London as a whole.

In the 1965 reorganization, the City of London, steeped in historical trappings, retained its own government and it is unlikely that such a unique structure will not continue to be protected.

## Demography

The most striking feature of London's demography is its declining population. The 1981 census figure of 6,713,165 is the lowest total registered since 1901, and compared with the 1971 figure of 7,452,346 represents a fall of 9.9 percent in a decade (Great Britain, 1982). In a wider context the metropolitan area also declined but only by 3.3 percent, to a total of 12,112,900, also indicating that the period of continuous growth is at an end. Within Greater London were marked differences in change in the two zones. Outer London declined from 4,420,411 to 4,215,187—4.65 percent; Inner London declined from 3,031,935 to 2,497,978—17.61 percent. This figure is the lowest total population for Outer London since 1931, and the lowest in Inner London since 1841. The three central boroughs, Camden, Kensington/Chelsea, and Westminster fell from 638,957 to 507,813, a change of 20.5 percent. The pictures are of a city being drained from the center. The City of London itself is an exception; it had reached an unprecedented low of 4,245 in 1971, but increased to 5,864 on the opening of one housing project—the upper middle-class Barbican Towers. Natural change (births minus deaths) has shown many small variations since 1945, but usually a gain. This gain hovered between 25 and 60 thousand but dipped sharply after 1964 until 1976 when deaths actually exceeded births. Positive change has risen gradually to 14,522 in 1982. Even this small component adds to the total change and indicates that in ten years there has been a net outward movement of 753,703.

Outmigration has been high in Inner London and less in the comparative stability of the suburban fringe. Here movement is between 5 and 10 percent, though this is not necessarily movement out of London, because a considerable amount is intraurban. But in Central London, change is very high, in many wards exceeding 20 percent. This figure reflects the temporary nature of residence in rooming districts, more the characteristic of a distinctive ecological region in London than indicative of underlying demographic trends in the city as a whole.

The decrease in population is paralleled by a lesser decrease in the number of households, from 2,651,815 to 2,517,641, a change of –5.1 percent. Fifty-eight percent of the population lives in households of one or two persons, and single-person households account for 26 percent of the total—well above the national average of 21 percent. One person in ten in London lives alone.

Population structure by age and sex, compared with Britain as a whole, shows relatively less in the age group 5-20, and a slight excess at 20-35, the kind of imbalance one would expect. As in the whole country, London's population is aging, with 17.88 percent of pensionable age (men over 65, women over 60), a figure which varies very little from zone to zone or even from borough to borough. But changes in some age categories are startling. During the intercensai period the numbers aged 0-4 decreased by –27.8 percent; 5-15 by –16.7 percent; 45 to pensionable age by –21 percent; and those over 75 increased by +10.7 percent.

As far as the immediate future is concerned, trends show that relatively small change is likely. In the next decade the decrease in population will be about 800 thousand compared with 700 thousand in the last decade, so the total population in 1991 will be about 6.7 million, compared with 6.8 million in 1981 (Hamnett and Randolph, 1982). This prediction suggests that London is approaching a relatively stable condition. There will continue to be a considerable contrast between Outer London, where natural growth may well more than balance the loss by outmigration by the late 1980s, and Inner London, which will show a continuing sharp net decline. Overall stability will mask an increasing internal imbalance. Trends also indicate that the number of one-person households will increase significantly with implications for the kind of housing being built.

Apart from population decline, the most important variable in London's demographic structure is the number and distribution of colored immigrants and their children. For simplicity and convenience I use the figure for native-born New Commonwealth[1] and Pakistan immigrants. As the main influx of migrants was in the 1960s, we now have a second and even third generation of people in London of colored or mixed parentage, but who are not differentiated by color or ethnic parentage by the census because they were born in Britain. The numbers registered are, therefore, very much smaller than they are in fact, and the distribution may seem relatively restricted because we have no indication of how far second-generation colored are moving into new localities (see table 3.1). The total increase 1971-81 was from 465,515 to 630,859, or +35.5 percent. Among the migrants the greatest increase has been among those from Bangladesh and Pakistan (+93 percent).

TABLE 3.1
Country of Birth of Population by Continent/Region—
Greater London, 1981 (in thousands)

| Continent/Region | N |
|---|---|
| Europe | 382 |
| Asia | 296 |
| Africa | 170 |
| Caribbean | 168 |
| Mediterranean (Gibraltar, Cyprus, Malta, Gozo) | 69 |
| United States | 22 |
| Canada | 10 |
| Middle East | 30 |
| Australasia | 27 |
| South and Central America | 11 |
| U.S.S.R. | 7 |
| Total population | 6600 |
| Born outside U.K. | 1200 |

SOURCE: Great Britain, Office of Population Censuses and Surveys (1982: Table 51).

The distribution of New Commonwealth immigrants makes a distinctive and familiar ecological pattern (Lee, 1977). In Outer London as a whole they account for 11.81 percent of the population, in Inner London for 19.37. Even these figures are extremely generalized, for in Outer London half the boroughs are considerably lower than the average figure, but two have very high proportions—Brent with 33.46 percent and Ealing with 25.41 percent—and even within these boroughs high figures are associated with only a few wards: in Brent four wards have over 50 percent and in Ealing one has 71.0 percent, another, 85.35. London's concentration is a broad belt around the city center, a belt which dominates Inner London and at times laps over into Outer London. This belt would certainly correspond to what the Chicago ecologists called the "zone of transition," where obsolescence, multiple occupancy, poverty, and lack of assimilation combine. The zone does not completely ring the center, being most conspicuously broken by a sector lying on a northwest axis having its origin in the rich West End and extending through fashionable Hampstead to the northwest suburbs. A similar axis goes west along the river. The westward concentration of New Commonwealth immigrants is extensive partly because of its accessibility to Heathrow, the present-day point of arrival (see figure 11 in Greater London Council, 1984a).

What little evidence there is suggests that second-generation immigrants have not become suburbanized to any particular degree, but rather have intensified the concentration in existing and, by now, well-established areas in north, west, and south Inner London (Lee, 1977). Wards of very intense concentration are still few in number, and in no sense can we distinguish ghetto conditions. But the need is felt, particularly by immigrants from India and Pakistan, to provide mutual support and form cohesive communities in which they can provide their own food, services, and places of worship. There is a voluntary acceptance of conditions approaching an "urban village" which gives a feeling of protection to new migrants, and which acts as a cultural cushion even for those who will eventually become culturally assimilated into the host community. Nevertheless the presence of extensive minorities of this kind forms a critical backgound to some of the problems of the Inner London area, discussed below.

## The Fabric of the City

London's different parts were an outcome of its historic growth; their characters reflect the nature of the urban environment at different phases of expansion. Thus much of Central London has areas and buildings of great historical significance, particularly in the City, with the Guildhall, Bank of England, and St. Paul's Cathedral, and in Westminster, the embodiment of the machinery of state for six hundred years, as witnessed by the Houses of Parliament and Whitehall, the royal palaces and parks.

Although much remains of historic Central London, particularly its monumental buildings and residential squares, its character has been changed by constant rebuilding and the development of new functions. Today it is an area dominated by office blocks—many of them high-tower blocks, retail shopping, theatres and restaurants, galleries, and public buildings. Most of this development is contained within what was Georgian London (pre-1830) but the expansion of office blocks has now spilled beyond in the north and west, so that the newest buildings are often cheek by jowl with the oldest, and the most sophisticated offices overlook the most ancient slums.

In general terms we can equate the remainder of Inner London with Victorian London (1840-1900) and its legacy of high-density, well-built terrace houses, most of them lacking what have now become essential facilities. The more prosperous houses of that period are given to subdivision and multiple occupancy, thus exacerbating the paucity of amenities. Obsolescence is, then, all pervading. But the low urban profile of this zone has been broken dramatically since the early 1960s by large blocks of apartments which all the Inner London borough authorities saw as the solution to problems of accommodation arising from war destruction and slum clearance. Theoretically the blocks were meant to maintain high densities while at the same time releasing land for recreation, and bringing the living standards of poorer people into the 20th century. Unfortunately they also created new social problems, as well as revealing faults in construction.

Outer London is the creation of post-1918 suburban expansion, dominated by the semidetached villa and epitomizing the Londoners' dislike of living in apartments unless forced to because they are very poor, or choose to because they are very rich; and their insistence on the privacy of a garden, or even a garden with no privacy at all. These desires result in a light density, and a general formlessness which pervades vast areas. The semidetached house first appeared in a London suburb in 1800, and can be thought of as the last stage in the devolution of the English country house, its minute lawn and garden the last vestige of an estate. The semidetached house has become the acceptable norm of the middle class, the ideal of all those below.

The housing contrast between Inner and Outer London is reflected too in tenure, the link between socioeconomic conditions and residence (Shepherd et al., 1974). Owner occupancy, 48.55 percent for Greater London as a whole, is 61.85 percent in Outer London, 27.29 in Inner London (see table 3.2). The privately rented sector has been virtually eclipsed apart for the rich West End and the rooming areas, and the vast majority of tenants are tenants to local authorities, dominating Inner London.

Housing apart, changes in the last two decades in London's urban environment and land use have been continuations of long-established trends. The Second World War had destroyed vast areas of Inner London, including a third of the City of London. Warehousing in 1949 was less than a million square meters compared with two million in 1939. Industrial space, at less than a million square meters had been halved, and offices decreased from 3.3 million square meters to 2.8 million. Housing had lost a third of its stock. But by 1970 warehousing had largely recovered its loss, offices had already surpassed replacement, and housing was beginning to replace the bulk of its losses. Industrial land use remained static. Between 1971 and 1980 office use increased by another 8.5 percent, warehousing by 7.5, housing by only 1.5 percent while industrial land use fell by a further 8 percent.

Industrial change will be dealt with below, but all of its most dramatic environmental effects can be seen along the river, which has also been consistently losing its functions as a major port. The historic expansion of the port was down river, from the Pool of London, where once all activity was concentrated, to new docks which could accommodate larger vessels and greater handling: the West India Docks (1802), London Docks (1805), East India (1806), Surrey (1855), Royal (1880), and Tilbury Docks (1886). The same procession has now seen abandonment and dereliction moving downstream, leaving the former Docklands an area of waste and speculation. With the exception of Smithfield meat market, wholesale handling of goods has dispersed from its central location. Retail shopping is still concentrated in the West End, and increased by the closing of stores on the eastern part of the shopping axis. Efforts have been made to consolidate the traditional location of local shopping in London, and one exception only weakens the general pattern and points to possible future dispersion and suburbanization of shopping: Brent Cross, located where the North Circular Road meets the M1 highway, is a response to the need for accessibility at a time when congestion is making a nonsense of the benefits of centrality; it is a completely enclosed and air-conditioned shopping center of 21.5 hectares with parking for 3,500 cars. Brent Cross provides two department stores and branches of several others, a supermarket, and 80 shops—a fragment of the West End in suburban London.

One last conspicuous change has been the increase in the number and size of hotels, encouraged by government aid to boost the tourist industry. The spate of new hotels also shows a shift westward as links with Heathrow become increasingly important.

TABLE 3.2
Percentage in Major Tenure Classes—London,
1971-1981

| Location | Owner-Occupier | 1971 Local Authority | Owner-Occupier | 1981 Local Authority |
|----------|----------------|----------------------|----------------|----------------------|
| Greater London | 44.4 | 14.9 | 48.6 | 30.7 |
| Outer London | 55.7 | 20.9 | 61.9 | 23.2 |
| Inner London | 19.4 | 30.3 | 27.3 | 42.8 |

SOURCE: Greater London Council.

The fabric of the city is held together by its transport system, its streets and railways and their ability to cope with the demands made on them by an extremely mobile population. Apart from the pressures of countless intra-urban movements, the main stress comes from the daily influx of commuters. Central London has a resident population of less than a quarter million, but a daily work force of well over a million, though this has fallen fairly consistently since 1971 when it peaked at 1.25 million.

Of those entering Central London daily, over 71 percent use British Rail or London Transports Underground system or both. Figures also show that the number of rail passengers has fallen, though most of this decline is within Greater London, and much less among long-distance commuters. During the last two decades the percentage using private transport has risen from 14 to 19. The remainder use buses. Private movement has increased in spite of mounting controls; perhaps it would be fair to say that it would have increased even more were it not for controls, or until congestion would have been intolerable. Traffic control has become increasingly sophisticated, with computer-controlled lights, bus lanes, and complicated one-way systems. Together with parking restrictions these measures have effectively kept London road traffic moving with no loss of speed and little environmental deterioration. But passenger traffic is only the lesser part of the total: most of it is freight, and little can be done to lessen the traffic or control the size of large trucks.

The obvious way to relieve the funneling which results from the basically radial road system is to create circular and transverse roads (a device first used in the 1750s with the first ring road, Marylebone Road). London's only clearly circumferential road for fast traffic is the North Circular Road, which sweeps through the western and northern suburbs. A tenuous route running south of the city is its counterpart, but more in the abstract than in concrete. A solution is in sight: a comprehensive ring system further from the city and in the green belt. This orbital road will soon be of motorway standard throughout and should prove a great relief to London from through traffic. The even more acute problems are in the center where the Greater London Council has made several proposals for ambitious ring roads. They would all require capital sums of astronomic proportions and would also involve resettlement of 100 thousand people whose houses would have to be destroyed. This plan is unlikely to materialize. Acceptable movement will continue to depend on traffic management and control.

# Changes in Occupation
## and the Industrial Base

Traditionally London has held a premier position in Britain in all aspects of city functions; it was a magnet for all activities and consequently for a flow of work-seeking migrants at all levels. In the last two decades, and particularly since 1971, the picture has changed markedly.

The number of jobs in London fell from 4.21 million in 1971 to 3.38 in 1982, a loss of .83 million or −19.7 percent. The number of service industries increased marginally from 2.69 million to 2.7, though this is forecast to fall a little. The number in manufacturing industries fell from 1.09 million to 0.63—a dramatic change of .46 million or −42 percent.

As far as industry is concerned, it would be a mistake to think of change or even decline as something new in London's history, though the degree of change is startling (Clout, 1978). Even in the late 19th century heavier industries had already moved out of the capital, and riverside activities were moving downstream. Changes now affecting London are more structural than spatial; to understand this one must look at changes in the economy as a whole (Keeble and Hauser, 1972). What is affecting London in particular is the weakening of those well-established industries specifically associated with Inner London. This area has a long tradition of specialized industrial quarters, most established in the last century, some before that (P. G. Hall, 1962; Martin, 1966). Best known are the clothing districts of East End and West End, the furniture industry northeast of the center, printing and publishing close to the City and Fleet Street, instrument making in Clerkenwell, and jewelry in Hatton Garden. All focused on a large number of small units inextricably woven into the fabric of the city. Many were geared to London's luxury markets, to immediate local demand, even to seasonal fluctuations. By the 1970s consumer demands for many goods had fallen; many small firms had gone out of business and were not replaced; and many succumbed to obsolescence as they failed to match modernization of processes with old buildings. A large number of factors, environmental and social, led almost inevitably to decline. The falling population meant not only that an immediate market had declined, but that there were also fewer in the work force: so much so that there was not a little in-commuting. The inability of small firms to cope with technological changes led to the breakup of family firms and loss of skills and know-how. Supplies of raw materials were moving away, and increasing congestion meant that workshops were losing their inaccessibility. All this had a multiplier effect (Leigh et al., 1983; Wood, 1978).

Not surprisingly, many manufacturers took advantage of moving their plants elsewhere, particularly to new and expanding towns which offered financial aid, new premises and plant, accessibility, and a pleasant environment. Between 1966 and 1974 the net loss in manufacturing industries because of the balance between opening and closing of firms of over 20 employees was 170,300, of which 88,200, or 51.8 percent were from Inner London. A total of

105,300 jobs went out of London—36,200 to assisted areas, 26,000 to overspill towns, the remainder elsewhere.

To some observers there seems to have been a "London factor" which produced a bias even in the structural changes of industries in Britain as a whole (Wood, 1978). Decline was more accelerated in many sectors. For example, between 1966 and 1974 clothing and footwear manufacture in London dropped by 38 percent compared with 10 percent nationally; engineering and electrical goods manufacturing dropped by 25 percent although there was little appreciable change in Britain as a whole. Food and drink manufacturing dropped by 28 percent compared with 3 percent nationally, paper and publishing by 24 percent compared with 3 percent nationally. In all these activities the Inner London area was proving the more vulnerable region, as its manufactories were the oldest, most restricted, least able to adapt to changing conditions. Technologically weak industries in obsolescent buildings suffer disproportionately, a problem aggravated by the difficulties of developing and changing in situ because of high costs and shortage of skills. Nor is the situation helped by the positive advantages offered elsewhere, as mentioned above. These, together with managerial preferences and regional policies, have aggravated the problem. Paradoxically, encouragement to increase manufacturing activities in assisted areas (deprived areas) in depressed parts of Britain was helping to create a new depressed area in the very heart of London.

Looking to the future, further decline seems likely, if not inevitable, particularly in view of the economic problems of Britain as a whole. It is unlikely that circumstances will encourage new investment. Some traditional industries are relatively strong and should survive, such as the clothing industry, printing, and publishing; but there seems little opportunity for craft industries generally. Outer London is now being hit in the same way as Inner London has been in the last two decades, for, paradoxically, when firms have moved back to Outer London with reinvestment and innovation, efficiency has increased resulting in shedding of labor. To a large extent both Inner and Outer London are now looking to government policy and aid, because borough initiative and finance is of necessity limited. There is a strong case for a policy based on London as a whole. Meanwhile, piecemeal developments in specific areas, some of which I mention below, offer only minimal alleviation.

In recent structural changes the service sector has continued to flourish, and London has shared fully in subsequent developments. Within the sector the office component has seen the more spectacular changes, with very obvious results in the urban landscape. The specialization of activities which first gave rise to the office became apparent very early in London. Even in the 18th century, the Bank of England, the Admiralty, the East India Company, and Somerset House were explicit and purpose-built. During the early 19th century insurance companies built their headquarters near Fleet Street and the Strand. In the City itself the great exchanges, coal (1849), stock (1854), and wool (1874), marked an expansion of corporate activity; and soon office streets like Victoria and Queen Victoria (both 1871) and Shaftesbury Avenue (1886) appeared. By 1891, 78 thousand clerks worked in London. In this century the City consolidated its primacy by extending to Kingsway (1904). In addition to commerce and finance, bureaucracy added its quota in a burgeoning civil service in Whitehall, London's County Hall, and several borough halls, and so did communications in Fleet Street and its associated publishing activities buildings, and in Broadcasting House, heralding the beginning of the transactional society.

New and massive buildings are the manifest signs of increasing office activity, but they can also exploit the potential of existing buildings which undergo a change in their use. In London the function of residential squares of the 18th and 19th centuries changed dramatically with the wholesale suburbanization of the former residents. Their large houses were particularly suited for small firm and professional activities, not least because they were architecturally attractive, an advantage to those selling their skills to discriminating clients. By the postwar period few of London's squares were residential: size, flexibility, and the need for personal contact make the houses ideal for professional use. Perhaps the best example is the once-residential Harley Street, which now houses over nine hundred doctors in consulting rooms which have a different kind of multiple occupancy than that usually met in city centers (Pound, 1967).

Office building is another matter: it means producing a commodity to meet specific needs, built to last, and subject to planning control. Office production has created a new industry, property companies, whose number grew from 25 before the war to 150 by the office boom of the early 1960s (Jones, Lang, and Wootton, 1980). Control has been implicit in all office-planning proposals for London since 1944, but it was fairly relaxed until the London County Council took firm control in 1956, when 400 thousand square meters of office space were being built per annum. Much of this building was, of course, replacement for war damage, but it led to a boom which peaked in 1962. The figure of 400 thousand square meters was repeated for five years. In an effort to stem the demand, an advisory body (Location of Offices Bureau) was set up to advise companies on office location and point them away from London (Manners, 1977). This trend was within a general policy of decentralization put forward by Abercrombie (1945), which lay within the broader guidelines of earlier policies on redistribution of population in Britain. Results were modest. In 14 years, up to its dissolution, two thousand firms and 150 thousand jobs moved, but a marked feature was the reluctance of firms to move far from the capital. In the first three years 46 percent remained within the Greater London area, and only 24 percent moved more than 80 kilometers away (P. K. Hall, 1972).

The spate of office building continues. Between January and June 1983 development began on .3 million square meters, 50 percent more than the growth in 1982; and 22 percent of this amount was pre-rented. True, half the figure represents refurbishment, but the total space is still considerable. The total space under construction in mid-1983 was over a million square meters;

and at that time well over a million square meters was on the market in Central London—8.6 percent of the total office stock.

As far as occupational structure is concerned, the shift in the economy toward a postindustrial period is clear (Evans and Everseley, 1980). With its traditional role in the tertiary and quaternary sectors, it is not surprising that London is in the forefront. The numbers employed in offices is constantly high, 1.7 million jobs in 1966 and 1.75 in 1982, contrasting with the decline in manufacturing. This situation does little, however, to provide work for those who lose it in other occupations, simply because Inner London can provide only a small proportion of the skills required in office jobs—hence the immense inflow of daily commuters. The bulk of office jobs are in the lower-skilled, lower-paid categories. See table 3.3.

## Some Special Characteristics of London

Two characteristics make London unique and contribute to its well-being today. Both are intrinsic and long standing. The first is relative location. For five centuries London has been a node of international communication, and particularly a focus for transatlantic traffic. For the last few decades this activity has centered on Heathrow Airport. Air passenger traffic at all London's airports grew phenomenally from about 2 million in 1950 to a peak of 30 million in 1973. Two-thirds of this is handled at Heathrow; Gatwick deals with 5.7 million; Luton and Stansted share 3.4 million. London airports handle 80 percent of Britain's total air traffic. Since 1968, enormous energy has been expended in nominating a third major site, but the pressure is still on Heathrow, where developments may well point to a future in which the airport itself and its associated hotels will become the center of transactions of all kinds, especially if links with the city cannot cope efficiently with traffic flow. The site itself could be used not only as a transit point, but as a venue for business meetings, symposia, and exhibitions. By the mid-1970s Heathrow had seven hotels with greater capacity than 200 rooms each, and a further ten have been added since then.

The second distinctive characteristic of London is its history and the consequent attraction for tourists. Of tourists who come to Britain, 95 percent visit London, and 75 percent of them stay in the central area. Peak numbers were in 1978, when 8.4 million visited London, but this figure dropped to 7.2 million in 1982. In that year, however, tourism brought in £2 billion, or 8.2 percent of London's gross domestic product. Directly and indirectly tourism supplies a quarter of a million jobs, of which 130 thousand are in catering (London Tourist Board, 1982).

A Development of Tourism Act in 1969 gave an enormous impetus to hotel building and improvement schemes by giving substantial grants per room. The result was a boom period. Although several new hotels have been built in the traditional center of the West End, there has also been a conspicuous

TABLE 3.3
Jobs—Greater London, 1951-1982
(in millions)

| Category | 1951 | 1966 | 1971 | 1976 | 1982 |
|---|---|---|---|---|---|
| Manufacturing | 1.55 | 1.30 | 1.09 | 0.83 | 0.63 |
| Offices | 1.56 | 1.70 | 1.65 | 1.78 | 1.75 |
| Others | 1.18 | 1.43 | 1.47 | 1.10 | 1.00 |
| Total | 4.29 | 4.43 | 4.21 | 3.71 | 3.38 |
| Manufacturing as percentage of total | 36 | 29 | 26 | 22 | 19 |
| Offices as percentage of total | 36 | 38 | 39 | 48 | 53 |

SOURCE: Greater London Council (1984b).

movement westward of hotel provision, partly to serve the terminal of Heathrow. The one outstanding exception is an 826-room hotel in St. Katherine's Dock, near a new World Trade Centre and very near the Tower of London. The riverside location and nearness to the Tower and resources of the City compensate for an otherwise East End location.

It would be tedious to recount the attractions of London to the visitor. Most are historic and many connected with transatlantic links. But in addition to monuments of the past and their associations must be added London's attraction as a retail shopping center, especially for luxury and specialty goods, and its value as a center for the arts. Its 40 or so regular theaters, crowned by a national theater, 2 opera houses, and several concert halls are the solid core of its entertainment. Recently a cultural center (The Barbican) has been opened in the City of London: its theater, concert hall, and museum may well be a countermagnet to the West End.

In view of the intrinsic value of the fabric of London it is not surprising that the city is extremely conscious of the need for historic preservation. One aspect has been its reluctance to change the aspect and character of small shopping centers—and residential areas—like Hampstead and Blackheath. Equally jealously kept is the character of its 18th- and 19th-century squares and its historic buildings. London has 350 preservation areas and 30 thousand listed buildings. Since the beginning of the century the London County Council, and later the Greater London Council, has had a team of specialists whose task it has been to safeguard its historic heritage. It has more than a thousand buildings in its care. Considering the financial return in tourism, this investment is sound.

It is appropriate to cite one specific case of preservation and the way in which change can be accommodated within a comprehensive plan. Departure of the wholesale fruit and vegetable market from Covent Garden left an extensive area of land which by virtue of its central location could have been ripe for office building: but it was also part of a considerable community of people. A planning team representing two boroughs and the Greater London Council put forward a plan in consultation with the local people. Much of the property was purchased; the result is an area of considerable rehousing, including new homes for 15 hundred, modest office accommodation, but, most important, a revitalized shopping area focusing on the early 19th-century market and retaining all its essential features. The Covent Garden project is the best example of capitalizing on a historic site and providing a new focus for tourism as well as resuscitating an area which had been in severe decline.

From the last point of view the refurbishing of St. Katherine's Dock near the Tower of London may be deemed less successful. There were outcries from the local borough of Tower Hamlets at the building of the hotel and trade center referred to above, and four hundred private houses, when there were six thousand families on the borough housing lists waiting for accommodation. The St. Katherine's project is part of a continuing controversy on the projected use of derelict land along the riverside in London which I refer to later.

## Global Aspects

During the last 20 years London has maintained its position in the international urban system by virtue of its share in international transactions. In this period the scale of internationalization of industrial, commercial, and financial activities has increased dramatically, and London has capitalized on its geographical location, expertise, and traditions. The square mile of the City in particular, but also the West End, has attracted a massive share of the multinational market. The new confidence of the City may well be symbolized by the tower of the Nat West bank, 600 feet high and by far the tallest building in London, built from profits one-third of which came from overseas. It seems as if most of the world is anxious to locate in this magic spot.

According to Blandon (1983), all but four of the hundred largest banks in the world have a branch in London. European banks have long-established links and in 1961, 100 foreign banks were represented. By 1983 the figure was 460, 391 directly represented and 69 indirectly. Seventy countries are represented and their banks have a total of 38 thousand employees. Most of these banks employ between 50 and 400 people, but the Bank of America has a staff of a thousand.

Banking and its associated activities find a number of advantages in locating in London. Although communications now make physical accessibility less important, being at a node in international air traffic is a great asset. Another plus is that much international transaction is in English. In addition London, exploiting a long tradition, supplies an expertise, not only in finance, but in all associated professional activities. One of its great advantages, too, is its compactness. The concentration of activities, however much inertia plays a part, is partly because face-to-face transactions are still thought important, in part because of the status associations of the City. Land use in the City is very specialized and highly selective. Prestige dictates that what property dealers call "prime banking areas" are in the immediate vicinity of Threadneedle Street; a further "acceptable banking area" is at the south end of Moorgate. Consequently the pressure is extremely high in a very restricted area, and higher density makes for higher office blocks. Outside the banking world there is more lateral growth, though most of it tries to maintain a perceived connection into the city. The alternative is development in the West End (King, 1985).

Competition, particularly for prime City sites, leads to very high rents, largely offset by low labor costs; on the whole London compares very favorably with all major business centers. A third of the thousand largest companies in Britain have headquarters in Central London, and 70 percent of the hundred largest (King, 1985).

London is also a major center for core activities relating to peripheral development: it supplies the expertise for overseas development, particularly in the developing world. It has a very large number of engineering firms, planners, architects, transport experts, et cetera, much of whose work is overseas. In addition, its universities and colleges have an extremely high ratio of foreign students. An increasing industry, to which over 80 establishments are now dedicated, is teaching the English language.

To the extent to which all these activities are open to change from without and have to be gained in competition with other world cities, this particular sector of the economy can be said to be vulnerable. On the other hand the economy also has strength in knowledge-based industries—the quaternary sector—already well established: not only education, but publishing and printing (now the largest single manufacturing industry), broadcasting and television, and such enterprises as Reuters' new multimillion-pound global economic information service.

## Inner London

A fundamental part of London's life which has emerged from the above account is that the initial convenient division of the city into Inner and Outer London reflects a fundamental division warranting the recognition of an inner city problem. We can dispense with those parts of Inner London which comprise the City and the West End. The City is a thriving, largely 20th-century financial machine: the wealth-producing area which has too few permanent inhabitants to warrant a socioeconomic analysis. The West End is the wealth-consuming area par excellence, the bureaucratic and shopping area, again largely divorced from the relatively few people who live there. For the rest, Inner London is a city within a city, and a socioeconomic analysis might be a tale of two cities.

I have already outlined some of the characteristics of Inner London in terms of activity and affluence. Environmentally it has a 19th-century inheritance with all the consequences of obsolescence. Continued decay has led to a crumbling fabric within a mixture of land uses: warehousing, manufacturing, railway yards, gas works, institutions, and derelict land all provide the warp and woof of the environment, and lack of open space, noise, dirt, and traffic pollution add further familiar ingredients. The physical disadvantages of the fabric of residential areas are aggravated by overcrowding, the need to share amenities, the high proportion of children. The social component is dominated by high unemployment, lack of job opportunities, social tensions of ethnically mixed communities, and the residents' inability to escape these conditions. The contrast with Outer London, a relatively prosperous, stable, mobile suburban community in a favored environment is clear-cut. It is probably true that society in Britain as a whole is becoming more polarized in spite of all efforts to improve the conditions of poorer elements. Disraeli first brought attention to the two nations, and it has long been a habit to equate them with North and South: in the last two decades they have been reflected in Inner and Outer London.

Between 1971 and 1981 Inner London lost half a million people and even more jobs. One house in three is unsatisfactory in some respect and over a quarter of a million people are on local authority housing lists. The bulk of London's colored immigrants share the conditions.

Let us consider unemployment first. In 1980 the former plateau of unemployment in London as a whole—about 150 thousand—soared dramatically to 386 thousand. A new basis of registration revised this figure to 315 thousand, but by 1983 even on this new basis the figure was 375 thousand. Comparative percentages of unemployed persons are:

|  | 1971 | 1981 | 1983 |
|---|---|---|---|
| Greater London | 4.7% | 9.0% | 10.34% |
| Outer London | 3.7% | 7.0% | 7.94% |
| Inner London | 6.1% | 12.3% | 14.40% |

These figures should be seen against national averages and the structural changes which have produced them. But the point being made here is the relative impact of the changes on Inner London, which now compares in unemployment with the depressed areas of the peripheral regions of Britain (figure 21 of Greater London Council, 1984a). Unemployment is a critical element in identifying the pathology of Inner London. We have already seen how jobs in the thriving office sector are largely taken by people living outside the area. Nor is the Inner area uniformly affected: there are boroughs and wards of boroughs in a much more critical situation than the average figure suggests. In Tower Hamlets, for example, unemployment is 19.21 percent. In 12 Inner London wards the figure is more than 40 percent, and the ward of Spitalfields in Tower Hamlets records 63 percent. Averages became academic: the perceived problem is tied to the worst conditions, so that within a uniformly depressed area are festering sores of unemployment which form the basis of a deep-seated malaise.

Housing is a closely associated problem in Inner London. Owner occupation accounts for 27.29 percent of tenure in Inner London, compared with 48.55 percent in Greater London as a whole. Owner occupancy is much lower in some boroughs: in Tower Hamlets it is only 4.59 percent. Local authority housing accounts for 42.77 percent in Inner London but is much higher in, for example, Southwark (64.8 percent) and Tower Hamlets (81.75 percent) (figure 31, Greater London Council, 1984a). Even so, municipal housing by no means meets the total need, and 250 thousand are on the waiting list for new houses. In addition in any one year the estimated number of homeless families is about 20 thousand. In the owner-occupied and rented sectors deterioration of stock is taking its greatest toll. It is calculated that a quarter of a million Inner London houses are unfit for habitation and half a million in urgent need of repair. One in four is in some way unsatisfactory.

Problems have been aggravated in municipal housing areas, not only by normal aging and deterioration which demand considerable resources, but in addition by the deficiencies of high-rise blocks giving rise to immense and unforeseen problems. Demand for housing after the devastation and lack of building during the Second World War and efforts to clear slums in the 1950s and 1960s coincided with the introduction of unit construction and the erection of blocks and slabs of apartments. This effort was a shortcut to provide maximum living space at high density, though not necessarily at low cost. Structurally, many of the blocks have proved suspect, and a section of one collapsed after a gas explosion (1968), revealing flaws in construction which cast doubts on the safety of scores of blocks, in addition to dampness and condensation resulting from the same techniques. Many of these blocks now face demolition.

Social problems arising from living in blocks were no less worrying. Inefficient elevators and overcrowding produce great hardship, and many communal areas are centers of delinquency and vandalism. Families feel isolated and threatened, and young mothers in particular cannot cope with lack of play areas. The break from the traditional two-dimensional life of the Londoner was not a success, in spite of the visual appeal of some block department schemes (e.g., Roehampton). Too many blocks have been the focus of social and economic problems. Some believe that ownership would produce the responsibility to counter the disadvantages and there are some experimental schemes in which tenants have become owners.

The second social phenomenon which by now has become part of Inner London life is New Commonwealth immigrants. I have already referred to their number and distribution. They replicate a classic pattern familiar in North American cities but which emerged very suddenly in London during the 1960s. Historically the city has seen successive waves of immigrants, but assimilation has never been an insurmountable barrier. This situation was even true of the Jewish movement into the East End at the end of the last century; although orthodox Judaism still identifies a concentration in this original hearthland, very large numbers have moved from this poor zone and enjoy a suburban existence in the northwest sector of Outer London. But migrants of the last two decades, West Indians, West Africans, Indians, and Pakistanis, are highly visible, easily stereotyped, and because of color, very difficult to assimilate. The Asians have an ambiguous view of the problem. While many of them are extremely eager that their children take full opportunities in education in order to prosper in Britain, they also insist on maintaining their own language, religion, and mores. That schools are trying to accommodate these ideas points to a future society which will be pluralistic.

Most of the colored community live in old and often obsolete houses, many in multiple occupancy and in conditions of grave social disadvantage: their unemployment rate is double that of whites in the same localities. Incomes are low and discrimination high, and young colored people in particular have been very severely affected by lack of work.

The Race Relations Act of 1976 sought to ensure equality of opportunity and elimination of discrimination, but some feel that we now need positive discrimination to improve immigrants' position and meet their special needs, a view held strongly by the Greater London Council. But the critical and immediate problem is the perception of Inner London's white population of this issue. It is almost inevitable that tensions at times are very high, and not unexpected that at least one serious riot has taken place in London (Scarman, 1981).

Inner London, then, is beset with problems; loss of people, jobs, and morale; aggravated by social and racial tensions which only long-term economic, legislative, and educational measures can hope to overcome. The immediate need is for alleviation, but it is difficult to see where limited aid can best be used.

## Planning and Policies

The foregoing description of the physical, social, and economic structure of London and changes in it in the last two decades or so must now be seen in terms of overall postwar planning and policies. Policy has largely stemmed from and been dominated by the Abercrombie Plan of 1944. The zones of London which have been the framework of this chapter are Abercrombie's (1945). In the simplest terms he saw a pre-1914 Inner London that was overcrowded and obsolete, and the physical destruction of the war encouraged him to think of radical changes in population distribution. Suburban London was stable and must remain so. A green belt must delimit the entire city. Beyond this would be growth, not in an urban sprawl but in discrete communities. The balance of the entire metropolitan area was to change by decanting people and jobs from the overcrowded center and recreating communities in the countryside beyond.

The lowering of Inner London's densities to 336 per hectare entailed an overspill of one million people to be rehoused by public action and possibly another quarter million to privately built housing. Resettlement would be in new towns, finite communities of 50-60 thousand, more or less self-sufficient and having many of the characteristics of Ebenezer Howard's garden city. Eight such towns were designated in 1947 and three more added later, all between 35 and 60 kilometers from the center of London. Low density (six to seven houses per hectare), traditional two-story houses, landscaping, industrial zoning, built-in social facilities all produced an environment admired worldwide. Later target populations grew, and by 1981 their combined total population was 900 thousand, well on the way to the original overspill envisaged. The new towns have attracted industry and maintained considerable self-sufficiency, in some cases to a very high degree.

Two unforeseen consequences of the overspill policy had important effects for Londoners (P. G. Hall, 1984). Initially those who moved to the new towns had specific manufacturing skills, necessary for the development of newly attracted light manufacturing in the new towns. This selective process meant that the least qualified and poorest elements in Inner London did not take part in the move. This situation led to the intensification of socioeconomic problems in the inner area while it contributed to the alleviation of problems of those who did move.

Second, planners underestimated the continuing dynamic nature of population changes when overspill had been taken care of. The southeast region itself was a powerful attraction to people in the remainder of Britain, and the new towns became magnets for people outside London. Also the new towns, based on a young initial population, were initiating considerable natural growth. In addition, the entire area beyond the green belt was proving very attractive to speculative builders meeting the needs of those who, on their own initiative, wished to leave London or wanted to live near the city. Abercrombie had envisaged the zone beyond the green belt as increasing from 2.6 million to 4.2 million, and 90 percent of this in new towns, but by 1981 the population was 5.4 million. The 1961-81 increase in this zone was a third of the total increase of the United Kingdom. Newcomers are not all long-distance commuters. Rather, the new pattern of settlements points to an emerging polycentric urban pattern in the southeast. By the mid-1960s even, the government was encouraging decentralization by establishing growth points beyond even the metropolitan belt. Swindon was elevated to a new city, Northampton and Peterborough greatly enlarged, and an entirely new city, Milton Keynes, was established about 75 kilometers from London.

Most of the population redistribution in the outer metropolitan area and beyond has been spontaneous, and controlled elements have been partly overshadowed. Nevertheless, a new element was successfully established—the new town—and the green belt had been preserved. The latter seems more or less inviolate. Although the idea predates Abercrombie, it was he who thought of it as a distinctive and permanent element, and it was confirmed in the 1947 Town and Country Planning Act. The green belt was subject to much discussion in the 1970s, partly because such an attractive piece of land, several kilometers wide, was inviting to speculators—partly because some thought its control was too negative and it was not accessible enough to provide the amenities it was meant to preserve. Pressure from speculative builders has increased suddenly, encouraged by government policy on private initiative. The green belt is largely agricultural (70 percent) and well wooded (12 percent), only 6 percent being residential and 6 percent used for recreation.

The final 6 percent includes a variety of uses such as mineral working, industry, and transport.

Post-Abercrombie planning has confirmed his broad outlines of the shape and size of the capital. The first South East Study (1964) emphasized the need for counter-magnets to relieve continuing pressure, and the 1967 Strategy for the South-East and subsequent reviews (U.K., Department of the Environment, 1976; U.K. House of Commons, 1978) gave substance to these ideas. Above all these studies confirm that London must be seen in a regional context (Standing Conference, 1981). The interaction between economic and demographic processes in the city itself and in the outer metropolitan region has dominated planning policy. Ultimately these must be seen in a national context. In the 1960s it seemed probable that Britain was moving toward a regional framework, and that a new planning tier would be set up between local and national. The economic councils of these regions, however, were dissolved in 1979, and regionalism is a weak component of contemporary policymaking (P. G. Hall, 1980).

The argument for the comprehensive view is clear and was in part embodied in the idea of a Greater London Council; but in fact its powers are restricted. Each borough is its own planning authority, and the Greater London Council is limited to some control over land use, transport and highways, office and industrial development, and some aspects of housing. The Greater London Council has produced a development plan (GLDP) (1969; Great Britain, Department of the Environment, 1973), but as a planning authority it is rather emasculated. In addition the GLDP has become increasingly irrelevant in view of the pressing issues of the 1970s and 1980s. Like all plans it was based on an assumption of growth, but London was facing a period of decline. The immediate issues were unforeseen and acute. Industry was fading, unemployment soaring, job losses were vast. Office space was increasing faster than needs demanded. The problems of housing were more acute than ever. Deprivation was acute, particularly in the inner zone (P. G. Hall, 1981). Policy issues seemed concentrated on balance and social equity, and access to services, for example bus fares, became a major issue.

There were some successes, such as the fares policy, preservation of the green belt, development of some strategic centers and some local environmental schemes. But there were also conspicuous failures, partly because some solutions were simply not in the purview of the Greater London Council, such as inability to check the fall in employment, to control housing location, to reduce traffic. The greatest single dilemma facing the Greater London Council was resolving the conflict between community needs and changes related to London's role as an international city. One of the greatest obstacles in implementing social improvements was London's inability to raise revenue to meet the demands, particularly in the depressed inner zone, whose boroughs raise an average of only £45 million in a total rate base of £2 billion. To compensate for this there was a gain of £261 million in urban programming grants, but a further massive loss in rate support grants from the central government (£864 million) in 1983.

The Greater London Council saw its role as delegating as much development control as possible to the boroughs, and restricting itself to influencing that small proportion of the 40 thousand or so annual planning applications which affect London as a whole, or which have implications for strategic policies. The council plays a major role in office location decisions and in public inquiries regarding future use of derelict land; it safeguards land for housing as far as it can; it integrates traffic modes. It is also the local planning authority for action areas such as Covent Garden.

But on the whole, the Greater London Council bemoans its impotence to tackle comprehensively problems at the heart of London's planning. It has seen the initiation of piecemeal developments, not always in sympathy with its own aims. For example, the central government initiated a scheme to revitalize London's dockland. The older parts of the dockland, from which activities have long migrated down river, is London's major area of dereliction. Reference has already been made to St. Katherine's Dock, but its rehabilitation as yacht basin, hotel, and trade fair center was seen as indicative of policies which would give the river an upper-class frontage and give priority to office space and luxury apartments. Below this area, however, dockland comprised 2,250 hectares (encompassing the old Royal Docks and Surrey Docks south of the river), which was derelict, isolated, had only a sparse population, and was ripe for development. A comprehensive study begun in 1971 made suggestions seen as biased to middle-class intrusion and high capital cost (Falk, 1981).

The Dockland Development Corporation (1981) is built on the same model as those corporations which successfully initiated the new towns. Its aim is to attract new activities as well as to stimulate housing. An industrial park, trade center, workshops, and superstore are already under construction. But plans for an airport have been very controversial, as have plans for linking it directly with the business center of the City. Within the Dockland development scheme the government has designated a Free Enterprise Zone in which development will be largely freed from bureaucratic constraints and even from some financial burdens until 1992. Efforts are concentrated on environmental reclamation and laying down an infrastructure for future housing and development.

It is too early to assess even the possibilities which lie ahead in Dockland. Some think designation of the area was made on false premises. No one is denying that there was a wasteland there, but whether it has the distinctive character of a unique region can be questioned. On social grounds it would be very difficult to define. Some assumptions behind the designation have created a major planning dilemma on how far its future should be decided by the local people. Plans for the future will be delicately poised on the conflicting needs of London and local inhabitants, or central government decisions versus community reaction.

Elsewhere in London the Greater London Council has promoted community projects in areas under threat of urban change or suffering the disadvantages of being in a deprived zone. One aim is to reassert the feeling of distinctiveness associated with so many localities in London, and to reawaken neighborhood pride. Sixteen such communities have been recognized, and since 1982 about 90 projects funded, including better social facilities, support for local organizations, and environmental improvement schemes.

## The Future

London is unlikely in the year 2000 to be vastly different from the London of today, partly because some characteristics pointing to continuing radical change are balanced by inertia and in particular because in no way can the existing infrastructure of so large a city be ignored either technologically or financially. London of the past 20 or even 50 years will in a large measure dictate the London of the next 20 years.

London will continue its role as capital city, with no indication that either devolution of political power or decentralization of essential functions will alter the role. It will also, in all probability, capitalize on its role as a leading financial city in the world urban system. Everything indicates a continuation and even an expansion of this function. It may seem that in these two very important roles the future of London is assured.

Yet there are no indications that London is moving toward a solution of its major problems. The first is the changing balance of its economic base. London will continue to reflect the structural changes in British society, essentially the swing away from traditional manufacturing industry and an increasing emphasis on service industries, and the emergence of a transactional society. We have seen that on the negative side London may even suffer disproportionately compared with the rest of the country in this transition, but it is also in a favorable position to extend its control of the expanding activities of the future. London's major problem will be that the emerging postindustrial structure will be excessively constrained by the infrastructure of an outmoded classical industrial city. In no way can changes in the city's fabric keep pace with the demands of new activities, except at the risk of grave disruption, unless far-reaching policy and planning initiatives are promulgated. Rather, congruence will diminish, aging and obsolescence inevitably increase, congestion become worse. Even the perception of these problems in a time of growing awareness and increasing demands for high environmental standards will become much more acute.

In response to these urban pressures the counterattractions of suburbanization, or even of semiruralization will be very great indeed, further increasing the disparity between center and periphery. The strategic plan for the South East (U.K., South East Joint Planning Team, 1970) projected a stabilization at seven million for Greater London by 1991. The figure is well below that already, and stabilization is likely at 6.7 million. Similarly its estimate for the Outer Metropolitan Region for 1981 was too high (6.4 compared with 5.4 million) as well as that for the London Region (13.7 compared with 12.1). By the year 2001 it is unlikely that Greater London will grow, and the Outer Metropolitan Area and the London Region will grow less than anticipated. But the South East Region as a whole will probably grow toward 20 million. It is the region, and the proximity of London's advantages that will attract, not the city itself. Decanting of the better off will continue, and this can only mean greater social polarity. The richer will continue to move out, the poorer to remain. There is some indication of a counter-movement to this, as a small number of wealthier people have already moved back to Inner London. In many cases they have reestablished themselves in some of the formerly middle-class squares and terraces vacated by rehousing the poor. Such gentrification has transformed the urban environment in many small and scattered neighborhoods. We have also seen that prime sites such as the river frontage are being developed for prosperous buyers. The most comprehensive scheme for the return of the rich is that of the Barbican, three 40-story blocks housing 65 hundred on the very edge of the City. These very high-density apartments are rented, and their attraction is enhanced by the inclusion of an arts center, a school of music, and a museum. Nevertheless, most of the apartments are no more than pieds-à-terre for those who live much of their lives well outside London, and perhaps the very name Barbican unwittingly suggests the defensive ghetto-like aspect of these blocks. They do no more than highlight polarization.

Consolidation of ethnic enclaves, albeit not of ghetto proportions, will further emphasize the deprived nature of the inner zone. So although we may envisage a London as prosperous as ever, it will have even greater problems of unequal distribution of wealth. Its international role and eminence in an emerging transactional society may well be in conflict with the needs of London's own people. In the heart of the city, houses will have to fight for existence beside offices, and people outside such activities will have to fight for jobs. Beyond the green belt open space will fight for its existence against the enroachment of an increasing population, who through modern technology may well carry on with their work in semisuburban conditions. In all this we may be having a preview of the decline of the city as we knew it, where accessibility and centrality were the same thing, and the gradual emergence of a more dispersed city, perhaps even polynucleate in character.

If this is the case, then the gravest criticism of Greater London as an administrative unit is not that it is unnecessary, but that it is too small. A tier of responsibility which relates to the entire region, to promote its growth strategically within acceptable environmental standards, and to ensure a greater equity of wealth, will probably be a more pressing need in the future than it is now.

## Note

1. New Commonwealth countries are those which have achieved self-government within the Commonwealth since 1945.

# Bibliography

Abercrombie, Sir Patrick. 1945. *Greater London Plan, 1944*. London: His Majesty's Stationery Office.

Blandon, Michael. 1983. "The Foreign Banking Community Grows." *Banker* 133 (Nov.):111-12+.

Clout, Hugh D., ed. 1978. *Changing London*. Slough, England: University Tutorial Press.

Coleman, Alice. 1980. "Death of the Inner City: Cause and Cure." *London Journal* 6:1-22.

Coppock, John Terence, and Hugh G. Price, eds. 1964. *Greater London*. London: Faber & Faber.

Damesick, Peter, 1979. "Offices and Inner-Urban Regeneration." *Area* 11:41-47.

———. 1980. "The Inner City Economy in Industrial and Post-Industrial London." *London Journal* 6:23-35.

Dolphin, Philippa, Eric Grant, and Edward Lewis. 1981. *The London Region: An Annotated Geographical Bibliography*. London: Mansell.

Donnison, David V., and David E. C. Eversley, eds. 1973. *London: Urban Patterns, Problems and Policies*. Beverly Hills, CA: Sage.

Evans, Alan W., and David E. C. Eversley. 1980. *The Inner City: Employment and Industry*. London: Heinemann.

Falk, Nicholas. 1980. "London's Docklands: A Tale of Two Cities." *London Journal* 7:65-80.

Great Britain. Department of Employment. 1973. *Greater London Development Plan: Report of the Panel of Inquiry*. London: Her Majesty's Stationery Office.

Great Britain. Office of Population Censuses and Surveys. 1982. *Census 1981. County Report, Greater London*. Part 1. London: Her Majesty's Stationery Office.

———. 1984. *Census 1981. Results for Greater London and the Boroughs. Statistical Survey 19. Greater London Council*. London: Her Majesty's Stationery Office.

Greater London Council. 1969. *Greater London Development Plan: Report of Studies*. London.

———. 1984a. *London Facts and Figures*. London.

———. 1984b. *Planning for the Future of London*. Issue Paper no. 2. "Employment and Industry." London.

———. Planning Committee. 1981. *Report and Review of Office Policy in Central London*. London.

Hall, John M. 1974. *London, Metropolis and Region*. Oxford, England: Oxford University Press.

Hall, Peter G. 1962. *The Industries of London since 1861*. London: Hutchinson University Library.

———. 1963. *London 2000*. London: Faber & Faber.

———. 1980. "The Life and Death of a Quasi Quango: The South East Planning Council." *London Journal* 6:196-206.

———. 1980. *The World Cities*. 3d ed. London: Weidenfeld & Nicholson.

———, ed. 1981. *The Inner City in Contest: The Final Report of the Social Science Research Council Inner Cities Working Party*. London: Heinemann.

Hall, P. K. 1972. "The Movement of Offices from Central London." *Regional Studies* 6.385-92.

Hamnett, Chris, and Bill Randolph. 1982. "How Far Will London's Population Fall? A Commentary on the 1981 Census." *London Journal* 8:95-100.

Hillman, Judy, ed. 1971. *Planning for London*. Harmondsworth, England: Penguin.

Hoare, Anthony G. 1975. "Foreign Firms and Air Transport: The Geographical Effects of Heathrow Airport." *Regional Studies* 9:349-67.

Jones, Lang, and Wootton. 1980. *Offices in the Centre of London: A Special Report*. London.

Keeble, David E., and D. P. Hauser. 1971. "Spatial Analysis of Manufacturing Growth in South-East England, 1960-1967. I. Hypothesis and Variables." *Regional Studies* 5:229-62.

———. 1972. "Spatial Analysis of Manufacturing Growth in South-East England, 1960-67. II. Methods and Results." *Regional Studies* 6:11-36.

King, A. D. 1984. "Capital City: Physical and Social Aspects of London's Role in the World Economy." Brunel University, Middlesex, England.

Lee, Trevor, 1977. *Race and Residence: The Concentration and Dispersal of Immigrants in London*. Oxford Research Studies in Geography. Oxford, England: Clarendon Press.

Leigh, Roger, David North, Jamie Gough, and Karen Sweet-Escott. 1983. *Monitoring Manufacturing Employment Changes in London, 1976-81.* 2 vols. London: London Industry and Employment Research Group, Middlesex Polytechnic.

Lomas, Graham, M. 1975. *The Inner City: A Preliminary Investigation of Current Labour and Housing Markets with Special Reference to Minority Groups in Inner London, July 1974.* London: London Council of Social Services.

London Tourist Board. 1982. *Tourism in London.* London.

Manners, Gerald. 1977. "Office Policy: National and Regional Perspectives. The Strategic Role of LOB." *Town and Country Planning* 45:25+.

Martin, John Edward. 1966. *Greater London: An Industrial Geography.* Chicago: University of Chicago Press.

Norton, Alan. 1984. "Metropolitan Government Abroad." In *Governing London. Papers and Proceedings of the Conference arranged by members of the University College London and London School of Economics and Political Science: April 6th 1984.* London: University College and the London School of Economics.

Pound, R. 1967. *Harley Street.* London: Joseph.

Rasmussen, Steen Eiler. 1967. *London, the Unique City.* Cambridge, MA: MIT Press.

Rhodes, Gerald W. 1970. *The Government of London: The Struggle for Reform.* Toronto: University of Toronto Press.

Runnymede Trust. 1975. *Race and Council Housing in London: Census Returns Examined by the Runnymede Trust Research Staff.* London.

Salt, John. 1978. "Population and Employment." In *Changing London. See* Clout, 1978.

Scarman, Sir Leslie George. 1981. *The Brixton Disorders 10-12 April, 1981.* Cmnd Paper 8427. London: Her Majesty's Stationery Office.

Sharpe, L. J. 1984. "The Debate on Metropolitan Government: The Functional Dimension." In *Governing London. See* Norton, 1984.

Shepherd, John, John Westaway, and Trevor Lee. 1974. *A Social Atlas of London.* Oxford, England: Clarendon Press.

Smart, G. 1984. "Governing London: Metropolitan Planning." In *Governing London. See* Norton, 1984.

*South-East Study, 1961-81.* 1964. London: Her Majesty's Stationery Office.

Standing Conference on London and the South East. 1981. *South East Regional Planning in the 1980s.* London: Greater London Council.

Thomas, David. 1970. *London's Green Belt.* London: Faber.

Thomas, Ray. 1982. "Stability for London." *Town and Country Planning* 51:98-99.

Travers, Tony. 1984. "Local Government Finance in Greater London." In *Governing London. See* Norton, 1984.

United Kingdom. Department of the Environment. 1977. *Inner Area Studies. Liverpool, Birmingham and Lambeth: Summaries of Consultant's Final Reports.* London: Her Majesty's Stationery Office.

United Kingdom. Department of the Environment. South East Joint Planning Team. 1976. *Strategy for the South-East. 1976 Review: Report and Recommendations.* London: Her Majesty's Stationery Office.

United Kingdom. House of Commons. Standing Committee on Regional Affairs. 1978. *The 1976 Review of the Strategic Plan for the South East.* 2d sitting, July 26, 1978. London: Her Majesty's Stationery Office.

United Kingdom. South East Joint Planning Team. 1970. *Strategic Plan for the South East; Report by the South East Joint Planning Team.* London: Her Majesty's Stationery Office.

Wood, P. 1978. "Industrial Changes in Inner London." In *Changing London. See* Clout, 1978.

# 4

# Tokyo

## Hachiro Nakamura and James W. White

Tokyo, called Edo in Japan's feudal period, was the seat of the Shogun's government and in his day comprised more than one million people—the world's largest city. Social turmoil during the Meiji Restoration of 1868 reduced the population nearly by half, but with the transfer of the national capital from Kyoto to Edo, now renamed Tokyo, the lost population was gradually regained. By the time Tokyo was classified as a *shi* or city in 1889 in accord with the new municipal code adopted in the previous year, the city's population already had reached 1.4 million.

Above the municipal level of government, the new Meiji regime established prefectures (*fu* and *ken*) to replace the fiefs of the feudal era. In the year Tokyo was officially recognized as a city, the population of Tokyo Prefecture totalled 1.63 million. When the Tama district, previously part of neighboring Kanagawa Prefecture, was annexed to Tokyo in 1893 the prefectural population rose to 1.79 million. Until World War II the population of both the prefecture and city of Tokyo continued to grow, interrupted only by the devastation of the Great Kanto Earthquake of 1923 which killed over 100 thousand persons (see table 4.1).

Outward dispersion of the city population was already underway by the 1910s, when electric streetcars and suburban railways were introduced in Tokyo. The dispersion gained speed along with the industrial boom accompanying World War I and accelerated further as families and enterprises opted for the suburbs rather than return to the earthquake-ravaged central city in the mid-1920s. With suburbanization, outlying towns and villages rapidly gained population, and many of these communities were merged into the city

TABLE 4.1
Population—Tokyo, 1920-1980 (in thousands)

| Year | Tokyo Prefecture | Ward District |
|------|------------------|---------------|
| 1920 | 3,669 | 3,358 |
| 1925 | 4,485 | 4,109 |
| 1930 | 5,409 | 4,987 |
| 1935 | 6,370 | 5,897 |
| 1940 | 7,355 | 6,779 |
| 1945 | 3,488 | 2,777 |
| 1950 | 6,278 | 5,385 |
| 1955 | 8,037 | 6,969 |
| 1960 | 9,684 | 8,310 |
| 1965 | 10,869 | 8,893 |
| 1970 | 11,408 | 8,841 |
| 1975 | 11,674 | 8,647 |
| 1980 | 11,618 | 8,352 |

SOURCE: Tokyo-to Somu-kyoku Tokei-bu Jinko Tokei-ka (1975: 68-69); Japan, Prime Minister's Office, Statistics Bureau (1982c: 58-59).

of Tokyo in 1932, as 23 new wards (ku) were added to the 15 preexisting ones.

This pattern of urban growth was short-lived, however. During World War II the national government decided that wartime exigency demanded more efficient administration of the national capital, and the city of Tokyo was abolished, with the ward area simply becoming one part of a new Tokyo Prefecture or Tokyo-to). Thus the jurisdiction of the Tokyo prefectural government came to comprise, as it does today, the ward area (reduced from 38 to 23 ku after the war) plus the old Tama district, which today contains an additional 26 cities, 7 towns, and 8 villages.

Partly because of these administrative changes, many Tokyoites are now unable to define the formal boundaries of Tokyo, and indeed the metropolis is a difficult entity to define. Tokyo Prefecture includes a considerable rural mountainous area in western Tama and a chain of small islands in the Pacific Ocean several hundred kilometers to the south. In neither are there any appreciable urban settlements, and most Tokyoites would not associate them with the giant city they live in. On the other hand, the giant city sprawls not only into much of the Tama district, but also into parts of three neighboring prefectures—Kanagawa, Saitama, and Chiba—thus constituting a conurbation which includes the cities of Yokohama, Kawasaki, and Chiba, each approaching or exceeding one million people.

Thus "Metropolitan" or "greater" Tokyo is a combination of Tokyo Prefecture and parts of three other highly urban prefectures, although the rural areas are far larger geographically than the urban. In analyzing the metropolis, however, we should not ignore these rural areas, since statistics pertaining to them are very often important to an understanding of the dynamics of the entire metropolitan area.

Over and above metropolitan Tokyo there also exists, by virtue of the National Capital Region Development Law, a wide region comprising

Tokyo Prefecture and seven surrounding prefectures designated as the "national capital region." As industrialization and cityward migration gained momentum and Tokyo encountered serious urban problems in the 1960s, the national government felt that a wide region had to be covered by any relevant planning area. A regional approach to problem solving was adopted, for example, in dealing with the development of water resources in the mountainous northern part of the national capital region to meet Tokyo's increasing demands.

"Tokyo" thus may be defined—and is, by different people—in at least five different ways, in order of increasing geographic size: (1) the 23-ward central district; (2) Tokyo Prefecture; (3) the conurbation area of continuous metropolitan settlement and land use, which includes parts but not all of (4) the 4-prefecture metropolitan region; or (5) the national capital region. Of these the most appropriate one for discussing Tokyo, especially in an ecological sense, is the conurbation. The problem here is how to demarcate the boundary of Tokyo so defined, since statistical data are usually collected on the basis of administrative, not ecological, units. Because the Tokyo prefectural government provides the largest amount of data of the sort relevant to this chapter, the most frequent references will be made to Tokyo Prefecture and the 23-ward district within it. Accounts of other defined areas will be also made where appropriate and where data are available.

## Population

Because of the damage and wholesale population evacuation of World War II, the population of Tokyo Prefecture fell sharply from 7.35 million in 1940 to 3.49 million in 1945, with a corresponding drop in the ward area from 6.78 million to 2.78 (see table 4.1). While the short-run effects of the war included poverty, privation, and near famine, recovery of the lost population was so rapid that by 1953 the prewar peak population was exceeded, as was the prewar peak population growth rate. This pace continued unabated, and the prefecture grew to 8, 9.6, and 10 million in 1955, 1960, and 1965, respectively, while the ward area grew from 6.9 to 8.3 to 8.8 million. Japan at the time had embarked on a course of high-speed economic growth, stimulating the cityward migration of workers from rural areas in response to worker shortages and higher wages in cities like Tokyo, Osaka, and Nagoya. Increasing numbers of youths who moved to large cities to enter colleges and universities as a result of Japan's economic growth were also a contributing factor. Japan's population, which was less than 40 percent urban in 1950, was almost 70 percent urban by 1965. Since Tokyo is the capital of Japan, not only local but national authorities also paid particular attention to the endless swelling of Tokyo's population and held it imperative that policies be developed to cope with its overcrowding, urban sprawl, pollution, and other attendant problems.

Tokyo's population growth began to slow in the mid-1960s, and a Ministry of Health and Welfare demographer, named Toshio Kuroda, discovered that net migration into central Tokyo had become negative. He dubbed this trend the "U-turn" phenomenon, and numerous observers contended that it demonstrated that young people who had migrated to Tokyo attracted by the splendor and opportunities of urban life often became aware of its problems and limitations and returned to their native communities where life was more comfortable, if not as affluent. This theory was widely accepted presumably because it was congruent with the antiurban values of the general public who were pleased to know that young people had grown disillusioned and had awakened to the evils of urban life.

The U-turn theory is still popular, especially among most government officials and urban affairs specialists. It is not, however, empirically tenable today, if in fact it ever was. Even in the peak period of internal migration in the 1960s many of those departing central Tokyo moved outward only to the suburbs, and even those returning to their home prefectures were more likely to settle in a city there than go all the way back to their village of origin. Thus a J might provide a better fit than a U to the trend under way. And if one adduces data from metropolitan Tokyo (Tokyo Prefecture plus its three adjoining prefectures) one again sees that suburban dispersion better describes the migratory process.

In the first place, although Tokyo Prefecture lost almost 1.6 million people in the birth-to-14 and 20-to-39-year-old age groups between 1965 and 1975, the three other prefectures gained over 1.4 million people in the same cohorts (changes due to death are negligible at these ages). Moreover, in terms of social increase alone, the negative net migration flow out of central Tokyo from 1967 onward is comprised of a continuing net inflow from outside metropolitan Tokyo and a massive outflow only as far as the suburbs.

Second, when one returns to the age-specific data, one finds several phenomena:

(1) Young people in their teens and 20s are still moving into Tokyo and the three neighboring prefectures.
(2) When they are 5 to 15 years older, those who entered Tokyo start to leave, but only for the three neighboring prefectures or farther west in Tokyo prefecture, and almost always within commuting distance of central Tokyo. Presumably becoming married and having children compels them to find more spacious housing for their families.
(3) A decrease in Tokyo's child and young adult populations and an equivalent increase in its three neighboring prefectures suggests that children born in Tokyo are taken from the prefecture by their parents when they move outward.
(4) This situation confirms our suspicion that the migration at work is of a J rather than a U shape, and amounts to little more than the expansion of the Tokyo conurbation into neighboring prefectures.
(5) Young people who originally migrated directly, as single individuals, into the three prefectures continue to live there, now with spouses and children. Subsequently they are joined by many of their peers who originally migrated directly to Tokyo and later left there with their families, thus contributing to a tremendous increase of population in the three prefectures, from a total of 7.4 million in 1955 (8.1 percent of the national population) to almost 18 million (15 percent of the total) in the early 1980s.

(6) Since the young population is still increasing in Tokyo and the neighboring prefectures, albeit at a decreasing rate (see table 4.2), one cannot expect the demand for land and housing in the metropolitan area to abate in the near future.

Thus, the exodus from Tokyo implies not a U-turn but the dispersion of population beyond the boundaries of Tokyo Prefecture. Consequently, the metropolitan area as a whole has come to be home to an ever-growing population (see table 4.2), which in 20 years grew from less than a fifth to nearly one-fourth of Japan's total population by 1980. The rate of growth peaked in the late 1960s; the subsequent drop may be largely attributable to stagnation of Japanese economic growth following the oil shock of 1973, since migration rates are correlated closely with economic growth rates. But even though the Japanese economy has largely recovered from the 1970s, the days of double-digit economic (and metropolitan population) growth are probably gone for good.

One final confirmation of the argument made above lies in the commuting patterns of the metropolitan Tokyo work force. We asserted above that housing is the main reason for the exodus from central Tokyo, and that the movers retain their downtown jobs, moving only so far as is still within commuting distance of them. As seen in table 4.3, Tokyo Prefecture had a 1980 daytime population 2 million persons greater than its resident population; the 23-ward area's daytime population was 2.4 million persons greater. And table 4.4 reveals that except for a small fraction, these commuters come from the three neighboring prefectures whose size has been growing during these years.

## Internal Differentiation

The demographic processes discussed above are not occurring evenly throughout the city. Since a better knowledge of Tokyo requires examination of internal spatial differentiation, let us look now at population in the central ward district.

By 1960, when authorities were beginning to worry about Tokyo's continuing growth, census data already revealed population losses in two of Tokyo's three innermost wards. The depopulated area has spread outward since that time; the 1980 census disclosed depopulation taking place in 19 of the 23 wards, a magnitude unseen in any of Asia's other giant cities. This loss created uneasiness in local government. Some public facilities became idle because of the shortage of users but the amount of public expenditure required for their maintenance remained the same. This situation held true especially for primary schools which have been losing students. Some inner-city neighborhoods are nowadays virtually deserted at night and tasks such as sweeping alleys and maintaining street lights, previously taken care of by neighborhood associations, are likely to be abandoned. These problems are

TABLE 4.2

Population (in thousands) and Rates of Increase (in percentages)—
Tokyo Prefecture and Metropolis, 1960-1980

| Location | 1960 | 1965 | 1970 | 1975 | 1980 |
|---|---|---|---|---|---|
| Tokyo Prefecture | 9,684 | 10,869 | 11,408 | 11,674 | 11,618 |
| Kanagawa | 3,443 | 4,431 | 5,472 | 6,398 | 6,924 |
| Saitama | 2,431 | 3,015 | 3,866 | 4,821 | 5,420 |
| Chiba | 2,306 | 2,702 | 3,367 | 4,149 | 4,735 |
| Total | 17,864 | 21,017 | 24,113 | 27,042 | 28,699 |

Rate of Increase

| Location | Social | | | | Natural | | | | Total | | | |
|---|---|---|---|---|---|---|---|---|---|---|---|---|
| | 1960-65 | 1965-70 | 1970-75 | 1975-80 | 1960-65 | 1965-70 | 1970-75 | 1975-80 | 1960-65 | 1965-70 | 1970-75 | 1975-80 |
| Tokyo Prefecture | 4.8 | -2.8 | -5.0 | -5.0 | 7.4 | 7.8 | 7.3 | 4.5 | 12.2 | 5.0 | 2.3 | -0.5 |
| Kanagawa | 20.6 | 14.1 | 7.4 | 2.2 | 8.1 | 9.4 | 9.5 | 6.1 | 28.7 | 23.5 | 16.9 | 8.2 |
| Saitama | 17.3 | 19.0 | 14.5 | 6.0 | 6.8 | 9.3 | 10.2 | 6.4 | 24.0 | 28.2 | 24.7 | 12.4 |
| Chiba | 11.5 | 17.2 | 14.5 | 8.1 | 5.7 | 7.4 | 8.8 | 6.0 | 17.2 | 24.6 | 23.2 | 14.1 |
| Total | | | | | | | | | 17.6 | 14.7 | 12.1 | 6.1 |

SOURCE: Japan, Prime Minister's Office, Statistics Bureau (1982b: 15-19).

TABLE 4.3
Daytime Population—Tokyo Prefecture and Ward District,
1955-1980 (in thousands)

| Location | 1955 | 1960 | 1965 | 1970 | 1975 | 1980 |
|---|---|---|---|---|---|---|
| Tokyo Prefecture | 8,332 | 10,200 | 11,752 | 12,655 | 13,343 | 13,481 |
| Ward District | 7,323 | 8,971 | 10,040 | 10,433 | 10,709 | 10,602 |

SOURCE: Tokyo-to Kikaku Hodo-shitsu Chosa-bu (1983: 4).

TABLE 4.4
Changing Size of Population Commuting to Tokyo Prefecture,
1965-1980 (in thousands)

| | 1965 | 1970 | 1975 | 1980 | Rate of Increase | | |
|---|---|---|---|---|---|---|---|
| | | | | | 1965-70 | 1970-75 | 1975-80 |
| Total | 921 | 1300 | 1682 | 1929 | 7.1 | 5.3 | 2.8 |
| From 3 adjacent prefectures | 872 | 1244 | 1612 | 1840 | 7.4 | 5.3 | 2.7 |
| Kanagawa | 332 | 454 | 573 | 645 | 6.5 | 4.8 | 2.3 |
| Saitama | 317 | 464 | 594 | 660 | 7.9 | 5.1 | 2.1 |
| Chiba | 223 | 326 | 444 | 536 | 7.9 | 6.4 | 3.9 |

SOURCE: Tokyo-to Kikaku Hodo-shitsu Chosa-bu (1983: 4).

especially serious in the three innermost wards—Chiyoda, Chuo, and
Minato—the first to undergo depopulation.

In Tokyo, this situation does not imply urban decay, for depopulation does
not leave the poorest people in the central city, as is the case in the United
States, and most of the depopulated wards harbor more daytime population
than nighttime population. This phenomenon occurs because the innermost
wards embrace Tokyo's central business district (CBD), and indeed, land
prices in these wards are the highest in the metropolis. In the CBD growing
numbers of public and private offices, luxurious retail shops, first-class
department stores, hotels, theaters, and restaurants have essentially displaced
residential land uses. In the past, two- to four-story buildings combining
homes and shops or miniscule factories predominated there, but these
buildings have been remodelled to convert rooms once used for residence into
space for sales or have been wholly replaced by high-rise buildings able to turn
a profit commensurate with the high land prices and CBD location. Those
who previously lived in those buildings have moved to the suburbs and
commute to their former home areas, now transformed to commercial use.

Thus, in Tokyo's innermost area, the proportion of workers whose
residences are also their working places is the smallest of all the wards. The
combination of decreasing nighttime population and increasing commuting
has created greater discrepancies than in any of Asia's other giant cities: the
daytime population of Chiyoda ward is 170 times greater than its nighttime
population; for Chuo ward it is 79 times greater and for Minato, 35 times.
Although the CBD contains the Ginza, Tokyo's most distinguished shopping

district, its outstanding characteristic is the predominance of office activities: more than half of all jobs in this area are related to office work.

While this land-use pattern has led to a very sparse nighttime population in central Tokyo, a second group of wards surrounding the innermost area has the metropolis's highest population densities, although it too has undergone continued loss of resident population. Because of the spread of office activities from the innermost wards, some wards to their north, west, and southwest now provide a high proportion (about 40 percent) of Tokyo's office-related jobs. Another characteristic of this outlying area is the relatively high proportion of sales-related jobs. Most notable in this respect is the development of three centers of sales, service, and entertainment enterprises in Shinjuku (Shinjuku ward), Shibuya (Shibuya ward), and Ikebukuro (Toshima ward), which have become subcenters, if not rivals, of the CBD and the Ginza. Development of these three districts may be attributed to locational advantage: each is situated at a traffic hub on the western periphery of central Tokyo where commuters must transfer among subways, trains, and buses. As Tokyo's demographic center of gravity has moved westward, the number of persons passing through their stations daily has grown into the millions. Recently Shinjuku became noteworthy for a concentration of skyscrapers built on the former site of a once-rural water filtration plant and an attendant proliferation of office activities.

An additional notable downtown district is Asakusa in Taito ward, east of the innermost wards. Taito ward is part of the old downtown, or *shitamachi*, part of Tokyo, long associated with small-scale mercantile and industrial activity and lower-middle-class life-styles as opposed to the "uptown," or *yamanote*. Uptown Tokyo includes the belt of western wards extending from Setagaya and Meguro in the southwest around to Nerima and Toshima in the northwest, associated with higher social strata and more tertiary industry. Asakusa, in particular, has long been a popular business and entertainment center for factory and shop workers from the shitamachi; given the spread of office activities in this direction, also from the innermost wards, Asakusa too embraces more daytime population than nighttime, though to a lesser extent than the CBD.

The wards lying farthest out from the central area provide more residential than working places, so the population tends to be smaller in the day than at night. The tendency toward depopulation prevalent in the central wards is at work here also, with four exceptions: Koto, Katsushika, and Edogawa in the East and Nerima in the northwest. As noted, the western yamanote wards count among their residents more people of comparatively high class in terms of education and income than in the eastern wards. Retail activity is greater in the yamanote, and so consequently is the number of sales-related jobs. In the remainder of the wards in the south, east, and north, jobs are more likely to involve factory work. In the north and east (the Joto and Johoku areas) factories are usually small-scale and often combined with residences; here the proportion of live-in workers is higher than in other wards. Factories in the southern (Jonan) area are much larger and often owned by Japan's leading industrial concerns.

Much differentiation is thus visible within the ward district of Tokyo. When the 23 wards are compared with the Tama district as a whole, further differences are observed. The Tama district is essentially a bedroom for the ward district, although some suburban cities such as Musashino and Tachikawa have larger daytime than nighttime populations. The outward relocations of factories which began before the war and has continued ever since has made some cities and towns in Tama into industrial suburbs. Even so, the district has absorbed so many movers from the ward district that it is typically residential throughout. In the 1980s some suburban municipalities recorded decreasing population, but up until then all the municipalities except those on the westernmost rural fringe witnessed an increase while almost all wards declined. Commuters from Tama to the 23-ward area almost tripled in numbers between 1960 and 1980 (see table 4.5), and this flow has grown not only from Tama but from adjacent prefectures as well. Table 4.5 reveals clearly the economic and demographic influence exerted on metropolitan Tokyo by its central core.

Because the exodus from central Tokyo has been primarily in search of housing, residential areas for commuters have spread throughout the metropolis. As Tama has filled in, more recent movers from central Tokyo are forced to find housing in ever more distant places. If metropolitan Tokyo is divided into concentric zones ranging from 10 to 50 kilometers, and we examine the differential population increase in each zone, we find that the zone of highest increase (the suburban "doughnut") has been moving farther and farther outward with time (see table 4.6). Although the rate of increase has diminished recently, Tokyo's commuter hinterland is still the biggest of all Asia's giant cities.

A preponderance of industrial jobs characterizes such wards as Edogawa, Adachi, Koto, Katsushika, Arakawa, and Ota; these areas are low in clerical occupations (most common in the three-ward CBD and Shibuya, Bunkyo, Shinjuku, and Meguro in the yamanote ring), and in sales and service jobs, which are more concentrated in the yamanote (Toshima and Shinjuku), notably in its outermost edge (Setagaya, Suginami, and Nakano wards). The Tama district includes some distinctive cities both industrial (Musashimurayama, Inagi, Akishima, Ome, and Hino) and white-collar (Tama and Kiyose and, strung along the busiest commuter rail line in suburban Tokyo, Kunitachi, Koganei, Kokubunji, and Musashino). The clearest trend visible, albeit a most modest one, is the trend toward more agriculture as one moves down (westward) along the list of municipalities.

Appendix 4.1 fills out this picture with data on the daytime, nighttime, and home-employed populations of Tokyo Prefecture. The daytime-nighttime population ratios clearly differentiate the sources and destinations of the commuter work force, and the contrast between the large-scale industrial development of Shinagawa and Ota wards and the small-scale enterprises in the equally industrial north and east (Arakawa, Adachi, Nerima, and Katsushika) may be seen in the proportion of the work force employed in their own residences. Except for the long-standing cities of Musashino and Tachikawa noted above, all of Tama has a smaller daytime than nighttime

TABLE 4.5

Workers and Students in Ward District during Daytime Hours,
by Residence—Tokyo Area, 1960-1980 (in thousands)

| Residence | 1960 | | 1970 | | 1975 | | 1980 | |
|---|---|---|---|---|---|---|---|---|
| | N | % | N | % | N | % | N | % |
| Ward District | 1811 | 69.4 | 4929 | 72.2 | 4762 | 66.6 | 4613 | 63.6 |
| Tama District | 190 | 7.3 | 441 | 6.5 | 509 | 7.1 | 532 | 7.3 |
| Prefectures | | | | | | | | |
| Kanagawa | 231 | 8.9 | 518 | 7.6 | 642 | 9.0 | 703 | 9.7 |
| Saitama | 193 | 7.4 | 509 | 7.5 | 654 | 9.2 | 714 | 9.8 |
| Chiba | 145 | 5.6 | 363 | 5.3 | 499 | 7.0 | 597 | 8.2 |
| Other places | 38 | 1.5 | 67 | 1.0 | 80 | 1.1 | 97 | 1.3 |
| Total | 2608 | 100.0 | 6827 | 100.0 | 7147 | 100.0 | 7256 | 100.0 |

SOURCE: Tokyo-to Somu-kyoku (1965: 156-75); Sori-fu Tokei-kyoku (1972: 655-58); Japan, Office of the Prime Minister, Bureau of Statistics (1977: 13-115–13-119); Japan, Prime Minister's Office, Statistics Bureau (1982a: 13-125–13-130).

population—fully 30 percent of Komae's population leave the city each morning for work. The Joto and Johoku areas also are net providers of commuters, but on a smaller scale, and they share the innermost area's trend toward depopulation.

## Transportation

Tokyo's massive increases in commuters has created unprecedented traffic burdens. As the Japanese economy has grown, Tokyo has accommodated an ever-growing number of high school and university students whose use of trains and buses during heavy commuting hours has worsened the traffic problem. Improvements in Tokyo's excellent transportation system are largely offset by growing demands, so Tokyo is notorious for its overcrowded trains and buses, especially during rush hours, and for its almost nonstop traffic jams.

Tokyo has been accessible by national and private rail lines since the 19th century. In addition to five radial routes from central Tokyo, the recently privatized Japan Railways has a loop line around the ward district. Other private railroads connecting suburban areas or adjacent prefectures have not been allowed to extend their routes into the area within the loop line, where streetcars and buses were the major means of public transport.

Routes of prewar railroads have continued to the present with only minor extensions. Since junctions of the loop line and private railroads (such as Shinjuku, Ueno, and Ikebukuro) were terminals of the private lines, passengers from the suburbs desiring to enter central Tokyo had to transfer to alternate carriers at these stations, which were consequently always crowded with people. Taking advantage of the captive throngs of passengers, entrepreneurs in the terminals' neighborhoods developed flourishing shopping districts, some of them becoming the subcenters of Tokyo already discussed.

TABLE 4.6

Differential Rate of Population Increase in Concentric Zones—
Tokyo Area, 1955-1980 (in percentages)

| Radius from Tokyo's Center | 1955-60 | 1960-65 | 1965-70 | 1970-75 | 1975-80 |
|---|---|---|---|---|---|
| 0-10 km | 13.4 | −1.4 | −6.5 | −6.5 | −6.3 |
| 10-20 km | 29.8 | 25.3 | 11.9 | 6.2 | 2.1 |
| 20-30 km | 22.7 | 40.4 | 31.6 | 22.5 | 9.2 |
| 30-40 km | 15.4 | 37.0 | 43.6 | 29.7 | 14.2 |
| 40-50 km | 3.3 | 14.9 | 19.6 | 22.1 | 16.1 |
| Total 50-km radius zone | 18.6 | 19.7 | 15.9 | 12.7 | 6.4 |

SOURCE: Sori-fu Tokei-kyoku (1983: 101).

In the immediate postwar period Japan lagged far behind the West in the technical level of her automobile industry. The level gradually improved and a type of small automobile—cheaper to operate and better fitted to the narrow and winding streets of Japan—was produced, which eventually acquired a competitive position with imported cars in the domestic market. Coincidentally, in the late 1950s Japan was launching its rapid economic growth to which the automobile industry development was a contributing factor. Car ownership in Tokyo rose sharply in the latter half of the 1960s, creating formidable congestion. Before the Tokyo Olympics in 1964 construction began on an overhead express highway which now crisscrosses Tokyo and extends out into the hinterland; improvement of road conditions is continuous; and presently no one may possess a car in Tokyo unless there is an off-street parking place for it. Nevertheless, the incessant increase of automobiles in Tokyo prevents relief from traffic congestion.

Floods of automobiles in the Tokyo streets became an obstacle for efficient streetcar operation because their tracks were very often occupied by automobiles and the resulting inefficient streetcar operation in turn further aggravated traffic problems. Rising manpower costs triggered by Japan's economic growth also exerted pressure on continuing operation of streetcars: personnel costs exceed those of buses because streetcars require a driver and conductor, doubling operator expenses. Thus, since 1963, streetcars have been gradually abolished with only a short route now remaining on the fringe of the yamanote.

The streetcar's replacement—which is coming to eclipse the bus also—is the subway, whose development in Tokyo has been truly remarkable: ten lines are presently in operation, with new lines and extensions of existing ones on the drawing boards. Subway routes not only run through the interior area of the loop line where private railroads have not been permitted but extend far into outlying suburbs and in some cases into neighboring prefectures. Subway lines frequently merge with private and national rail lines, so that commuters may leave the CBD by subway and then transfer to the national rail network merely by crossing a platform. Also notable is that within the interior district there is a subway station within five to ten minutes' walk of almost any place, suggesting the medium's paramount importance for public transportation in the area of heaviest daytime population flow. Buses and taxis are available, but congestion has diminished the efficiency of both, and the growing demand for transportation has been largely met by subways, and, for suburban commuters, private rail lines (see table 4.7). Crowding on subways and trains (riders as a percentage of capacity), which reached 250 percent for the principal lines during rush hours in the mid-1960s, was ameliorated by the subway-building boom of the 1960s, but it has never fallen below 200 percent and in the early 1980s, as construction has fallen off and population growth continues, it seems to be worsening again.

## Tokyo's Role as a Giant City

The position of Tokyo in its Japanese context has already been seen demographically—over one-quarter of all Japan's people live in the metropolitan area. Economically, too, Tokyo, is a giant: the prefectural gross product accounts for over 15 percent of Japan's GNP. The work force is concentrated in three areas—manufacturing, wholesale and retail trade, and communications and services. The manufacturing work force, by a slim margin the largest, is concentrated in small factories on the eastern side of the city and large ones in the south and southwest. The Tama district, as noted, also includes a few industrial suburbs. Products of the small factories include apparel, leather goods, toys, furniture, personal ornaments, footwear, and sundry goods; the large factories produce steel and metal products, general machinery, electric machinery, transportation equipment, precision instruments, and so on. Publishing and printing industries have been attaining remarkable growth in the inner area of the ward district, along with production of fashion ladies' wear. Because Tokyo, especially in the CBD, is where up-to-date knowledge and information are most available, the city offers a locational advantage for publishing and haute couture and for some products of small factories producing fashionable leather goods and footwear and luxury accessories.

The most rapidly growing economic sector is that of services, in contrast with the gradual waning of manufacturing. The decline of manufacturing as a whole, in spite of the previously mentioned flourishing industries, is in part attributable to the shrinkage of heavy and chemical industries after they peaked in the 1950s and the prefectural government subsequently made it essentially impossible for additional factories of this type to be built in Tokyo.

Before World War II factories were likely to be located at the periphery of what was then the built-up area where land was inexpensive and transportation was available. Thus, the eastern and southwestern outlying areas of today's ward district, then not under jurisdiction of the city of Tokyo, were turned into a factory zone. Eastern-area factory sites were cheap (because the land was low and swampy) and convenient (because the rivers and canals running through them were navigable). The southwestern area also abounds with lowlands; its locational advantage lies in easy access to sea transport on

TABLE 4.7
Daily Average Shares of Transit Facilities in
23 Special Wards–Tokyo, 1960-1980

| Transit Facility | Fiscal 1960 | | | Fiscal 1980 | | |
|---|---|---|---|---|---|---|
| | Passengers Transported (in thousands) | Percentage of All Passengers | Index | Passengers Transported (in thousands) | Percentage of All Passengers | Index |
| National railways | 5,447 | 36.0 | 100 | 6,821 | 30.3 | 125 |
| Private railways | 3,414 | 22.6 | 100 | 6,543 | 29.0 | 192 |
| Subways | 865 | 5.7 | 100 | 5,377 | 23.8 | 622 |
| Streetcars | 1,914 | 12.7 | 100 | 120 | 0.5 | 6 |
| Buses | 2,259 | 14.9 | 100 | 2,077 | 9.2 | 92 |
| Taxis | 1,230 | 8.1 | 100 | 1,620 | 7.2 | 132 |
| Total | 15,129 | 100.0 | 100 | 22,558 | 100.0 | 149 |

SOURCE: Tokyo Metropolitan Government (1984: 113).

Tokyo Bay, close to the ports of Tokyo and Yokohama. Over time, the southwestern zone expanded into neighboring Kanagawa Prefecture, making Kawasaki—which lies between Tokyo and Yokohama—Japan's largest industrial city. After Kawasaki was fully occupied, the industrial zone stretched inland, giving rise to the present-day industrial suburbs of the Tama district.

Factories, like population, have shifted outward from Tokyo since World War II, and new factories have tended to be located in adjacent prefectures or even further out in the northern part of the national capital region. Thus, while the economic giantism of the capital region persists, that of Tokyo Prefecture has declined, at least in the number of factories, industrial workers, and value of national industrial shipments.

Associated with the dispersion of factories from Tokyo is the growing proportion of workers in the services and commerce—the gross product is greater in the service sector, but Tokyo's predominance is clearest in commerce: it accounts for almost one-third of all sales in the entire country. The share of the retail trade has declined, perhaps reflecting the dispersion of consumers to Tama and neighboring prefectures. Wholesale trade has fared much better, which may be proof of the city's advantage as a strategic point for wholesale business with its availability of up-to-date information, proximity to the financial heart of the country, and position at the center of the distribution network for northern and central Japan. Because wholesale firms—especially those involved in international trade—must keep informed of current trends in production, consumption, market quotations, financial affairs, governmental trade policies, and the like in Japan and abroad, they may prefer to locate their offices in Tokyo, especially in the CBD. Thus, Tokyo's innermost three wards contained in 1982 one-third of the firms, half the workers, and five-sixths of the value of annual wholesale trade in metropolitan Tokyo.

Noteworthy among wholesale firms are the *sogo shosha* or general trading companies, comprehensive merchandising, purchasing, marketing, and brokering firms dealing in every conceivable commodity, whose geographical sphere is worldwide, with offices in all major foreign cities. Constituting only .03 percent of Tokyo's wholesale firms and with only 4.2 percent of the employees, the sogo shosha account for 40 percent of Tokyo's wholesale trade and 80 percent of the total sales of all the sogo shosha in Japan. Tokyo offices of the sogo shosha are all located in the three-ward CBD.

In addition to commerce, services in general—and especially business-related services—are an important part of Tokyo's economy. Industrial and business machine leasing, information services, data processing, market research, and business consulting show marked development, while personal services such as laundries, barbers, furniture repair, and movie theaters, and social services like broadcasting, education, and clinics and hospitals are declining. As with wholesale trade, business-related services flourish in Tokyo's inner area.

The discussion above gives one some idea of the structure of Tokyo's economy, but not the full picture of Tokyo's role in the national arena. Turning to this role, we may first note the tendency—despite some dispersal— of corporations to locate in Tokyo, and the extraordinary overrepresentation of Japan's biggest corporate headquarters there (see table 4.8). With roughly 10 percent of Japan's population, Tokyo Prefecture has roughly one-quarter of all main offices, and of enterprises listed on the Japanese stock market 94 percent locate their main offices in the prefecture, especially in the innermost three wards (69 percent) and secondarily in adjacent wards (15 percent). Of those 2,555 firms with 30 or more branch offices nationwide, 1,051 (41 percent) are based in Tokyo.

Main offices located in Tokyo find it an advantageous place for purchase of stock and sale of products because of ample chances to obtain information from governmental sources and business circles and to raise funds for their businesses. Tokyo also has all necessary means of communication and transportation. With these advantages main offices concentrated in Tokyo control branch offices and factories all over Japan.

As the seat of the national government, Tokyo functions as Japan's control center so far as the public sector is concerned, and given the *dirigiste* qualities of Japanese government it serves the same function for the private sector as well. Tokyo is the nerve center of Japan, equipped with all the complex machinery needed for economic coordination, direction, and control. Because Tokyo is such a nerve center, the wholesale trades prefer to be located there. Manufacturing firms find it more effective to locate their managerial functions in this nerve center and their factories on the periphery or in more distant hinterland.

Because offices have been increasing in this way, business-related service industries have developed to reinforce office work. Similarly, financial agencies have accumulated in Tokyo since financial operations are integral to office work, especially that of main offices. As service industries and financial agencies also require office space, the increase of total office space is remarkable in the ward area, in contrast to the shrinking of factory sites: office

TABLE 4.8
Concentration of Head Offices of Private Corporations—
Tokyo, 1967-1977 (in percentages)

| Size of Firm (Capitalization) | 1967 | | 1977 | |
|---|---|---|---|---|
| Over $500,000 | 44 | | 41 | |
| $500,000-$5 million | | 41 | | 30 |
| $5 million-$25 million | | 55 | | 52 |
| Over $25 million | | 60 | | 58 |
| Under $500,000 | 27 | | 24 | |
| All private corporations | 27 | | 24 | |

SOURCE: Tokyo Metropolitan Government (1981: Tables 1-9).

floor area in the ward area increased from roughly 25 hundred acres in 1965 to roughly 75 hundred in 1978, while the number of factories declined from about 8 thousand (on some 125 hundred acres of land) in 1965 to 45 hundred (on 75 hundred acres) in 1977.

Since main offices are staffed with executives, managerial, and clerical workers, Tokyo has the highest proportion in Japan of these jobs among its working population. To these operations a high level of professional knowledge and technical expertise is indispensable; not surprisingly, the average level of education of Tokyoites is higher than that of Japanese living anywhere else. And much of this education is obtainable right in Tokyo, home to over one-quarter of all Japan's colleges and universities and almost half of all college students. The resulting population is a voracious consumer of printed material, which has led to the proliferation of publishing and printing industries in the central city. Writers, musicians, actors, painters, photographers, scientists, and lawyers gravitate to Tokyo also, implying that Tokyo's nerve center function is not confined to the political and economic spheres. Specialists of all kinds consider Tokyo the center of their cultural, political, and academic work and organizations such as labor unions also locate their headquarters there.

Perception of the external world is indispensable to the function of the nerve center, and Tokyo plays an outstanding role in Japan's external affairs. Ambassadors to Japan are stationed in Tokyo, especially in Minato ward, one of the innermost three, and Japanese offices of international organizations are also located in Tokyo. Tokyo has grown in international importance; it houses more foreign business firms, hosts more international conferences, and receives more foreign visitors than any other Japanese city. Tokyo is the gateway for foreigners who come into contact with Japan, and the home base for Japanese firms which launch their business abroad (as exemplified by the sogo shosha), and has wholly eclipsed such previously important international centers as Kobe and Yokohama.

Thus Tokyo's most, and increasingly, important function is as a nerve center. This function became especially salient in the late 1960s. There was no discernible pattern of change in the occupational structure from 1950 to 1965; from 1965 onward, however, white-collar jobs increased in proportion and absolute number in contrast to the decline of blue-collar jobs (table 4.9). We have already seen how office space grew and factory space shrank. Tokyo has been, and will continue to be, increasingly specialized in its administrative and nerve center function, thereby displacing even more factories and changing outlying areas into dormitory suburbs for white-collar workers who commute to downtown jobs.

## Urban Problems

Tokyo is not immune to the plethora of urban problems which beset other giant cities of today. Depopulation in the inner area and overcrowding of public transportation facilities have been discussed already. Air and water pollution, garbage disposal, congestion and noise of road traffic, and increased traffic accidents are just a few among other growing problems. Countermeasures have been taken, but in many areas appreciable progress has not been attained yet.

Nevertheless Tokyo has been free from some of the problems plaguing other giant industrialized cities. For instance, no massive unemployment occurred in Tokyo even after the oil shock of 1973, though as late as the 1950s full employment had been but a utopian dream for many Tokyoites. In those days Tokyo was still dotted with slums inhabited mostly by ragpickers, day laborers, and the like. However, with subsequent growth of the Japanese economy these slums have all been eradicated except for one neighborhood inhabited by about seven thousand day laborers. And even this "slum" would be hardly recognizable as such to visitors from the Third World—or from New York City. The unkempt streets, decaying or makeshift housing, and large numbers of able-bodied but idle working-age males characteristic of slums elsewhere are rarely seen; a Tokyo slum would strike most foreign visitors as closest in appearance to a stable but lower working-class neighborhood back home.

The racial/ethnic problem, often combined with urban poverty in large cities of the West, is almost absent in Tokyo because of the ethnic homogeneity of the Japanese population. There is a Korean minority, which disproportionately occupies the lower socioeconomic strata, but they account for less than 1 percent of Japan's population. There are also the *burakumin*, descendants of the outcast class in the feudal period. Like the Koreans, they have long been the object of social discrimination and economic deprivation, but they too account for only a few percent of the total population. They are found mostly in southern Japan and in recent years the national government has devoted considerable sums to improving their position. They still constitute an urban problem where found, but the real problem now is the attitudes of majority society, and their situation constitutes a national, not distinctively Tokyo, problem.

One of the most striking characteristics of Japanese urbanization over the last century has been the ability of neighborhoods to maintain cohesion and

## TABLE 4.9
### Working Population by Kind of Job—Tokyo Prefecture, 1950-1980
#### (in thousands)

| Occupation | 1950 N | 1950 % | 1955 N | 1955 % | 1960 N | 1960 % | 1965 N | 1965 % | 1970 N | 1970 % | 1975 N | 1975 % | 1980 N | 1980 % |
|---|---|---|---|---|---|---|---|---|---|---|---|---|---|---|
| Total | 2343 | 100 | 3346 | 100 | 4552 | 100 | 5444 | 100 | 5647 | 100 | 5597 | 100 | 5650 | 100 |
| White Collar | 704 | 30 | 969 | 29 | 1381 | 32 | 1808 | 33 | 2088 | 37 | 2238 | 40 | 2397 | 43 |
| Professional and technical | 172 | 7 | 248 | 7 | 329 | 7 | 380 | 7 | 523 | 9 | 563 | 10 | 673 | 12 |
| Managers and officials | 113 | 5 | 183 | 6 | 214 | 5 | 294 | 5 | 398 | 7 | 376 | 7 | 392 | 7 |
| Clerical and related | 419 | 18 | 539 | 16 | 837 | 20 | 1134 | 21 | 1167 | 21 | 1299 | 23 | 1332 | 24 |
| Blue Collar | 907 | 39 | 1311 | 39 | 1920 | 42 | 2129 | 39 | 2020 | 36 | 1797 | 32 | 1659 | 29 |
| Transport and communications | 72 | 3 | 122 | 4 | 189 | 4 | 243 | 4 | 236 | 4 | 222 | 4 | 220 | 4 |
| Craftsmen, production process, and laborers | 834 | 36 | 1187 | 35 | 1729 | 38 | 1885 | 35 | 1783 | 32 | 1574 | 28 | 1438 | 25 |
| Mining | 1 | 0 | 2 | 0 | 2 | 0 | 1 | 0 | 2 | 0 | 1 | 0 | 2 | 0 |
| Sales and Service | 588 | 25 | 935 | 28 | 1149 | 25 | 1424 | 26 | 1469 | 26 | 1509 | 27 | 1536 | 27 |
| Sales | 359 | 15 | 552 | 16 | 658 | 15 | 857 | 16 | 875 | 16 | 922 | 17 | 951 | 17 |
| Service | 183 | 8 | 328 | 10 | 430 | 9 | 498 | 9 | 519 | 9 | 510 | 9 | 513 | 9 |
| Protective service | 46 | 2 | 56 | 2 | 60 | 1 | 69 | 1 | 75 | 1 | 76 | 1 | 72 | 1 |
| Agriculture, Forestry, and Fishing | 148 | 6 | 131 | 4 | 101 | 2 | 77 | 1 | 56 | 1 | 40 | 0 | 34 | 0 |
| Unclassifiable | 6 | 0 | 0 | 0 | 2 | 0 | 7 | 0 | 13 | 0 | 13 | 0 | 24 | 0 |

SOURCE: Tokyo-to Somu-kyoku Tokei-bu Jinko Tokei-ka (1974: 97-99); Japan, Prime Minister's Office, Statistics Bureau (1982c: 51).

NOTE: Totals and subtotals underlined.

organization in the process. Almost every neighborhood in Tokyo has an association (*chokai* or *jichikai*) which assists the government in the coproduction of municipal services, promotes neighborhood solidarity, and acts to articulate neighborhood demands to the government. These associations are also integrated with fire and police protection agencies, and their prevention of social atomization and active observation of local goings-on contributes to the astonishingly low crime rate in Tokyo. If one excludes from this crime rate traffic offenses, crimes of negligence, and political crimes (demonstrations are concentrated, not surprisingly, here in the nation's capital), the remaining violent crime is so infrequent that Tokyo is perhaps the safest giant city in the world.

In some other areas, though, Tokyo is still far from approximating the standards of cities in other advanced countries. In terms of open space, especially public parks and playgrounds, and sewage facilities, Tokyo still lags far behind. Not only Tokyo, but Japan as a whole, cannot solve this sewage problem in the foreseeable future. When urban growth began in the early stages of Japanese modernization, vaccines were already available remedies for infectious diseases, and the Japanese had long recognized the relationship between clean drinking water and health. Besides, human excrement was a valuable source of fertilizer. Therefore a sewage system was not recognized as indispensable and its construction was delayed. Partly as a result, Japan is now more plagued with water pollution than most other advanced countries. In Tokyo, almost 75 percent of the population of the ward area occupy sewered dwellings, but in Tama the proportion is still below 50 percent.

Another serious problem in Tokyo is the amazing rise of land prices and the attendant housing problem. In Japanese cities land prices, perhaps in association with overall economic growth, started to rise sharply in the late 1950s until they outstripped the rise in commodity prices. If 1955 is taken as a base year with an index of 100, commercial land prices by 1983 had risen to 5,100 and residential land prices to 3,100. And the situation is worse in Tokyo than elsewhere—land prices in Tokyo are 50 percent higher than the next most expensive city, and if one can bear up under an hour-plus commute (each way) one may work in the Tokyo CBD and live in Yokohama on land at half the Tokyo price (table 4.10). Sumitomo Shoji, a leading sogo shosha, found in 1983 that the price of a condominium-type residence in Tokyo is $1,763 per square meter compared to $858 in London and $686 in New York. And these prices are directly reflected in rents: a 1981 survey disclosed that Tokyo had the highest rents of all Japan's 47 prefectures—and Kanagawa, Saitama, and Chiba ranked second, third, and seventh.

Since many Tokyoites who want to buy a house cannot afford to even if they rely on a bank loan, their alternatives are to continue living in a rented home, be content with buying a small house on a site of 100 square meters or less, or move to Tama or a neighboring prefecture where they can buy a bigger, less expensive house but spend a longer time commuting on a crowded train. Nevertheless, the housing situation has improved over the years (table 4.11),

TABLE 4.10
Comparison of Land Prices in Major Cities—Japan, 1983

| City | Yen (in thousands) | $ (= ¥255) | Index |
|---|---|---|---|
| Tokyo Ward District | 356.2 | 1397 | 100.0 |
| Kyoto | 233.4 | 917 | 65.5 |
| Osaka | 221.6 | 869 | 62.2 |
| Kawasaki | 206.2 | 808 | 57.9 |
| Yokohama | 182.0 | 714 | 51.1 |
| Kobe | 170.2 | 667 | 47.8 |
| Nagoya | 141.1 | 553 | 39.6 |
| Kiroshima | 118.7 | 463 | 33.3 |
| Fukuoka | 91.2 | 358 | 25.6 |
| Kitakyushu | 63.0 | 247 | 17.7 |
| Sapporo | 56.9 | 223 | 16.0 |

Land Price per m²

SOURCE: Tokyo-to-Kikaku Hodo-shitsu Chosa Bu (1984: 99).

in terms of living space per capita and proportion living in dwellings with baths; those having to share common toilets now number only about one in ten. But the percentage of Tokyoites owning their homes or living in rented single-family homes seems to have undergone little change.

Of Japan's prefectures, Tokyo thus has the worst housing conditions. Tokyo still has large numbers of wooden apartment buildings; many are poorly built; only about half provide private toilets; and all are firetraps. Although such housing was home to less than 30 percent of the population in the late 1970s, in the outlying wards of Suginami, Ota, Toshima, and Nakano this proportion approached or exceeded 40 percent. And even where redevelopment occurs, it often leads to greater population densities: the proportion of landholdings of 100 square meters or less is increasing in the ward area as landowners subdivide their lots, often erecting new single-family dwellings standing less than one meter apart. Under such conditions the family garden is usually restricted to a few flower pots on a windowsill.

As noted above, this housing situation is intimately related to commuting time and the family finances of Tokyoites. More house for one's yen is bought in the coin of commuting time and paternal absenteeism. According to a traffic census by the Ministry of Transportation, more than half the commuters to Tokyo's three innermost wards spend more than one hour commuting; those commuting more than an hour and a half rose from 18 percent to over 20 percent between 1975 and 1980, and homeowners commute farther than renters. Given the typical 5.5-day work week, most commuter husbands see their families essentially one day each week. And home ownership strains the pocketbook as well as the family. In 1983 the average annual net income of a worker in Tokyo Prefecture was 4.32 million yen ($18,000), but it could easily cost five times that to buy even a small house in a rather distant place. A larger house might cost seven times the average income. A survey by the Tokyo Prefectural government disclosed that the average worker, if a homeowner, allocates more than 900,000 yen ($3,750) per year, or

TABLE 4.11
Changing Housing Conditions—Tokyo Prefecture, 1965-1980

| Characteristic | 1965 | 1970 | 1975 | 1980 |
|---|---|---|---|---|
| Percentage living in own house | | | | |
| Tokyo | 39.6 | 38.5 | 38.4 | 42.3 |
| Ward district | 38.3 | 37.3 | 37.1 | 40.9 |
| Percentage living in rental house | | | | |
| Tokyo | — | 43.7 | 43.8 | 40.8 |
| Ward district | — | 47.0 | 47.0 | 43.8 |
| Living space per head (m²) | | | | |
| Tokyo | 6.8 | 8.5 | 10.2 | 11.8 |
| Ward district | 6.7 | 8.3 | 10.2 | 11.8 |
| Owned houses | 8.8[a] | 11.2[b] | 13.0[c] | — |
| Rental houses | 5.8[a] | 7.0[b] | 8.3[c] | — |
| Percentage of houses with bath | — | 45.5 | 55.9 | 64.7 |
| Percentage living in wooden apartment houses | — | 30.6[b] | 27.6[c] | — |
| Percentage living in apartment houses with common toilet | — | — | 16.4 | 12.4 |

SOURCES: Tokyo-to Somu-kyoku Tokei-bu Tokei Chosei-ka (1984: 48, 106, 110); Japan, Office of the Prime Minister, Bureau of Statistics (1974: 4-5; 1979: 2-3).
a. 1968.
b. 1973.
c. 1978.

some 20 percent of annual net income, to the payment of housing loan installments.

It is interesting to speculate on the relationship between Japan's housing problems—the small living space and the large debt incurred in buying it—and international trade. Japan is often accused of substantial excesses of exports over imports and stimulation of domestic consumption has been suggested as a remedy. However, those who must be content with very small dwellings have only limited space for furniture and other durable goods and certainly are unlikely to buy, for example, a seven-foot-tall American-made refrigerator. And the oppressive scale of home loan payments and relatively low availability of such loans necessitates very high rates of saving, and consequent limitation of one's consumer behavior. Imposition of income tax on the presently tax-exempt interest from small savings accounts would make them less attractive, but the present trend toward deregulation of interest rates on larger accounts means that savings-minded Japanese are unlikely to stock up on foreign—or domestic—durables beyond their current habits. Nevertheless, as we shall see, housing is today a major focus of those who wish to reduce Japan's trade surplus.

But perhaps Tokyo's most acute, and most intractable, problem today is only potential: her extreme vulnerability. This vulnerability has two sources—one seismic and one technological. The potential for earthquakes such as the one which levelled much of Tokyo in 1923 is unabated today; indeed, Tokyo is overdue for a big quake. And should one strike, the immensely greater size of the population concentration, the number of (at best hypothetically safe) skyscrapers with their glass facades, elevated highways, subways, wooden buildings, gasoline-filled vehicles, and landfill areas makes the prospect fearful indeed. Every neighborhood in Tokyo has posted maps showing the nearest open space, but given the paucity of parks in the city these are woefully inadequate. In the 1970s, one magazine published a map of Tokyo with shaded areas showing the extent of damage expected in case of a quake equivalent to that of 1923. Large areas east of the CBD, home to thousands of people, were left unmarked. These were filled land reclaimed from Tokyo Bay, and in the event of such a quake they would simply sink into the Bay. Therefore detailed explanation of their fate was unnecessary.

The technological vulnerability of Tokyo stems from its modernity—and thus the exquisite interdependence of its parts—combined with the size of its commuter network and the precise scheduling of the trains which alone make the commuter-based metropolitan economy work. During the rush hours surface trains and subways run at two- to three-minute intervals, *on schedule* (one cannot set one's watch by them exactly but they are seldom more than 30 seconds late). Extremely cold or hot weather, strikes, or (as seen in the fall of 1985) sabotage by political radicals are capable of stranding literally millions of commuters in motionless trains or stations so overcrowded that being pushed off the platform is a likely danger. A well-planned general transportation strike could paralyze the metropolis, its efficacy enhanced by the horde of commuters who would decide to try to get to work in their cars. And there is precedent for riots by enraged commuters, their stoic endurance worn through by the wilfulness of the strikers responsible for their imprisonment.

Neither of these possibilities is a daily concern for most Tokyoites, who leave them more in the hands of concerned and well-trained (but untested) disaster planners, the flexible and conciliatory managers of the Tokyo transportation system, and the Tokyo police. But they suggest that despite all the best efforts of government policymakers and the frequently beneficent collective effects of the unregulated decisions of millions of Tokyoites, the whims of nature and the inevitable consequences of technological advance create potential problems beside which the smallness of one's apartment pales into insignificance.

## Political and Public Policy Responses

Since the mid-1950s public concern has increasingly focused on the tremendous concentration of population in and the consequent urban sprawl of Tokyo. One of the first governmental responses was the National Capital Region Development Law (1956), the underlying assumption of which was that a wide region had to be covered by any policies aimed at coping with Tokyo's urban problems. The law's main purpose was to prevent further sprawl by designating the area surrounding the existing built-up area as a greenbelt zone where urban land use was not permitted.

In view of the stubborn protests of people in the greenbelt zone against the prohibition of urban land use and further concentration within the zone of population which was greater than expected, the law was revised in 1965 and plans designed to comply with it were amended three times. Consequently the area earmarked for the green belt was designated a suburban development area to permit orderly urban development compatible with preserving as much green land as possible. For the population expected to migrate into the existing conurbation, satellite towns were to be provided outside the suburban development area to absorb the population who might otherwise move into the inner urban area. Thus, conditional urban development was permitted in the former green belt zone; it was intended to suppress anything that would lead to further concentration of population in the suburban development area and existing conurbation—the ward district, eastern Tama, the metropolitan areas of Kanagawa Prefecture including Yokohama and Kawasaki, and parts of Saitama and Chiba prefectures.

The National Capital Region Law is a copy of England's new town policy, but whether it has attained success is questionable. In the suburban development area, population has increased sharply in spite of the original population control policy. In the depopulating central district local authorities are anxious to restore lost population. Such authorities are, for example, opposed to the present restrictions on factory construction. The restrictions were originally regarded as a beneficial measure to prevent excess population growth and pollution, but when growth came to a standstill local authorities feared that the measure would lead to urban decay.

The Tsukuba "academic garden city" was a satellite town constructed as part of the capital region development plan. A number of governmental research institutes and a national university were relocated to Tsukuba, formerly a village in Ibaraki Prefecture about two hours from Tokyo by train and bus. The population originally estimated for the new city was 100 thousand, but after more than a decade its population remains at 35 thousand despite about $6.5 billion of government funds (at $1 = Y155—actually over $8.8 billion at current prices) invested so far in research institute and public facility construction and road connections to Tokyo. The ward area population has not shrunk appreciably as a result.

Nevertheless, the law and associated plan had seemed appropriate at a time when the Tokyo prefectural population was growing apace, exceeding 10 million in 1962. Tokyo's overcrowding seemed an evil requiring an all-out effort for its remedy. Eight separate plans for reconstructing or relocating the capital were announced by different specialists. Although Tokyoites were enjoying the ample benefits of economic growth, the costs were also becoming clear in the turgid, opaque waters of polluted rivers and smoggy, ear-rending din of cars in the streets. Increasingly, popular expectations of the good life in an affluent society and the gritty reality of life in Tokyo became contradictory. The discrepancy between the objective, environmental standard of living and these subjective expectations led to pervasive discontent with urban life which, in turn, triggered hypersensitivity to environmental affronts. In the late 1960s and early 1970s many Tokyoites became preoccupied with costs of economic growth but, in sharp contrast to an earlier day, almost dismissed its benefits. Widespread discontent sparked efforts to reorganize a political system seen as hand-in-glove with the polluters and developers now blamed for urban ills of all sorts, and the electoral result was the frequent replacement of conservative mayors and governors by those supported by leftist political parties and environmentalist citizens' groups throughout the Tokyo metropolitan area.

These new leaders unanimously emphasized participatory democracy and promised to pay attention to local public opinion, especially in the area of development and environment. Their policies were usually congruent with the social perceptions of their discontented constituents, but even their policies sometimes failed to meet the endlessly ascending level of expectations, and the mushrooming protest movements which had ushered in the era of reformist mayors did not recede. These movements usually argued that the evils of urban life which had grown more and more serious were caused by the insatiable industrial concerns which had concentrated in Tokyo to maximize their profits at the sacrifice of the public interest, but they were equally willing to take up parochial cudgels against the local construction of facilities like garbage treatment plants which some might see as part and parcel of the public interest.

Public opinion warmly supported their arguments—or at least the antibusiness ones. Antiurban observers welcomed both the protest and the U-turn theory, holding them as proof of the healthiness of young people who had become aware of and were rejecting in two different ways the evils of urban life. In this social milieu vigorous protest arose against construction of high-rise apartment housing in the ward area and spread to the suburbs as well. Japan is a land of frequent earthquakes, and until techniques for earthquake-proof building were developed, high-rise housing had been nonexistent. Construction in 1968 of a building over 100 meters high in the core of the CBD, however, demonstrated the feasibility (presumed, since Tokyo has had no major quakes since then) of earthquake-proof architecture, and was followed by a series of 30-story or higher skyscrapers, plus hundreds of smaller, 4- to 10-story apartment buildings.

These smaller buildings became the focus of controversy, since some of them, built like everything else in Tokyo cheek by jowl with their neighbors, deprived neighboring one- or two-storied houses of sunlight. Protests were organized demanding "sunlight rights," and the courts have accepted these demands as part of the people's entitlement to "the minimum standards of wholesome and cultured living" as stipulated in Article 25 of the Constitution. In response, the Tokyo prefectural government led by Governor Ryokichi Minobe designated a large part of the ward area as a "first-class residential area" where construction of any building more than 10 meters high is prohibited. Many cities in the metropolitan area also set forth restrictions on the height and shape of high-rise buildings, and imposed quotas of open space around them.

In the 1960s and 1970s the average number of stories in buildings in the ward area was less than two; by 1983 it was still only 2.5, and only 10 of the 23 wards exceeded this figure—the highest average was Chiyoda's 6.1. Even so public opinion, urban planners, and social scientists generally favored these prohibitions, assuming that taller buildings were the work of avaricious private developers sacrificing public welfare to the pursuit of profit. The assumption about motivation is doubtless correct, but few seem to have considered the welfare of those who watched their children grow up practically as strangers, and who commuted for hours and hours every day because of the unavailability of housing close to the city center.

Housing is in short supply in most cities in the Tokyo conurbation. Developers' prospects for profit were thus excellent, giving concerned local governments considerable leverage in dealing with them. Many stipulated guidelines for developers, compelling them to provide such facilities as roads and parks, plus contributing to the municipality a certain amount of money or developed land. The argument behind the guidelines was that municipalities had to allocate sizable chunks of their budgets to delivering services to the people who would flow into the newly-developed residences.

Despite the accuracy of the argument, the ultimate effect of developers' contributions was not as foreseen. The contributed money and land were not always used to benefit the new residents concerned; indeed, they were sometimes used for other space or budgetary needs of the government, and thus did not enhance the environment commensurately. Moreover, these guidelines led to higher land prices, since developers could hardly be expected not to pass on their expenses to customers. Doing so was not difficult because of demand, and it contributed to the spiral of land prices in Tokyo. Public opinion, however, favored the guidelines, alleging that such measures controlled land development, population growth, and destruction of green space.

In the reformist climate of the late 1960s and 1970s no consideration was given to future housing demand, which might have been foreseen had the age structure of the Tokyo population been examined. Because the largest age cohort in 1970 was in its 20s, one could foretell that in several years when these young people married and had children they would need more living space, creating a huge demand for housing. There would be no such demand if the U-turn theory were true, but in fact this young population remained right in metropolitan Tokyo while local governments in the area were reinforcing measures which helped to suppress the supply of housing. The resulting imbalance between demand and supply led to the continuous rise of land prices. Because the U-turn theory was accepted with no careful examination, authorities had not been aware of the consequences noted.

Local governments alone were not at fault, however. National land policy also inflated land prices. A number of zoning codes and prescribed building standards enacted in 1968 and 1970 made development difficult and expensive, and in the early 1970s Prime Minister Kakuei Tanaka's plan to "remodel the Japanese archipelago"—in effect, a national land-use plan based on a highly integrated national communication and transportation network—convinced many landowners that nationwide growth was imminent and higher land prices justified. Legal measures were taken when prices rose so sharply that land became a major object of speculation in the first half of the 1970s. A new, heavy tax imposed on the sale price of land was designated to prevent outrageous markups, but it reduced the profitability of land sales to the point that there was little incentive to put undeveloped land on the market. The National Land Use Law of 1974 was the culmination of these measures. While preventing disorderly urban land use on one hand, the measures were designed on the other to restore the balance between demand and supply by suppressing excess demand which was stimulated by unscrupulous developers who attempted to buy up as much land as possible for the profit to be gained from resale.

Furthermore, policy inconsistency played a role. Though the Japanese economy has shown a remarkable growth, its course up to the present has not been straight upward. Several times, Japan has fallen into temporary economic stagnation. At such times the government has tended to increase housing loans, both governmental and private, in the expectation that the construction thereby stimulated would have a multiplying effect, revitalizing productive activities in many fields of Japanese industry. Such construction, however, presupposed a corresponding increase in the supply of land for housing, but, in the main, little change was made in the land policy. The inevitable consequence of a growing amount of money flowing into a housing market characterized by limited availability of land was simply a further rise in the price of land.

Thus, national and local governments tried to control private developers. Because governments failed to accept the movement of a free market, no consideration was given to the possibility that developers were trying to buy up land because they anticipated a huge demand in the face of the corresponding short supply of land. A more appropriate alternative would have been to stimulate supply, not suppress demand. As with high-rise buildings, authorities failed to recognize a process which was the inevitable consequence of meeting a huge demand for housing with limited available land, although developers had indeed profited from their buildings and their sales.

In Japan arable, and thus habitable, land accounts for only some 15 percent of the total land area—picture, if you will, half of the population of the United States packed into California or, more precisely, into 15 percent of the land area of California. Procuring a sizable parcel of land is difficult because there is little to begin with and because petty landownership is very common, especially in the cities. With such limited land, high-rise building construction would have been an effective alternative to increase dwelling space and open space after the development of earthquake-proof building technology. But housing planners in Japan are apt to introduce models of Western countries disregarding differences in the amount of habitable land. Moreover, educated Japanese are relatively trusting in local government and much more critical of national government, given their political preference for decentralization. Accordingly, they often fail to examine local government policy closely.

The factors discussed above combined to discourage residential development in metropolitan Tokyo, despite the huge demand for housing which authorities lamented, but did little to meet. The result was an extraordinary rise in land prices, increased fractionalization of holdings (if people could not afford 100-square-meter plots, perhaps one could find buyers for 80-square-meter ones), and—where government did step in—massive complexes of depressingly uniform apartment blocks in distant suburbs. Thus, in Tokyo's regional planning and land policies we find little appreciable success. Underlying factors—antiurban feelings and ignorance of the free market—have been the basis of urban planning which legitimated governmental intervention in urban affairs. In view of the housing tastes of Tokyoites we should reevaluate these assumptions, previously taken for granted. Governmental intervention was largely intended to meet the political preferences of the maximum number of citizens, but the free market appears to have been the best indicator of majority housing preferences, and an unregulated free market would most likely result in throwing a great quantity of underused land on the market and a massive movement of housing construction *upward* rather than ever farther *outward*. Overpopulation of a certain area has been taken as an evil of urban life by those with antiurban feelings (usually the same people who define *over*population), but they fail to recognize that areas become highly populated because many people prefer to live there as a result of conscious calculation of costs and benefits.

## Tokyo's Future

An examination of five- to ten-year-old plans, projections, and "visions" of Tokyo published by local and prefectural governments reveals that few of their forecasts have come true. Not wishing to fall into the same trap, we shall single out a number of factors which may influence Tokyo's future, but we make no predictions of their precise consequences.

### An Aging Society

The 1970 and 1975 population pyramids of metropolitan Tokyo exhibited the protrusion of the age cohort in its 20s, but in the 1980 pyramid there is no equally salient protrusion. As the population pyramid approximates the shape of a cylinder, we can expect a growing proportion of aged people, and in the near future. One cannot tell whether gentrification will emerge in Tokyo and to what extent if it does. Living in Tokyo's central area is very expensive and we cannot foresee the economic level of the prospective older people, although many will already own their own homes and thus be better able to live there than their incomes would suggest.

In any case, care for the aged will become a matter of great importance in future metropolitan policy. All public facilities will require remodeling for use by the elderly, and geriatric health care and the facilities necessary for it will require new budget allocations and probably tax impositions, which will constitute an increasingly heavy burden on the shrinking numbers of working-age people. Concurrently, industries catering to the needs of Tokyo's aged will flourish.

A further effect of the aging of Tokyo's population is the relatively diminishing proportion of the childbearing age group in the area's total population. Rates of use of facilities such as elementary schools will decrease and, as deaths increase and births decrease, the natural increase which alone sustains the population of Tokyo in the face of negative net migration will decrease, exacerbating the problems of depopulation already discussed.

## Urban Renewal

Prime Minister Yasuhiro Nakasone, in office from 1982 to 1987, was aggressive in launching urban renewal as a means of sustaining Japan's economic growth. Though the national government cannot afford to allocate funds for this purpose because of fiscal austerity, Nakasone tried to encourage the private sector to undertake urban renewal. Because a variety of regulations impede private sector undertakings, authorities have been reconsidering existing governmental interventions and investigating how to remove meddlesome regulations. Thus the bans on high-rise apartments and zoning codes restricting development of residential land are being reexamined. The Ministry of Construction has requested local governments to alter their guidelines for land development. Some planners and social scientists opposed Nakasone's policies facilitating urban renewal, and local governments have not as yet been inclined to comply with the request of the ministry. Since renewal policy is still in a formative stage, one cannot tell what will become of it.

So far as the inner area of the ward district is concerned, however, urban renewal has come to be widely recognized as a matter of immediate importance. The 1980s have seen a sudden, sharp rise in the price of land in this area because of the shortage of available space for office buildings. Figures referred to above indicate that total office space in the ward district, especially in the three central wards, has been on a constant increase in the past. Had due attention been paid to this indicator of demands, one could have predicted a shortage of land for office buildings (and hence of office space) if no measures were taken to meet the increasing demand. However, only after the sharp rise in land prices did the government decide to increase the supply of available land by converting a number of places which have been sites of warehouses, dockyards, cargo depots, and the like into land for office buildings in collaboration with the private sector. At the same time, the government is working out measures to control the purchase of land on the free market, thus still sticking to the policy referred to above. Zoning codes for first-class residential areas are not going to be changed in the outer area of the ward district, although it is feared that land prices in this area, too, will rise soon. It is expected that approval for construction of high-rise apartment houses will lead to rising land prices, but no proposals have been made to repress the rise through more effective use, and hence increased supply, of available land.

## Multinuclear Urban Cores

The present governor of Tokyo Prefecture, Shun'ichi Suzuki, has proposed a long-term plan called "My Town Tokyo," a policy concept projecting an optimum future for Tokyo in the 21st century. One of his ideas is creating a multinuclear urban structure. This policy implies building a city with multiple nodes in which homes and jobs would be in close proximity. It also means restricting further concentration of business functions in the central district and decentralizing and relocating them to the yamanote subcenters and Tama. It is not certain whether this idea will be successfully implemented. Artificial building of a city or a core area often fails, as did the grandiose Tama New Town project and, to a lesser extent, Tsukuba, if the individualized forces of the free market—which respond negatively to inconvenience, insufficient infrastructure, and architectural regimentation—are not fully considered. Unfortunately, planners in Japan (like most nations) are more inclined to control the market than to use it.

## Relocation of the Capital

When population increase in Tokyo was very rapid, around 1960, relocation of the capital of Japan was discussed, but without reaching any conclusion, and the idea was eventually dropped. More recently, government officials and cabinet members have intermittently toyed with the idea. Some agencies have indeed been relocated, for example to Tsukuba, but construction of a new capital is a monstrous venture requiring enormous effort and cost. Thus, the probability of such a move is almost nil, unless Tokyo should first encounter some devastating natural disaster.

In the more modest case of Tokyo Prefecture, however, relocation is an imminent reality. The prefectural government owns several dozen acres at the site of the disused water treatment facility previously mentioned, west of the principal subcenter of Tokyo, the booming loop line terminal of Shinjuku. The present plan is, over the next several years, to move the prefectural government from its present, cramped locations near Tokyo Station in the CBD to this new site, further spurring the westward movement of Tokyo's demographic and economic center of gravity. Commuting to the new CBD from ever more distant parts of the capital region will be feasible, and the already less favored eastern sections of the metropolitan area will face another handicap. But since not everything will move west along with the government, the overall effect of the move will probably be to disperse the metropolis' functions to a beneficial degree.

## Population

For the first time since World War II, the 1980 census recorded a decrease in the population of Tokyo Prefecture. Since the population of the entire metropolitan area was still on the increase, the decrease in the prefectural population indicated the trend of demographic dispersion to outlying parts of the area. After 1980, however, the prefectural population again began to increase. Without observing further population change for some years to come, one cannot identify the factors underlying the newly emerging trend. One provisional interpretation is that with land prices rising sharply even in neighboring prefectures, a growing number of young people have given up, perhaps forever, on buying homes in outlying areas and decided to stay in Tokyo in rented homes.

And although the population of the entire metropolitan conurbation is still increasing, the rate of increase has slowed. Some specialists predict the end of giant cities and the coming of an age in which regional and provincial cities will predominate, fueled by a popular desire for and administrative imperative toward decentralization. Even so, internal migration is associated with several factors, the most prominent being the national rate of economic growth. The slowdown of population movement coincided with the economic stagnation of Japan in the 1970s, and it is quite possible that the metropolitan area will undergo additional population increase with recovery of the economy. On the other hand, Japan's rural areas today do not contain a large surplus population available to move to larger cities, because of declining rural birth rates and the almost universal availability of nonagricultural employment in rural areas: today, most country people may have an urban life-style without moving to the city.

## Telecommunications

Recent progress in telecommunications has been remarkable as communication satellites, fiber optics, cable television, teleconferencing networks, and the CAPTAIN (character and pattern telephone access information network) system demonstrate. If the day comes when two widely separated persons can get in touch as if they were in direct contact and the cost is negligible, offices in Tokyo may no longer find it beneficial to be situated in the inner areas of the ward district where land is costly and traffic is congested. If such a trend should accompany the planned facilitation of physical movement around the metropolitan area—a bridge-tunnel "crossway" across the mouth of Tokyo Bay is planned which will marvelously assist in this— then the prefectural government may well be able to implement the idea of a multinuclear city. For the time being, however, what is heard seems to be largely hyperbole or wishful scenarios about the prospective uses of telecommunications. Since technology and application are still evolving, we must reserve definitive conclusions about the impact of telecommunications on metropolitan life and form.

## The International Situation

In Japan's modern history, cities in the coastal area facing the Pacific Ocean have tended to attain greater size than those in other areas. Because of a scarcity of natural resources Japan found it imperative to import raw materials and export processed industrial goods. The appropriate sea routes extend into the Pacific, and this coastal area was thus favored with opportunities for urban growth. Thus the coastal pattern of city development—especially in the 12-prefecture megalopolitan belt between Tokyo and Osaka, which is home to fully half of the nation's people—has been affected by Japan's international position and the same may be assumed for the future.

Regarding Japan's prospective international situation, attention must be given also to East and Southeast Asia in particular. It is very likely that many countries in this region will make considerable industrial progress, even if the pace differs from one country to another—indeed, South Korea, Hong Kong, Taiwan, and Singapore are already referred to as the "NICs", or newly industrialized countries, of Asia. Given Japan's predominant role in the foreign trade of every country in Asia, it is possible that Tokyo will come to play the role of nerve center of the entire region, thus reinforcing its international functions. This would in turn have an impact on Tokyo's internal structure, though we cannot predict precisely what this will be. Presumably, beyond that noted above, the CBD will have to provide even more office space, with an increase in skyscrapers and heavier traffic congestion or overflow of offices to outlying wards. Additional rail, highway, and airport construction may be necessary, with all the neighborhood protest and political conflict such projects entail in Tokyo today. And the overall strengthening of the Japanese economy, with an attendant rise in the value of the yen, will make life increasingly difficult for the already strained small-scale manufacturing areas north and east of the CBD, much of whose output is destined for export. In a democratic society, one cannot expect such hardship to go unprotested.

Another possible consequence with political ramifications is the advent of a multiethnic metropolitan society. Except for Yokohama and Kobe, no Japanese city has ever experienced the presence of a large foreign community—especially a largely Asian community—and whether Japan in general or Tokyo in particular will be able to adapt without significant ethnic conflict remains a lively but open question.

But it may be that the most significant international impact on Tokyo's form and content will come through the agency of Japan's trade partners and, although involving governmental-level conflict, it could conceivably bring a solution to the residential needs of Tokyoites which their own efforts alone cannot. Japan is presently under the gun, as never before, to rectify her gross trade imbalances with all of the other advanced industrial democracies. One of the keys to rectification, as seen by the Japanese and other governments, is stimulation of domestic demand overall, in the hope that this will (1) create less dependency by the domestic economy on exports and (2) create more consumption of imported goods as well. And one of the cornerstones of the stimulation plan is a ten-year, $100 billion-dollar-plus public works program, including the $2 billion Tokyo Bay Crossway. The program, like the renewal plans noted, relies primarily on the private sector, to which loan guarantees, low tax rates, tax exemptions, and multiyear tax writeoffs will be offered. And one of its major focuses is housing, to which end relaxation of zoning regulations, building height restrictions, and land-use limitations is either being advised to local governments or carried through by cabinet ordinance.

As noted in the context of urban renewal, such grandiose plans may come to naught. Prime Minister Nakasone stepped down in late 1987, and one cannot predict that his successor, Noboru Takeshita, will be equally committed to the idea. Local governments may again drag their feet, and environmental groups litigate delay. But there are reasons to think that now there is an impetus to substantive change in the legal framework of metropolitan form and the relative weight of the public and private sectors in the Tokyo arena.

In the first place, Nakasone's party was largely behind him, which increases the likelihood that Takeshita will continue the policy. Second, the wave of reformist mayors and governors which coalesced with the environmental movement in the 1960s and early 1970s to limit the vertical transformation of the metropolitan area has largely receded—most notably, the governor of Tokyo is today a conservative. Third, there is precedent for the intrusion of international imperatives into the planning of Tokyo: the largest planned program of change in the city's history (in contrast to the unforeseen changes brought by earthquake and war) was the creation of a vast building complex and a superhighway system in preparation for the 1964 Olympics. And finally, urban development now has a constituency with a power no domestic agency can match: Japan's trade partners, whose amity and cooperation are utterly essential to Japan's economic survival. For political reasons, Japan's trade imbalance simply cannot grow indefinitely.

Does all this mean that Tokyo will be offered up, with the support of developers and would-be homeowners and over the protests of environmentalists, on the altar of international commercial accord? Not necessarily. But if, a generation from today, Tokyo is a multinuclear metropolis with larger numbers of citizens living near their jobs in high-rise buildings, and if land use is far more economically and residentially rational than it is today, then Tokyo will in all likelihood be a giant city "of the world" in ways of which its citizens are quite unaware.

## Selected Bibliography

Allinson, Gary D. 1979. *Suburban Tokyo: A Comparative Study in Politics and Social Change.* Berkeley: University of California Press.

Bestor, Theodore C. 1985. "Tradition and Japanese Social organization: Institutional Development in a Tokyo Neighborhood." *Ethnology* 24:121-35.

———. 1987. *Japanese Urban Life.* Stanford: Stanford University Press.

Dore, Ronald P. 1958. *City Life in Japan: A Study of a Tokyo Ward.* Berkeley: University of California Press.

Economic Planning Agency. 1972. *New Comprehensive National Development Plan (revised).* Tokyo.

Fukuchi, Takao, Masahiro Chuma, and Makoto Yamaguchi. 1983. "Interregional Economic-Demographic Model of the Tokyo Region." *Socio-Economic Planning Sciences* 17:329-44.

Glickman, Norman J. 1979. *The Growth and Management of the Japanese Urban System.* New York: Academic Press.

Hanley, Susan B. 1976. "Population Change and Urban Logistics in Tokugawa Japan." Paper presented for the Workshop on the Japanese City, Mt. Kisco, NY, April.

Honjo, Masahiko, 1975. "Tokyo: Giant Metropolis of the Orient." In *World Capitals: Toward Guided Urbanization,* edited by H. Wentworth Eldredge. Garden City, NY: Anchor Press/Doubleday.

Japan. Economic Planning Agency. 1972. *New Comprehensive National Development Plan (revised).* Tokyo.

Japan. Executive Office of the Statistics Commission and Statistics Bureau of the Prime Minister's Office. eds. 1949—(annual). *Japan Statistical Yearbook.* Tokyo: Statistics Bureau, Management and Coordination Agency. [Editors and publishers vary.]

Japan. Office of the Prime Minister. Bureau of Statistics. 1974. *1973 Housing Survey of Japan.* Vol. 3. *Results for Prefectures and Cities. Part 13: Tokyo-to.* Tokyo.

———. 1977. *1975 Population Census of Japan.* Vol. 4. *Commutation. Part 1: Division 2 Kanto.* Tokyo.

———. 1979. *1978 Housing Survey of Japan.* Vol. 3. *Results for Prefectures. Part 13: Tokyo-to.* Tokyo.

Japan. Prime Minister's Office. Statistics Bureau. 1982a. *1980 Population Census of Japan.* Vol. 5. *Commutation. Part I: Place of Work or Schooling of Population by Sex, Age and Industry, Division 2 Kanto.* Tokyo.

———. 1982b. *Population of Japan: 1980 Population Census of Japan.* Abridged Report Series no. 1. Tokyo: Japan Statistical Association.

———. 1982c. *Population of Tokyo-to: 1980 Population Census of Japan.* Abridged Report Series no. 2, Part 13. Tokyo: Japan Statistical Association.

———. 1984. *Statistical Indicators on Social Life.* Tokyo: Japan Statistical Association.

Japan. Office of the Prime Minister. Bureau of Statistics. Quinquennial. *Population Census of Japan.* Tokyo: Japan Statistical Association.

Kornhauser, David. 1976. *Urban Japan: Its Foundations and Growth.* London: Longman.

Kuroda, Toshio. 1973. *Japan's Changing Population Structure.* Tokyo: Ministry of Foreign Affairs.

McKean, Margaret. 1981. *Environment Protest and Citizen Politics in Japan.* Berkeley: University of California Press.

Otomo, Atsushi. 1984. "Geographical Distribution and Urbanization." In *Population of Japan.* Country Monograph Series no. 11. New York: Population Division. Economic and Social Commission for Asia and the Pacific. United Nations.

Seidensticker, Edward. 1983. *Low City, High City: Tokyo from Edo to the Earthquake.* New York: Knopf.

Smith, Henry D., II. 1986. "The Edo-Tokyo Transition: In Search of Common Ground." In *Japan in Transition from Tokugawa to Meiji,* edited by Marius Jansen and Gilbert Rozman. Princeton, NJ: Princeton University Press.

Sori-fu Tokei-kyoku, ed. 1972. *Showa 45-nen Kokusei Chosa Hokokusho. Dai-6 kan: Tsukin, Tsugaku Shukei Kekka, sono 2: Jugyochi, Tsugakuchi. Dai-1 bu: Higashi Nihon.* Tokyo: Sori-fu.

———. ed. 1983. *Showa 55-nen Kokusei Chosa Monogurafu Shiriizu no. 3: Nihon Jinko no Chiiki Bunpu to Henka.* Tokyo: Nihon Tokei Kyokai.

———, ed. 1984. *Showa 55-nen Kokusei Chosa Monogurafu Shiriizu no. 6: Tsukin, Tsugaku Jinko.* Tokyo: Nihon Tokei Kyokai.

Steiner, Kurt, Ellis S. Krauss, and Scott C. Flanagan, eds. 1980. *Political Opposition and Local Politics in Japan.* Princeton, NJ: Princeton University Press.

Tanaka, Kakuei. 1973. *Building a New Japan: A Plan for Remodeling the Japanese Archipelago*, translated by Simul International. Tokyo: Simul Press.

Tokyo Metropolitan Government. 1978a. *An Administrative Perspective of Tokyo, 1978*, edited by Liaison and Protocol Section, Bureau of General Affairs. TMG Municipal Library no. 1. Tokyo.

————. 1978b. *City Planning of Tokyo*, edited by Liaison and Protocol Section, Bureau of General Affairs. TMG Municipal Library no. 13. Tokyo.

————. 1980. *Plain Talk about Tokyo*. TMG Municipal Library no. 15. Tokyo.

————. 1981. *Tokyo at a Glance. Facts and Figures 1981*. TMG Municipal Library no. 16. Tokyo.

————. 1982. *Tokyo Tomorrow*. TMG Municipal Library no. 17. Tokyo.

————. 1984. *Plain Talk About Tokyo*. 2d ed. TMG Municipal Library no. 15. Tokyo.

————. Annual. *Tokyo Statistical Yearbook*. Tokyo: TMG.

Tokyo-to Kikaku Hodo-shitsu Chosa-bu, ed. 1983. *Tokyo ni okeru Dai-toshi Juyo no Kozo: Hiru, Yakan Juyo no Konzaika to sore wo Kiteisuru mono*. Tokyo: Tokyo-to Seikatsu Bunka-kyoku Koho-bu Tomin Shiryo-shitsu.

————, ed. 1984. *Tokyo no Tokushitsu wo Kangaeeru*. Tokyo: Tokyo-to Kikaku Hodo-shitsu Chosa-bu.

Tokyo-to Somu-kyoku, ed. 1965. *Tokyo-to no Jinko ni kansuru Tokei Shiryo*. Tokyo: Tokyo-to Somu-kyoku Tokei-bu.

Tokyo-to Somu-kyoku Tokei-bu Jinko Tokei-ka, ed. 1975. *Tokyo-to no Jinko Tokei no Aramashi*. Tokyo: Tokyo-to Koho-shitsu Fukyu-bu Tomin Shiryo-shitsu.

Tokyo-to Somu-kyoku Tokei-bu Tokei Chosa-ka, ed. 1984. *Tokyo-to Shakai Shihyo*. Tokyo: Tokyo-to Seikatsu Bunka-kyoku Koho-bu Tomin Shiryo-shitsu.

White, James Wilson, ed. 1979. *The Urban Impact of Internal Migration*. Chapel Hill: Institute for Research in the Social Sciences. University of North Carolina at Chapel Hill.

————. 1982. *Migration in Metropolitan Japan*. Berkeley: Institute of East Asian Studies, University of California.

Witherick, M. E. 1981. "Tokyo." In *Urban Problems and Planning in the Developed World*, edited by Michael Pacione. New York: St. Martin's.

Yazaki, Takeo. 1970. *The Socioeconomic Structure of the Tokyo Metropolitan Complex*, translated by Mitsugu Matsuda. Honolulu: Social Science Research Institute, University of Hawaii.

APPENDIX 4.1

Census Data on Daytime and Nighttime Populations (in thousands)
and Percentage of Workers Employed at Home—
Ward and Tama Districts, 1980

| Location | Daytime Working Population | Population Density (per km²) | | Ratio of Daytime to Nighttime Population | Percentage Working at Home |
|---|---|---|---|---|---|
| | | Daytime | Nighttime | | |
| Ward District | | | | | |
| Chiyoda | 767 | 81 | 5 | 17.1 | 1.7 |
| Chuo | 619 | 65 | 8 | 7.9 | 2.9 |
| Minato | 574 | 36 | 10 | 3.5 | 4.5 |
| Shinjuku | 446 | 38 | 19 | 2.0 | 8.1 |
| Bunkyo | 167 | 28 | 18 | 1.6 | 15.4 |
| Shinagawa | 242 | 20 | 17 | 1.2 | 15.1 |
| Meguro | 130 | 18 | 19 | .9 | 18.6 |
| Shibuya | 285 | 31 | 16 | 1.9 | 8.4 |
| Toshima | 205 | 28 | 22 | 1.3 | 15.7 |

*Continued*

APPENDIX 4.1  Continued

| Location | Daytime Working Population | Population Density (per km²) | | Ratio of Daytime to Nighttime Population | Percentage Working at Home |
|---|---|---|---|---|---|
| | | Daytime | Nighttime | | |
| Taito | 257 | 34 | 19 | 1.8 | 18.6 |
| Sumida | 173 | 20 | 17 | 1.2 | 28.4 |
| Koto | 210 | 10 | 10 | 1.0 | 19.0 |
| Kita | 159 | 17 | 19 | .9 | 23.5 |
| Arakawa | 112 | 19 | 19 | 1.0 | 31.6 |
| Itabashi | 223 | 15 | 16 | .9 | 20.9 |
| Nerima | 172 | 9 | 12 | .8 | 30.0 |
| Adachi | 242 | 10 | 12 | .9 | 30.7 |
| Katsushika | 177 | 11 | 12 | .9 | 31.7 |
| Edogawa | 193 | 9 | 11 | .9 | 29.0 |
| Ota | 360 | 14 | 14 | 1.0 | 18.4 |
| Setagaya | 241 | 12 | 14 | .9 | 24.7 |
| Nakano | 115 | 17 | 22 | .8 | 25.5 |
| Suginami | 161 | 13 | 16 | .8 | 25.5 |
| Tama District | | | | | |
| Musashino | 59 | 13 | 12 | 1.0 | 16.1 |
| Tachikawa | 69 | 6 | 6 | 1.0 | 15.3 |
| Hachioji | 138 | 2 | 2 | 1.0 | 20.8 |
| Mitaka | 56 | 8 | 10 | .8 | 19.8 |
| Ome | 40 | | | .9 | 22.4 |
| Fuchu | 79 | 6 | 6 | .9 | 15.1 |
| Akishima | 37 | 5 | 5 | .9 | 16.8 |
| Chofu | 62 | 7 | 8 | .9 | 18.3 |
| Machida | 75 | 4 | 4 | .9 | 19.8 |
| Koganei | 27 | 7 | 9 | .8 | 21.3 |
| Kodaira | 53 | 7 | 7 | .9 | 19.2 |
| Hino | 46 | 5 | 5 | .9 | 14.1 |
| Higashi-murayama | 32 | 6 | 7 | .8 | 23.6 |
| Kokubunji | 25 | 6 | 8 | .8 | 21.3 |
| Kunitachi | 19 | 7 | 8 | .9 | 21.8 |
| Tanashi | 26 | 8 | 10 | .9 | 16.8 |
| Hoya | 22 | 8 | 10 | .7 | 29.2 |
| Fussa | 19 | 4 | 5 | .9 | 18.7 |
| Komae | 17 | 8 | 12 | .7 | 27.9 |
| Higashi-yamato | 18 | 4 | 5 | .8 | 23.9 |
| Kiyose | 16 | 5 | 6 | .8 | 25.5 |
| Higashi-kurume | 24 | 6 | 8 | .8 | 24.0 |
| Musashi-murayama | 24 | 4 | 4 | 1.0 | 16.4 |
| Tama | 18 | 4 | 5 | .8 | 17.7 |
| Inagi | 15 | 2 | 3 | .8 | 20.5 |
| Akigawa | 10 | 2 | 2 | .8 | 31.4 |

SOURCE: Sori-fu Tokei-kyoku (1984: 216-19); Japan, Prime Minister's Office, Statistics Bureau (1982c: 58).
NOTE: Data for Tama district include only cities; towns, villages, and islands are excluded. Daytime working population data include workers in unclassified jobs.

# 5

# Shanghai

## Rhoads Murphey

Although its beginnings as a regional port and trade center for the lower Yangtze valley go back more than a thousand years, Shanghai's major growth occurred and its present character was largely formed only after it was "opened" to Western trade and residence in 1843, after the first Anglo-Chinese War (Opium War) and the Treaty of Nanking in 1842. The treaty opened four other ports in south and central coastal China. More treaty ports were added in subsequent years, to a total of about a hundred by the time of the First World War, including nearly all of China's large cities except Beijing. Almost from the beginning of this system Shanghai grew more rapidly than any others in trade volume and population. Accurate figures are lacking, but by about 1910 Shanghai, at roughly 1.2 million, had become China's largest city, as it still is. Its dominance is unlikely to be challenged since it remains the country's premier port and largest single industrial center. As trade between China and the rest of the world increases, especially with trends since 1978, and as post-Mao national policy continues to favor "the four modernizations," which strongly emphasize industrialization, Shanghai's size and importance seem certain to grow still more, together with the problems urban size and industrial concentration bring.

AUTHOR'S NOTE: I am indebted throughout this chapter to my Chinese collaborator, James Chan, for his invaluable assistance as critic and compiler of much of the data cited on Shanghai development since 1981, including observations, interviews, and statistical compilations. He collected Chinese press reports in May 1984 and subsequently obtained them. Without Chan's help, this chapter would not have been possible in its present form.

For all its national and global importance as one of the world's ten largest cities, modern Shanghai has been since the 1840s an anomaly in China. Shanghai parallels the nature and role of similar cities in the developing world which grew under colonialism in some form and have remained major centers of foreign contact and foreign-style growth: Bangkok, Seoul, Calcutta, Bombay, Manila, Djakarta, Istanbul, Lagos, Cairo, Buenos Aires, and even to an extent Tokyo, Yokohama, and Kobe. In its semicolonial days, Shanghai was described as "in but not of China." Shanghai's connections were far more importantly overseas than with the rest of the country; its character, style, and even physical appearance were more Euro-American than Chinese. Most downtown public and commercial buildings date from 1900-35 and strongly suggest an American or 20th-century European cityscape. Shanghai's consistent economic emphasis on manufacturing, banking, overseas trade, education, and publication further distinguish it from most of the rest of China, where industrial centers grown up since 1949 acknowledge and depend on Shanghai's pioneering and leadership role. Shanghai remains the country's chief style center, a pacesetter in literature and the arts, fashion, ideas, technology, and innovation. Most Chinese regard Shanghai with a mixture of envy and resentment. For many it is still the most desirable place to live for the same reasons which draw people to big cities everywhere but accentuated in China by Shanghai's perhaps greater degree of uniqueness. Migration is strictly controlled in China in a long-standing effort to limit urban giantism; this may further heighten Shanghai's appeal as something only for a lucky few.

During the first three decades of Communist rule, Shanghai was periodically viewed officially as more problem than promise, an unwelcome leftover from a humiliating and resented semicolonial past. There was official talk in the early 1950s of dismantling Shanghai and distributing its factories and experts over the previously neglected rest of the country. Shanghai was never dismantled completely, but skilled workers, technicians, machine goods and tools, and some whole factories were reallocated from the city to aid in developing new inland industrial centers, and Shanghai's own growth was sharply restricted. Part of this policy resulted from an understandable recognition of China's heretofore lopsided industrial growth, with over a third of its modern industry (and nearly half its foreign trade) in Shanghai alone before 1949, and with the only other major industrial center in Japanese-controlled Manchuria. Both were peripheral to China, vulnerable to outside attack, smacked distastefully of a hated past from which China was now liberated, and seemed still to be creatures of their former imperialist masters.

National policy was also motivated in part by a generally antiurban bias, reflecting the rural roots of the Communist power drive, the role of cities as bastions of Kuomintang (Nationalist) Party power and reaction, and their morally corrupting nature in this context, where Chinese people were drawn into bourgeois values, pseudo-Western life-styles, and consequent counter-revolutionary behavior. Western-style and "modern" cities—notably the former treaty ports but increasingly most growing cities—were seen as soul-destroying, antipeasant (hence anti-Chinese), and filled with corruption, crime, hypocrisy, suffering, squalor, and pollution. China had been unique in the degree to which such attitudes were, at least for a time, reflected in official policy and acted on. That policy was not applied consistently, even before the beginnings of major change in 1978; even by 1956 it was recognized that breaking up or stifling Shanghai would kill the country's major golden egg-laying goose. But Shanghai's growth has been curtailed and its population increase kept small—a considerable achievement especially given continued national demographic and industrial growth.

Shanghai nevertheless presents China with a chronic dilemma, one unlikely to be fully resolved. It is too obvious that modern Shanghai was created under imperialist management. The city is China's major window on a world which the revolution long tried to reject and still seeks to filter or sanitize. Shanghai is the apotheosis of the capitalist path with its stress on production, trade, and the rise of new urban elites. The opposite side of China's and the revolution's heavily peasant coin, Shanghai is physically marginal to most of the country. The city remains, as it has for well over a century, a hot bed of new ideas, intellectual ferment, bourgeois life-styles, and other actual or potential counterrevolutionary trends. Shanghai has been a consistent trouble spot politically through the often chaotic years since 1949, as expected. It is equally obvious that Shanghai represents in important respects China's future, especially now that official policy has moved away from its radical Utopian phase toward a position which gives first priority to modernization and production, always Shanghai's priorities. The government seems willing to make less static about the consequent rise of new elites generated by if not essential to such a policy, which Shanghai has long specialized in doing.

Without Shanghai, and many similar cities which can use it as a model, China will never escape from peasant poverty. If in such an escape, the country loses its original revolutionary soul, as seems probable, present policy appears to regard that as an unavoidable price to pay, although it denies that such a result will take place and continues to try a variety of ways to prevent or minimize it. One suspects that this is in part whistling in the dark, in part sincere effort to preserve at least some aspects of egalitarianism, countrywide development, and national integrity rather than merely following Western or foreign trends. But willingness to reward initiative and performance, stress on modernization, and tolerance of new managerial/technical elites together with expanded facilities to train them are more important long-run determinants of the shape of China's future. Shanghai has represented in many Chinese minds the Satanic side of the West, the preeminent temple to Mammon, the sink of material and spiritual pollution. In the mind of an early Communist revolutionary:

[Shanghai] was a bitch-goddess who gnawed at their souls, scarring them brutally and indelibly. Life in the Concessions was comparatively safe . . . but the security which Tai enjoyed there galled and tormented him at the same time. He, a Chinese, was being protected from his fellow Chinese by the grace and scientific superiority of Westerners. . . . [The countryside] represented idyllic, uncontaminated China . . . There he saw Chinese living in a natural economy . . . while the workers of Shanghai were in hell (Mast, 1971:229, 247).

Shanghai is still the hope of the future, even for nationalist-minded, even for revolutionary Chinese who care about the material welfare of the masses. Without the power Shanghai represents, China is condemned to economic and technological backwardness.

## Population

Let us first discuss the city's population totals, since this is one basis of Shanghai's distinctiveness as China's largest city and the one whose growth has been most successfully controlled. Immediately after liberation in 1949 the city grew very rapidly as urban refugees and others displaced by anti-Japanese and civil wars or eager for new economic opportunity streamed into Shanghai. We have no accurate counts from that period, but the total urban population probably reached about 6.7 million by 1957, from a 1936 maximum of about 4 million and a 1949 total of about the same. The city was not heavily contested by the retreating Kuomintang forces and there was relatively little damage. Most of the few remaining foreigners had left by the time Communist armies entered Shanghai, and virtually all were gone by the end of the year, although at their peak in the 1930s they had numbered over 60 thousand. For more detail on this period and earlier ones, and population figures, see Murphey, 1953; for data since the early 1950s see Howe, 1981, especially pp. 218-21, and Murphey, 1977 and 1980.

Chinese statistics, especially before the 1982 census, are too unreliable to give a precise picture[1], but a further source of ambiguity is that since 1958 Shanghai's administrative area has included an extensive surrounding rural zone, intensively cultivated to feed the city and hence densely populated but hardly urban. Population totals since 1958 include this large area, roughly 80 percent of the 6,186 square kilometers and ten counties which form the municipality, as if it were part of the city. Shanghai is sometimes referred to in the foreign press as the world's biggest city because most sources lump the population of the large rural areas in with the genuinely urban inhabitants. Once this has been sorted out, and based on fragmentary reports in the Chinese press, it seems clear that *urban* Shanghai grew out of control in the 1950s from rural inmigration.

From 1959, after establishing strict controls and reassigning large numbers to the countryside, Shanghai's urban total was reduced to about 5.5 million (plus about 5 million in the rural parts of the municipality) and held at roughly that figure until the early or mid-1970s. Since then it has grown to a 1984 estimate of a little over 12 million, of which about 6.5 million live in the unambiguously urban area, including new satellite industrial towns around its periphery (but see below on uncounted "illegal" residents). Since 1959 migration within China has been rigidly limited and Shanghai, as the largest city, has been especially carefully monitored. Success in inhibiting the city's growth has also been because of the forced assignment of urban residents to "productive labor in the countryside," especially young middle-school graduates.

## *Rustication*

In the Chinese revolutionary context, and remembering its antiurban bias, rustication serves two purposes: it restricts city growth by siphoning off a significant part of the yearly increase, and it educates or reeducates urban people in "correct" attitudes, kindling or recharging their commitment to "serve the people" who are primarily in rural areas. Theoretically, at least some urban people, especially middle-school graduates, have acquired skills which may be useful in the drive to raise productivity, and their labor alone may be of some help. More stress seems to have been placed on the morally purifying virtues of rural labor for city-bred people, since cities breed bourgeois values, as the word itself implies. More practically, urban jobs are still not increasing as fast as even natural city population growth, even in Shanghai, and cities cannot keep up with the rising demand for housing and basic services. In the countryside, urban people sent down could somehow be fed, housed, employed at least after a fashion, and provided with commune-based forms of social security at little evident additional costs.

From the beginning of the revolution, manual labor has been glorified. This work was further institutionalized in 1957 in the practice known as *hsia fang* (sending down) which involved assigning urbanites—mainly intellectuals, elites, and white-collar workers—for a month or more each year to labor in the countryside. The practice still continues, although as a declining gesture despite the good sense it makes (see Lee, 1966). Even before hsia fang was institutionalized, a more permanent form of urban-rural population transfer was established with Shanghai as the major target. The *shang-shan hsia-hsiang* movement (up to the mountains and down to the countryside), launched in 1955 and still technically in effect, assigned urban secondary school and college leavers and graduates to rural areas en masse and permanently. Like nearly all programs, it was disrupted by the Great Leap Forward in 1957-58 and the difficult years which followed, but by 1966 about a million urban youths had been so transferred. During the 1966-76 decade of the Cultural Revolution, with all radical programs in high gear, a further 15 or 16 million followed, something like a tenth of China's total urban population and also close to the total annual population increase (for more detail, see Bernstein, 1977).[2] In Shanghai, with few exceptions (only children of a single or ailing parent, a few youths with good political connections), all members of all graduating classes between 1966 and 1976 or 1977 were shipped out in this program.

Most of these exurbanites ended up doing only manual labor, and their morale was generally terrible. To many of them, and to their parents, rural relocation seemed a waste of their training and supposed talents, certainly of their urban rearing and expectations of life-style and employment—which of course was part of the point of the official policy. Higher education was not an alternative for most; there were places for only tiny numbers (fewer than 1 percent). During those radical years the universities were virtually shut down most of the time with the limited places going overwhelmingly to those who had already done at least two years of rural labor. From the beginning of the shang-shan hsia-hsiang program there was predictable discontent, and more and more of the "sent-downs" began to find their way back, illegally, to Shanghai and other cities. This escape required forged or stolen travel permits and ration cards (most basic food and other consumer goods were until very recently rationed, as some still are), moving in with friends or relatives, and living by their wits. The press reported panhandling, mugging, and street gangs composed of such refugees living by petty theft and dealing in the black market in rationed or forbidden goods.

It is hard to guess the numbers of such people in Shanghai or estimate the overall size of the U-turn movement but it is certainly not trivial. In the changing political climate of 1978, 1979, and 1980 after the reascendancy of Vice Premier Dengxiaoping there were repeated mass demonstrations in Shanghai by "sent-downs" who had filtered back to the city and now demanded official permission to return, plus assignment to urban jobs. (See the accounts in *Chieh-fang Jih-pao* [Liberation Daily] and similar stories in French and English in Agence France Presse [Peking] and New China News Agency.) In May 1984 an official in Shanghai said in private conversation that about 600 thousand illegal "temporaries" lived in Shanghai, plus 200 thousand other illegal infiltrators from rural areas outside the municipality, most of whom have not been counted in any official censuses. These figures are well over 10 percent of the official total of about 6.5 million urban inhabitants and should probably be added to make the real urban population of Shanghai close to 7.5 million. Another informant in May 1984 said that about 450 thousand people have returned to Shanghai every year since 1979. In addition, many people who live in rural parts of the municipality work in the city during the day. Visitors to Shanghai increasingly see and are sometimes approached by youths attempting to sell things in the street, perform services, or cultivate contacts which can lead to urban employment or emigration.

As policy continues to put new emphasis on production and incentives, without the former level of concern about consequences for the revolutionary goals of egalitarianism, the mass line, and controls on growth of cities, urban employment may begin to grow somewhat more rapidly. It will be some time yet before growth is rapid enough to absorb all of each year's new urban age cohorts, let alone those millions in rural areas who would rather be in a city, most of all in Shanghai. Attractions of big city life are much the same in all developing countries, as they still are to some extent in developed ones. In China, as in most poor countries, the attractions are heightened by the clearly superior urban living standard, not only in the far wider range of food and consumer goods (including television sets, radios, and refrigerators) but also in medical and educational services.

Admission to postsecondary education is by stiff competitive examination, and university and technical school graduates are again expected to move into elite jobs. Most universities are urban, notably in Shanghai with 16 universities and colleges, about 40 technical training colleges, and many large research institutes. The designated key schools whose graduates are almost the only successful candidates for the postsecondary entrance examinations are also urban, with the major cluster in Shanghai. In China, education has always been seen as the means to status improvement, and such a connection is no longer frowned on. Partly because cities are the home of elites, urban wages and incomes in China are about three times the rural average, and wages are highest of all in Shanghai. For a detailed account, see Howe, 1973, and Whyte and Parish, 1984.

Shanghai's features are clearly desirable, regardless of one's political views. There is so much more to spend money on—big department stores, theaters, sports and cultural events, and now even beauty parlors, fashionable dress shops, and "places of evening entertainment." For those still in the earlier stages of upward mobility, there is the simple excitement and bustle of street life, museums, parks, markets, billboards (no longer merely political), and the heady atmosphere of any big city, especially one with the persistent character of Shanghai.

Shanghai's attributes seem very bourgeois, but the state has, however reluctantly, accepted the notion that incentives are essential to better economic performance and has agreed to provide them in the form most people want. Shanghai is still a far cry from Paris and life there is on a very modest scale. From an outside perspective, life is drab and most people in Shanghai are still poor by European, American, or Japanese standards; most housing is especially dismal and inadequate and consumer goods still few. But the lure of what Shanghai can offer is powerfully there.

One of the sacrifices made in the interest of maximizing productivity and accelerating the drive for modernization seems to be the former rigid controls on growth of the biggest cities, most of all Shanghai. Shanghai and other big cities are the most efficient industrial centers, with the highest technical level, productivity per worker, quality, and output, and it makes little sense to restrict them artificially in the name of revolutionary ideology. Young people who have drifted back from the countryside are now increasingly tolerated officially and referred to as "waiting for employment"; many have even been issued ration cards and housing assignments. There will probably be some form of official amnesty for most of the 800 thousand-plus "illegal" residents referred to above, while at the same time new jobs are, however more slowly, created by new economic growth. The period when Shanghai's growth was stalled by controls seems over, and a new period of possibly rapid growth beginning.

Shanghai will still have to earn its keep, although perhaps less lopsidedly than before. During the 1950s Shanghai provided about 18 percent of the national government's revenue budget, as it still does, but received only about 2 percent of central investment funds in return, as the city in effect subsidized the building of new inland industrial centers. During the same period Shanghai produced about one-fifth of the gross value of national industrial output, although in the critical category of machinery Shanghai's share was much larger (see Lardy, 1975a, b). Measured another way, Shanghai contributed over 14 percent of China's industrial output by value in the 1950s but received only 1-2 percent of central investment, in addition to exporting large amounts of machinery, some whole factories, and over 250 thousand skilled workers and managers (Howe, 1981:166). The rise of new and expanded inland centers has greatly reduced Shanghai's proportional dominance, but since about 1958 coastal and inland regions have shown approximately the same industrial growth rates; if one omits oil (primarily extraction rather than manufacturing), the old coastal areas have grown more rapidly (see Roll and Yeh, 1975). Such a trend seems likely to accelerate in the 1980s, with Shanghai in the vanguard. In this connection, the biggest new steel plant in China is currently under construction in Shanghai, at Baoshan on the Yangtze near its junction with the Huangpu, with imported Japanese equipment and technology. As of 1982, probably on the eve of major new industrial growth, Shanghai accounted for 11.4 percent of China's total industrial output (Shanghai Academy of Social Sciences, 1983:124).

Shanghai remains attractive as an industrial and commercial center not only because of its long-standing position in the forefront of economic and technological change and its pool of skilled workers, technicians, and high-level facilities. Shanghai became China's largest city and chief center of modernization in the 19th century primarily because of its unrivaled geographical advantages. Situated on the Huangpu River, a tidal tributary emptying into the lower end of the Yangtze estuary, Shanghai's water transport connections by coastal and riverine routes throughout the Yangtze system, which drains the largest and most productive share of the country, are unmatched and have long made it the cheapest assembly and distribution point for raw materials and finished goods. Since 1910 railways and later motor roads and air routes have also focused on Shanghai, but the low cost of water transport, the city's central location on the coast, and its Yangtze system connections have made it China's premier port and leading commercial and industrial center. These advantages were clouded from 1949 to 1978 when antiurban, antitreaty port, and specifically anti-Shanghai policies inhibited the city's growth in favor of a better distributed pattern of development, but the advantages remain and seem now likely to be more fully pursued.

Further growth of this already huge urban mass will create serious problems, which, as in most big cities, seem inadequately anticipated or planned for. In the rest of this chapter, I survey those problems as the city's growth and industrial development move into what may be a more rapid phase.

## Site Deficiencies

Shanghai has some serious site deficiencies associated with the flat deltaic alluvium on which it rests, and chronic troubles with using the small, winding, and silted Huangpu as the harbor. Poor soil and surface drainage have had to be contended with at every period of the city's growth. Minor flooding is still frequent. With the naturally high water table, heavy rains characteristic of the summer monsoon, and onshore winds combined with high tides, result in drainage system backup in many areas, causing expensive damage and inconvenience. Sewage and waste disposal is an even more serious, related problem, and the urban area has almost no vertical relief, most of it being only a few feet above mean high tide in the Huangpu. As urban and industrial growth have drawn ever more heavily on local groundwater reserves by pumping, the water table in many areas dropped enough to produce ground subsidence in the unconsolidated alluvium, with disastrous results for roads and buildings. In some areas the water table dropped below tidal levels and was invaded by salt water, poisoning the aquifer. Large or multistory buildings must be based on concrete rafts on which the structures float in the jelly-like subsoil, adding significantly to costs.

These are familiar problems in many cities which have found lowland or deltaic sites advantageous for other reasons, as has Shanghai. They can be adjusted to, but only at additional expense, and often, as with depletion of groundwater reserves and consequent subsidence or salt invasion, only incompletely or by trade-offs. Shanghai before the 1840s was surrounded and crisscrossed by an intricate network of tidal creeks and canals which served for drainage and transport. As the city grew, more and more of these water bodies were filled in or covered over, worsening the flooding problem. Officials eventually realized that embankments were necessary (such as the famous Bund along the Huangpu) and land for roads and buildings would have to be raised by landfill, a solution which now protects most of the urban area from all but the worst flooding. One major stream remains, the Woosung River (formerly Soochow Creek), emptying into the Huangpu in the heart of the downtown core and still heavily used by smaller boats since it is navigable inland as far as the once dominant delta city of Soochow.

Perhaps the worst aspect of the site is the harbor. Like so many ports which grew originally in the age of sailing ships, Shanghai has long outgrown its harbor as the size of the most economical carriers has mushroomed. The river at Shanghai is a little wider than the Thames at London but carries a much heavier silt load. The major docks at Shanghai are about 20 kilometers upstream from the river's mouth at Wusong in the Yangtze estuary, along a winding course with several bends where silting continually builds up and is kept open to ocean shipping only by constant dredging. Sediment from the heavily silt laden Yangtze also enters the Huangpu with each flood tide. At Shanghai there is extremely limited anchorage, turning space, or docking space with deep water, and shoaling also builds up from silt brought down by the Woosung River. A daily tidal range of about two meters adds to the problems, although it does make the harbor accessible at high tide to ships drawing up to nine or ten meters. Deeper draft carriers and those too large to negotiate the awkward channels and limited space—most of the more efficient transocean shipping—use port facilities at Wusong, where the estuary is over 40 kilometers wide. Wusong offers inadequate shelter from winds, tides, and currents and limited dock facilities which often involve costly delays, in addition to expensive cargo shipment to or from Shanghai by rail, truck, and lighter.

Finally, the most intransigent problem lies at the mouth of the estuary where the Yangtze merges with the sea and drops its heaviest silt load to form a vast offshore bar. There is continuous dredging here too, but alongshore currents in the open ocean redistribute the silt, making this a Sisyphean task and keeping the largest carriers out altogether. These problems were sharply illustrated when it was belatedly realized that the new steel plant at Paoshan (see above), scheduled to use large amounts of imported iron ore, could not be served by the most economical ore carriers because of their draft. There and in the harbor along the Huangpu, much of the loading and unloading is done slowly and expensively by lighter from ships anchored in the stream, and delays are chronic.

Problems posed by the Shanghai site have been more than overbalanced by its wider locational access to and from the rest of the country and abroad by sea. Shanghai is far from unusual, especially among other ports such as Rotterdam, Hamburg, London, Calcutta, Bangkok, or New Orleans in illustrating the advantages of wider access despite disadvantages of similar site problems. If growth is to continue, Shanghai will have to provide far greater investment, as other deltaic cities have done, to mitigate those disadvantages. In addition to Wusong, two other ports now part of the regional Shanghai economic zone have been developed to ease congestion and handle larger ships. Ningpo, 160 kilometers south on the northeast coast of Zhejiang Province is largely free of silt and now has a large cluster of deepwater docks; it will be used for unloading some of the largest ore carriers to supply the new Paoshan steel plant. Nantung on the Yangtze more or less opposite Wusong has new docks and port transfer facilities. But at all three outports, despite new berths and container facilities, increasing traffic continues to outrun capacity, delays remain a major problem, and ships over ten thousand tons have difficulty entering or berthing. Press reports (China Daily, 1984b) show docking facilities at Shanghai adequate for 100 vessels but over 200 enter the port daily. Over a third of China's international cargo, 100 million tons in 1984, moves through Shanghai's overcrowded harbor, making it again, as before 1949, one of the world's ten largest ports.

Areally, the city has grown extensively from its pre-1949 base, despite official controls. The original Chinese walled city, still reasonably well preserved although its walls are gone, lay south of the Woosung River along the west bank of the Huangpu. Between it and the Woosung to the north, the first treaty port concession area began, with docks along the river and by the 1870s an embankment, the Bund, to protect the settlement from flooding. The earliest and still largest axis of growth was westward, away from the Huangpu; this is still the sector of major commercial arteries, shops, banks, offices, hotels, theaters, museums, parks, and bus lines. The Woosung River was a minor barrier to expansion but was soon bridged in several places and the downtown core spilled across it, including the main railway station and post office. Docks, wharves, warehouses, and (especially after 1895) factories and shipyards spread northward along the west bank of the Huangpu downstream, becoming by 1984 an almost solid industrial and shipping area all the way to Wusong, which was itself expanding southward as it increasingly served as the city's outport. Other more central industrial zones extended earlier patterns along both sides of the Woosung River westward, and southward along the Huangpu upstream from the old Chinese city, including another railway station with connections to the south. The chief central residential area developed in the old French and international concession areas, back from the river and the commercial sector and stretching out westward. The Huangpu has remained a barrier, despite the tunnel under it, but an industrial zone and associated dock and shipping facilities have continued to extend the late 19th-century ribbon growth along the east bank.

## Urban Development and Services

Areal growth since 1949 has expanded these patterns but has also taken the form of new industrial satellite centers in a rough ring around the central city and articulated with a new ring highway, Chungshan Road, as well as with the ring rail line connecting the north and south stations around the western edge of the downtown area. All cities in China are supposed to follow a functional zoning pattern, distinct areas of different categories of land use and industrial plants, with interdependent industries grouped together. Housing and service facilities for workers associated with each are to be located adjacently. This ideal plan is derived to some extent from earlier Soviet styles of urban development but seems to have been carried farther in China, for new growth.

Shanghai was already a big city before 1949 and it was difficult to reconstruct it except on a very limited basis. The new outlying satellite centers, mainly around the northern, western, and southern edges in previously rural areas, could be integrally planned, with housing units and workers' communities grouped around a new industrial cluster. There are really two such patterns, the first begun in the mid-1950s around the perimeter of the older urban area, and the second, more recent, as spatially separate towns of from 50 to 250 thousand population farther out, between 10 and 40 kilometers from the main urban area. There are now six "near-suburban" industrial districts of the first sort and six more distant suburban industrial townships of the second sort. Both kinds of complexes house interdependent or complementary industries (electronics, chemicals, steel, etc.), and in both most housing is within walking distance of work place, usually in large and often dreary blocks, although sometimes in a cluster of smaller communities with buildings one might almost describe as town houses. Most new growth since the early 1950s has followed such functional zoning as much as possible and has helped to cut down on wasteful commuting, although for the city as a whole that still takes place on a massive scale (see below).

The new integrated developments are provided with a full range of retail shops and services: schools, health clinics or small hospitals, day care centers for the young children of working mothers, often a savings bank, a cinema, and recreational facilities. Over 150 such work/housing complexes have been built in Shanghai since 1949, including several redesigned areas in the preexisting city as well as several in each of the new satellite clusters.

Foreign visitors are usually taken to see one or more of the newer and higher quality housing units and may derive a misleading impression of how most Shanghai people live—better perhaps than most rural people, but the shortage of adequate housing and poor quality of most of it may be the city's biggest problem, after the even more dismaying challenge of environmental hazards, discussed below. Probably over half the present housing stock dates from before 1935, most of that in poor repair. There are few detailed or reliable statistics, but probably at least a third of all dwelling units lack piped water. Most tenants share toilet, bath, and often cooking facilities, limited or primitive as these usually are. Housing space per person is by Western standards submarginal, although this all represents little or no change from traditional Chinese circumstances and is adjusted to more easily in a culture long used to extreme crowding, lack of privacy, and spartan living conditions.

Most foreign studies suggest between three and four square meters of floor space per resident—half the figure even for urban Japan, let alone Europe or the U.S.S.R. Studies agree that there was a major decline, perhaps as much as 40 percent, in total housing stock from 1949 to at least the mid-1960s as collapsed or demolished structures were not fully replaced by new units (see Howe, 1968, 1978; Kang, 1966; Ma and Hanten, 1981; and White, 1978). Housing of acceptable quality is critically scarce, a problem now acknowledged publicly. At least 10 percent of the housing stock consists of temporary or jerry-built shacks, many inhabited by the large population of "illegals."

Shanghai has avoided a totally out-of-control housing crisis only because its growth was restricted from 1959 to the early 1980s. As more rapid growth resumes, housing must be given urgent attention, which the state and municipality have been slow to do during the decades of emphasis on industrial development when consumer goods, housing, and services were consistently sacrificed to concentrate investment on what was seen as primary growth. Especially with the large returned population of formerly rusticated youth, the bulk of the city's population is now between 20 and 35 years old, the peak marriage ages, with consequent new demand for housing. The common solution is for newlyweds to move into already cramped parental units. Almost no housing was built during the chaotic decade from 1966 to 1976. The change of political climate since 1978 has produced a new emphasis on light industry and consumer goods, together with new gestures and major plans in housing. But it is easier, quicker, and cheaper to produce more radios or electric fans, even more TV sets (now common—a new "opiate"?) than to produce new housing, especially in the gargantuan amounts needed merely to provide adequately for the existing population, let alone probable future increases. Housing uses desperately scarce capital goods in large quantities: cement, steel, glass, electric power, and of course transport to assemble these and other materials.

Shanghai may continue to confront a severe housing shortage, perhaps a chronic crisis, for some time to come. More recent figures from the Chinese press (*China Daily*, 1984a) give an average of 4.8 square meters of housing space per capita in Shanghai in late 1984, and mention goals for the seventh five-year plan (1986-90) of adding five million square meters each year, with the expectation that this would provide seven square meters per capita by 1990. Settlement is already very dense, estimated at about 100 thousand per square kilometer in the central urban area and about on a par with Tokyo, Hong Kong, or Calcutta (see White, 1981). For the urban area as a whole (about 150 square kilometers, although such measures are both arbitrary and changing as the urban area grows), with an estimated official population of about 7.5 million, there are roughly 50 thousand per square kilometer.

Urban services—water, heat, light, sewers, schools, and hospitals—also lag far behind need or demand, although Shanghai is better provided for than almost any rural area and probably most other big cities in China. Urban transport is, by Western standards, hopelessly overburdened, although again probably better in Shanghai than in most other big Chinese cities. Planned development came late and has far from eliminated separation of work and residence. Many buses are at least relatively new and many, especially downtown, are electrified. Buses are in woefully short supply, evident from the long waiting lines to board already jammed vehicles. Official figures show approximately five million daily round-trip "potential riders," 2.2 *billion* person-trips per year (six million per day), and in 1977 only 2,600 buses (Ma, 1979), although this had risen by 1982 to 4,189 "public vehicles" (Shanghai Academy of the Social Sciences, 1983:1254). The chief alternative is bicycles, by far the commonest mode of urban transport, but although the supply continues to lag behind demand (and there are complaints about the high controlled price) streets are often literally jammed with bicycles, and police at intersections direct an overwhelmingly two-wheeled traffic. For urban transport, Shanghai will probably have to put up with at least massive inconvenience for some time, especially as the city grows more rapidly. Buses can only crawl in chronic congestion. Road improvements and extensions do not keep up with rising volume, although new expressways are planned to three of the outlying satellite towns—a change more likely to worsen central area traffic than to relieve it. A subway is under construction, bisecting the city from southwest to northeast, scheduled for completion in 1990. In 1970 a traffic tunnel was finished under the Huangpu, and two more are planned (for details, see *China Daily*, 1984c). But these additions seem likely to be engulfed by faster growing demand, and to pour more traffic into the central area, as has happened elsewhere in the world. State planning has long since decided not to invest scarce capital or resources in private car production. There are a few, for official use, and the commonest model is made in Shanghai and carries the city's name, but automotive production in the big Shanghai plant concentrates on trucks and buses for the national and local market.

Industrially and economically, Shanghai was fundamentally reshaped after 1949, from its prerevolution emphasis on trade, banking, publishing, entertainment, education, light industry, and consumer goods into a dominantly heavy industry center, in keeping with the national pattern and its reiterated efforts to transform all cities, especially Shanghai, from a consumer to a producer mode. The original prominence of the private sector in Shanghai manufacturing also made it a target, but the city has basically altered its aspect in three decades of radical change. In 1949 heavy industries accounted for only 13.6 percent of its total industrial output by value, while textiles accounted for 62.4 percent and other light industries 24 percent. By 1976 the heavies had reached 53.4 percent and textiles and light industries together had fallen to 46.6 percent, both within a context of rapid overall industrial growth, especially after 1956 (Howe, 1971 and 1981).[3] Despite the apparent implications of official policy, Shanghai grew industrially during this period and up to the present far faster than China as a whole.

When the new Paoshan steel plant is completed, perhaps at the end of the 1980s, Shanghai will probably be the country's largest steel center as it has long been the biggest producer of ships, vehicles, machine tools and machinery, electronics, pharmaceuticals, and petrochemicals. Such a growth pattern may suggest caution about predictions of even more rapid growth in the future, but the expectation is based primarily on changes in national policies and attitudes since 1978, on Shanghai's demonstrated efficiency and high marginal productivity (industrial productivity per worker was estimated for 1982 as ten times the national average), and on the chronic builtup pressures for migration to the city and its manifold appeals, in a climate which may begin to relax some of the controls preventing such movement heretofore. Simply softening the wholesale rustication of youth and the tacit acceptance of returnees would retain within the city 50 to 400 thousand new age cohorts each year, depending on the degree to which policy is modified. In any case, Shanghai is moving into high gear to update or replace its older plants and pockets of outmoded technology, regain ground lost in these terms to newer industrial centers elsewhere in the country, and push for a new spurt of industrial growth.

## Unemployment

But will Shanghai's effort be enough to absorb the rising numbers or to pay the costs of new housing and other urban services? Official sources have recently become more open about acknowledging massive urban unemployment: there have even been mentions in the Chinese press of as much as 20 or 25 percent unemployment. Population clearly grew more rapidly than employment up to the imposition of controls in 1958-59. The lid on further inmigration and reassignment of some two million people may have kept unemployment minimal for a time, but it probably began to creep up again as some demographic growth resumed in the 1970s, and by 1980 had become the major problem the press was beginning to refer to. It is difficult to see what can be done about unemployment, a universal problem in all cities in the developing world, mitigated in China until recently only by the imposition of Draconian controls on movement. One can understand the state's reluctance to ease these controls, but also their bitter unpopularity and efforts to evade them. It is a genuine dilemma, one which the state will probably continue to hedge on, without making major or official changes in stated policy but with a greater tolerance for some evasion.

One sort of solution, becoming increasingly public, is to allow the previously despised service sector to expand. Street vendors, repair or other services, and small private establishments of most kinds were banned until about 1978 but have now begun to blossom again. One suspects that the major reason for official acceptance of this resurgence of "capitalist enterprise" is its ability to absorb unemployed labor, but it also enhances the convenience of daily living for urban residents who now may no longer have to stand in line for hours to obtain scarce goods or services, or be obliged to do the work themselves on top of long hours on their jobs. There has been a big increase also in small privately owned restaurants, food take-out places, and even hairdressing and beauty shops. Small workshops and minor industrial establishments survived the radical policy stage to a great extent, however, and as late as 1976 still composed about 90 percent of all industrial units, as had consistently been the case before 1949; only about 2 thousand of the roughly 30 thousand industrial enterprises, or of about 8 thousand plants designated as "factories," in Shanghai in the late 1970s employed over a hundred workers (Howe, 1981). Such smaller units have multiplied since 1978 as controls on private enterprise have been somewhat relaxed for smaller scale units which continue to provide jobs for otherwise unemployed or underemployed residents, often in or attached to households and using family or immediate neighborhood labor.

One hears less these days about the need to transform Shanghai from a consumer to a producer city. Such a transformation has to a large extent taken place with the rapid growth of heavy industry. But there is also new growth in light industry (see below) and other major service categories—publishing and education, in both of which Shanghai remains China's major center, and trade and finance, especially with the change of outlook which recognizes China's need to exchange with the rest of the world. Shanghai, handling some hundred million tons of waterborne cargo a year, has again become one of the world's major ports. All of these developments generate new jobs, but although almost all enterprises are far more labor-intensive than in the West, including handling and transport of raw materials and finished goods, loading, and unloading, it seems unlikely that employment can keep pace with probable population increases, except with the retention of extensive migration controls and perhaps also some reassignment of new age cohorts. Shanghai is a magnet for the would-be upwardly mobile—most people everywhere, after all—and most rural people would probably prefer to pull a hand cart or pick rubbish, or simply to take their chances, in Shanghai than to remain on a commune.

## Pollution

As China's biggest industrial center, and especially with its increased component of heavy industry, Shanghai has created the country's, and one of the world's, worst pollution problems. It is very difficult to measure this problem precisely. A national environmental protection law was adopted "in principle" in 1979, and Shanghai Municipality has its own supplementary regulations. In 1982 a new national Ministry of Urban and Rural Construction and Environmental Protection was established. But the law and local regulations tend to be very general, or deal in what must be seen as ideal prescriptions, such as prohibiting all uncontrolled or toxic discharges into the Huangpu and all pollutants from ships into the river. There are no clearly-stated standards, very few published measures of actual pollution, and only the most skeletal and inadequate monitoring system. The new ministry is an unwieldy bureaucratic hybrid formed by merging organizations supervising construction, surveying, and mapping, and it remains unclear how effective its voice will be in environmental protection (see Smil, 1984:171-72, 231-38).

Some steps have been taken, especially since 1978, to begin controls on the worst aspects of air and water pollution, primarily through a system of fines. A small but devoted group of Chinese scientists and environmentalists, particularly in Shanghai, is deeply concerned about the problem and pursuing studies and control plans. But the following excerpt from the Chinese press in March of 1980 may not be overly cynical:

Although an environmental protection code has already been promulgated, there is no way to implement it, and in actuality it can function only as propaganda. At present, even criticism is subject to permission, so why bother even to discuss the rest? . . . Capital for control of the "three wastes" (solid, liquid, and gaseous) is distributed by environmental protection organs, while the other concerned departments are only uninvolved observers. When they need funds they are very enthusiastic and vocal, but once they have their money they lose enthusiasm for controls . . .

Our present environmental protection system is weak and powerless . . . When other systems were established, the concerned departments of the Party Central Committee immediately informed the lower levels of their organizational system; environmental protection has been the only exception. None of the responsible bureaus or large and medium-size plants and mines have environmental protection organs. How can a limited number of environmental protection cadres tackle this huge task all alone? There are also no special capital construction funds for scientific research in environmental protection or for a network of monitors, so how is such work to be developed? (Liu, 1980; quoted in Smil, 1984:176-77).

Uncritical Western observers had earlier reported, especially between 1972 and 1978, great successes in controlling pollution: new model plants with emission controls and waste treatment facilities, segregation of heavily polluting enterprises away from densely settled areas and downwind, and massive recycling and waste recovery, using the vast supplies of low-wage or free labor organized in mass campaigns. The last seems to have been the most effective but there is no way to estimate how much of total toxic wastes may have been treated in this fashion—or how much of this activity was primarily because of poverty and scarcity, which have traditionally motivated Chinese to minimize waste. Neighborhood processing stations for household trash and garbage in Shanghai were said to have extracted over two hundred tons of usable oils for lubrication and two thousand tons of reusable rags and yarn (see Salter, 1976). A 1974 report in New China News Agency described how one chemical plant in Shanghai sent people out every day to collect fats and oils from the waste water of dishwashing and sewer systems of hotels and restaurants, recovering six hundred tons of fats between 1971 and 1973, enough for three million bars of soap (see Smil, 1984:177). However admirable, such methods seem an inefficient use of any but unemployed labor, and they deal (or dealt) with comparatively trivial problems or produced trivial recovery, while the major industrial sources of pollution went largely unheeded.

## Air Pollution

Admittedly, industrial pollution control is a problem for all poor countries. They urgently need industrialization, as fast and as cheaply as possible; anything which delays it or increases its cost may represent an unacceptable burden, at least in the early stages of growth. Pollution control or concern for the environment may appear to be a luxury affordable only by affluent economies. Poor countries, including China, also often lack the technology for effective pollution control.

China has an additional problem: it has extensive and reasonably low-cost coal deposits and is dependent on coal for about 70 percent of all its primary energy needs, almost certainly the highest figure in the world. Coal is of course the dirtiest of all major fuels, and in China very little of it is washed or sorted before burning, a process which adds cost but is common in developed countries. Coal burning in China is extremely inefficient and emissions are correspondingly heavy. Smil (1984:116-17) cites fuel conversion efficiency rates for Japan of 60 percent, for North America and Europe of between 40 and 55 percent, but for China of 30 percent. This figure applies to the heaviest users of coal, industries—including iron and steel, power plants, and chemical factories—which consume over two-thirds of the Chinese total and whose fuel efficiency is very low, and to transport and household users. Coal-burning steam engines still dominate the railways and haul the bulk of all freight, as they seem likely to do for years to come; new steam locomotives continue to be built. Houses are predominantly heated (minimally) and most urban cooking is done with coal dust briquettes, in especially inefficient stoves and with no provision for trapping or venting emissions.

The soot and sulphur dioxide put out by China's coal-burning sources are all too noticeable, especially in winter but also throughout the year. Smil (1984:118) cites a figure from one Shanghai site showing 1,822 tons per month of dust fall, primarily from coal combustion, a figure hard to approach in China or elsewhere. The fallout from uncontrolled coal combustion contains strong carcinogens when inhaled, and contributes importantly to chronic smog, referred to locally as "Yellow Dragon." Clear days in Shanghai have become rare, and the incidence of lung cancer has shot up.

In the West and Japan, improvement in air quality in previously smog-ridden cities such as London, Pittsburgh, St. Louis, and Osaka resulted, before the imposition of major controls, largely from conversion from coal to oil and gas as the major heating, power, and transport fuels in the decade following World War II. Conversion was followed in most countries by controls, including requirements for coal washing and desulphurization where it remained in use, its banning from certain areas ("smoke-free zones" as in London), and the beginnings of controls on other sources of air pollution, although current controls still lag far behind urgent need. China, with its new emphasis on "production first," cannot easily contemplate reducing or inhibiting use of coal, its cheapest and most plentiful fuel, or imposing controls which would slow growth and add costs.

In one respect, perhaps the only major one, China is fortunate in having so far refused (nor can it really afford) to enter the automobile age. Most big cities elsewhere have created frightening air pollution from the still increasing number of motor vehicles, which also produce dismaying traffic congestion. Shanghai is far from free of traffic problems, as already indicated; like most cities, it was not built for the automobile, or for the volume of traffic of other sorts which now clogs its streets and the bridges across the Woosung River. But although vehicle exhaust adds substantially to air pollution, in Shanghai it is a relatively minor contributor compared with most big cities.

Unfortunately the difference is probably more than made up by industrial and household emissions. We have no specific data on acid rain damage in Shanghai, nor measures of sulphur dioxide or particulate content of the air, but the air looks, smells, and feels as bad as the worst situation elsewhere. Clearly Shanghai air is worse than air in London, Tokyo, or even Los Angeles, in all of which fuel shifts and controls have made some impact. One now sees occasional motor scooters, and the number of cars and taxis is increasing. The state still owns and runs all large enterprises, and the bureaucracy at all levels is immense, so the number of various officials is huge, many of them with cars assigned for their use or shared. As development proceeds and more managerial and technical elites emerge, the number of official-private cars may pose a larger environmental problem. Meanwhile, truck and bus emissions, much of them diesel, are already a significant factor and will certainly become more so. Fortunately, temperature inversions in Shanghai are rare and there is usually some lateral air movement, predominantly winds from the north-northwest in winter and south-southeast in summer, which in this flat deltaic landscape meet with no obstructions; nor are polluted air masses trapped by any surrounding hills.

There have been some efforts to control air pollution. Since 1980 over half the stacks in the city's largest factories have been fitted with electrostatic precipitators (Smil, 1984:163), and all newly built plants are supposed to be so equipped. But Smil (p. 166) also estimates that Shanghai still puts out about 16 hundred tons of fly ash and 15 hundred of sulfur dioxide for each square kilometer of the urban area, figures which he suggests are unprecedented.

When the full complex of the new Paoshan steel plant is finished, its uncontrolled sulfur dioxide emissions will add about 25 percent to the present total, and prevailing winter winds will carry most of it directly to the central urban area, although by that time some additional controls may have been imposed. Work on the Paoshan plant, originally planned for completion in 1983, was delayed, largely for financial rather than ecological reasons, and contracts with Japan cancelled or renegotiated to provide for construction in two phases, the first of which was completed in 1985.

Perhaps also for state budgetary reasons, some 340 new industrial projects in Shanghai planned in the enthusiasm of 1978 have been cancelled or postponed since 1979 (Smil, 1984:163). Reflecting the post-1978 national shift in priorities, light and consumer goods industries have grown much faster than the heavies; by 1981 they once again dominated Shanghai's manu-facturing with 55.6 percent of the total output by value (Zhen and Chaun, 1982), to that extent lessening the increase in pollution. By 1984, Shanghai was said to produce 40 percent of China's consumer goods.[4] Zhen and Chaun stated that expenditures for new housing, public utilities, and environmental protection had been increased to 20 percent of new investment in the city, and that over three million square meters of new housing had been completed in 1980, a rise of some 40 percent over 1979 but still far from adequate for the current population and its own natural increase, allowing for demolition of old housing and the substandard quality of many units. Nor is it yet apparently possible to provide even new housing with adequate electric power, piped water, cooking fuel, or waste disposal systems. How much of the budget went or goes now for environmental protection, and with what effect, is not clear.

## Water Pollution

Shanghai's major environmental problem is probably pollution of its municipal water supply and the rivers flowing through it. The Huangpu may now be the most heavily polluted stream in the world, and it would appear that all aerobic and most other aquatic life had been killed by 1980.[5] The entire river area and the lower reaches of the Woosung River smell terrible, worse during dry or low water periods, as in winter, but bad at all times; there are chronic complaints from workers and residents. Most of the pollution is from industrial wastes, but less than 5 percent of the city's sewage is treated, and hence something like four million cubic meters of raw sewage enter the Huangpu daily (Smil, 1984:104). As the rudimentary sewerage system is extended to cover more of the city's area and housing, the need for treatment plants will become even more urgent. China has long been admired by Westerners for its recycling of human wastes as fertilizer, and most households in Shanghai, where flush toilets are uncommon, still have bodily wastes picked up by night soil contractors who sell their collections to communes and farms in the municipality. Despite the labor and inconvenience involved, this is obviously far preferable to dumping household wastes untreated into the drinking water supply.

The Huangpu is Shanghai's main source of water, but it has become essentially a chemical cocktail composed of raw sewage, toxic urban wastes, and huge amounts of industrial discharges. Eight pumping stations along the river alternate with sewage outlets and industrial effluent dumping sites; the stations treat the water with further additions of chlorophenol and other chemicals to make it potable, but hardly pleasant or safe in longer-run terms, since the treatment does not remove the growing amounts of lethal heavy metals and other industrial toxins. The normally high water table also means that the variety of toxins from industrial plants and local rivers seeps relatively quickly into the ground water and contaminates the wells which provide the balance of the city's water supply. Sampling tests in 1979 found 96 percent of surface water test sites contaminated with heavy metals (Smil, 1984:165). The falling water table and subsidence which had resulted from overpumping for expanding water demand as the city grew have been slowed or largely corrected by pumping used water back into the aquifers, but at the obvious cost of further contaminating the ground water. Smil reports that "major industrial enterprises working with mercury and cadmium have drastically reduced the release of these heavy metals into the rivers" (p. 163), but provides no further details. Given the tragic history of mercury and cadmium poisoning, as with Minamata and *itai-itai* disease, it is understandable that, as earlier in Japan, the first controls should be imposed on those particular toxins (see Murphey and Murphey, 1984). But it is not clear that mercury or cadmium, let alone other toxic heavy metal discharges, have been stopped in Shanghai. The total pollutant load dumped into the rivers and seeping into the ground water seems more likely to be still increasing as industrial and population growth proceed and pollution control seems clearly to lag behind.

A few factories have been shut down (most of them temporarily only) for severe violations of the new environmental regulations, including air and water pollution. When all but the smallest enterprises are state-owned, use of fines as the major means of enforcing such controls seems at best a flawed instrument. Plant managers must get approval and financing from the higher bureaucracy to install or construct pollution control equipment. From both points of view, fines may seem easier and cheaper, and may even be absorbed, without great effect on practice, as may any official directive. Meanwhile Shanghai began, early in 1985, a 43-kilometer canal to divert cleaner water from the upper reaches of the Huangpu for city use, after passing through a new filtration plant yet to be built. In addition, a separate project is planned, with World Bank aid, for completion between 1992 and 1996 to intercept wastes now discharged into the Woosung River and the Huangpu and convey them by underground pipes to the lower Yangtze estuary, where they will be aerated through jets and mixed with the massive and rapid flow of the Yangtze (Zhen, 1985).

There have thus been the beginnings of public and official recognition that environmental problems are a genuine threat to Shanghai's future, and the beginnings of some regulations, controls, and remedial projects. The problems are dismaying in their magnitude; the costs of adequate solutions and lack of adequate technology and trained people are equally dismaying, but the first steps, starting with recognition, have been taken. As in Japan and elsewhere, it may need a major human disaster to heighten awareness of the problems and their urgency before truly effective measures are adopted and their costs accepted.

## Future Growth

At a recent conference of economists and planners in Shanghai (Ningiun, 1984) there was some acknowledgement of pollution problems, but the major emphasis was on future growth. Continued stress on light industries was urged, especially high-tech industries, as well as more rapid development of the service sector in information, higher education, research, management, and finance. Some suggested first priority on the further growth of Shanghai's electronic industry, including the newly begun production of computer software, and further efforts to develop export industries (Shanghai already provides over 30 percent of China's exports). Meanwhile the city hopes to attract more overseas investors and joint enterprises, especially in high-tech lines, offering concessions like those in the "special economic zones": preferential tax treatment, credits, markets, tariffs, and rights to remit earnings abroad in foreign currencies (see Fan, 1984; Singang, 1984). Any shift away from heavy industry into other fields will help to slow the buildup of pollutants, but at the same time any growth in the city's economy or total size will increase the pollutant load and further overstrain housing, basic services, and transport. If Shanghai is to grow, it must make major efforts to reverse the present losing battle on all of these fronts.

Since 1949, or the end of the Pacific war in 1945, Shanghai has developed pretty much as most other urban centers of modernization have, despite what now appear as relatively minor and short-term deviations from the standard pattern occasioned by the policy swings of a revolutionary system pushing for a series of radical departures. If one were to forget the first 30 years of revolution and look only at Shanghai since 1979, there would be few surprises. The city's development would seem to be, as indeed it is, a logical projection from its situation as of the last prewar year of 1936 (the full-scale Japanese invasion came in August 1937, including a frontal attack on Shanghai), in keeping with the growth of most other commercial/industrial centers during the same period, especially those in the developing world. Shanghai's *relative* dominance of China's trade and manufacturing have declined with the growth of new inland centers, but that too could have been expected under any political regime, as could growth of heavy industry (previously lacking) in the 1950s and 1960s, the newer boom in light and consumer industries, the beginnings of a new rise in the service sector, and the current emphasis on high tech, research, education, information, management, finance, foreign trade, and the early stages of the computer revolution, all in keeping with global urban patterns.

What did revolutionary planning do to alter the city's character or growth? In the first respect, Shanghai is certainly less "bourgeois" than it was in 1936 or 1949, and far more egalitarian. Housing has probably worsened on average in per capita space and quality, but the lower extremes of poverty, squalor, and suffering have been ameliorated. The municipal government no longer has to clear human corpses from the streets, as they did to a total of 5,590 in the comparatively peaceful year of 1935, rising to 20,746 in 1937 (Shanghai Municipal Council, 1939:175). Most people now have some shelter and some material support from a state, municipal, work unit, neighborhood, or family safety net. The range of incomes, although still fairly wide as compared with the society's egalitarian ideology, has also been reduced at the high extremes and there are few opportunities for the acquisition, use, or display of wealth, unlike the situation before 1949. Working conditions for most are, like housing, austere, but child and indentured labor and the miserable factory conditions and sweat shops which disfigured prerevolutionary Shanghai are gone. Transport workers and dockers, no longer called "coolies," are protected by the same regulations which supervise working conditions and wages for other workers. The ricksha, a foreign introduction, has been abolished as degrading, although carts drawn by human labor are still common.

Health levels have certainly risen, one of the major accomplishments of the revolution in China as a whole. Former epidemics have been eliminated, and there is universal basic medical care for almost everyone at very low cost or free. The major health problem is almost certainly air and water pollution, already affecting the death rate adversely. Total average food intake and probably nutritional quality have risen somewhat in Shanghai, especially since 1978. Education through at least six grades is available free to all residents and literacy is relatively high. Some consumer goods are newly available to most people: radios, sewing machines, bicycles, wristwatches, some cameras, and now television sets and the first of major household applicances—washing machines and refrigerators. These and other consumer goods are more readily available to Shanghai people than to others elsewhere in China and help to increase the city's attraction.

Especially compared with most cities in other countries (except for Japan), Shanghai is relatively orderly and free from at least major street crime, although petty theft remains a problem and there are cases of white-collar crime such as larceny, fraud, and embezzlement. Security is provided by police, peoples' courts, and the supervisory and regulatory role of work units and neighborhood committees. Some recent cases of violent crimes by unemployed or underemployed youths who have been tried and executed for rape or armed robbery have received much publicity. While there does appear to be a small hoodlum group, such problems are the palest of reflections of their counterparts in many Western cities. These criminals worry many Chinese as possible harbingers of trouble which may become more pronounced as Shanghai continues to pursue modernization and moves away from the moral restraints imposed by the old society and the ideals and discipline of the revolutionary years, in pursuing personal material gain and what are still warned against as "bourgeois values." The new openness to the outside world may also continue to bring in "spiritual pollution," a price which China, Shanghai especially, may simply have to pay for the benefits of exchange, as most of the rest of the world has long paid.

The major impact of revolutionary planning on the city's growth has been twofold: it clearly restricted population growth for a critical two decades, and to a large extent successfully shaped what new growth has occurred in the form of planned satellite centers, first along the expanding edge of the older urban area between 1956 and the early 1970s, then in new integrated clusters of interdependent manufacturing, housing, and services farther out into the rural parts of the municipality. While such planned development probably slowed the increase of overcrowding and shortage of housing and services in the original urban area, it did contribute to those problems since none of the new centers could be completely independent and all continue to rely to some extent on facilities, transport, and other uses of the main urban area. Crowding seems to have grown, and since 1978 to be accelerating. Admittedly, like most Chinese and other Asian cities, Shanghai has always struck outsiders as impossibly crowded, in part no doubt because housing is so cramped and inadequate that most people spend much of their leisure time outside, shopping for food daily and often having to wait in long queues but also just walking around as a form of recreation. All but the main arteries seem virtually solid with people, enough so that even bicycles have a hard time. The critically limited park areas for a population of this size and density are chronically jammed, as are the former gentry gardens now open to the public, museums and industrial exhibits, and theaters and sports events, where there are no empty seats and one must often wait weeks for a ticket or stand in line on the chance of picking up a cancellation.

Despite these and other problems of overcrowding, people still want to live in Shanghai and use whatever means they can find to effect a move, especially from the surrounding countryside within 150 kilometers or so. Perhaps the most basic failure of the radical policies of the revolutionary decades was thus the persistence of attitudes in favor of city life over the peasant rural world, and the continued threat that material values will overcome revolutionary ideology. One wonders, however, how much of people's preferences are ideological and how much the simple and universal desire for what is seen as a better material life. Whatever the disadvantages posed by unemployment, bad housing (much of it worse and more crowded than most rural housing), inadequate urban infrastructure, and truly poisonous environmental pollution, Shanghai offers, par excellence for China, the glitter and promise of stimulation, variety, choices, and upward mobility—a brave new world indeed.

Shanghai has not so much reverted to its earlier character as it has persisted in it through three decades of repression and now seems more likely to blossom again more openly as the chief center of cosmopolitan and sophisticated culture, the home of artists, writers, intellectuals, and political mavericks as well as of rising managerial and technical elites, in touch with the wider world abroad and leading a formerly exclusionist China into a new reacquaintance with things foreign. This view of Shanghai is no doubt deeply disturbing to many revolutionaries; some may feel the conflict between ideological purity and the economic development which China so urgently needs, if only to serve the interests of mass welfare, another revolutionary goal. Soon after Communist armies occupied Shanghai in May 1949, an official statement declared that Shanghai must terminate its dependence on the imperialist economy. Shanghai represented the concentrated, typical expression of the colonial or semicolonial nature of old China's economy. Communist officials felt the city's past prosperity was founded on bureaucratic capitalism, not an independent, solid economic base. To the Communists, Shanghai could only become a truly prosperous new peoples' city by shedding its colonial or semicolonial status (*China Digest*, 10 Aug. 1949).

Colonial or semicolonial rule is gone, but global interdependence is one of the chief marks of and means to economic development. China has damaged its own material interests by cutting itself off from the rest of the world, impoverishing itself spiritually and culturally. Now that it is more willing to pursue the benefits of interaction, Shanghai will resume much of its former role. This is not to say that it will once again become the "sink of iniquity" it was widely said to be in its treaty port days. China will remain a tightly controlled socialist state and there seems little likelihood that the drugs, prostitution, bars, gambling, speculating, white slavery, kidnapping, and gangsterism which earned Shanghai its earlier labels, including the nickname "paradise of adventurers," will revive. What seems more likely to revive is Shanghai's importance as China's window on the world, along with its leadership role in modern high-tech development. Such a trend is bound to accentuate Shanghai's long-standing cosmopolitanism and sophistication in all respects. The city has been out of step with the rest of China since 1843, or in a longer view has initiated and led trends which then spread more slowly. During the Cultural Revolution years, and as it began to wane, in Shanghai one found the greatest ferment and the earliest efforts to reassert freer (or simply more urbane) styles, as in earlier periods of change such as the May 4th

Movement of 1919, the Chinese Renaissance which followed it, the stirrings of Chinese nationalism, and the emergence of a new literature. In the 1970s it was in Shanghai that the first gestures away from the Cromwellian austerity of earlier years were made, including the beginnings of color, variety, and style in personal clothing.

Shanghai has not come full circle. The revolution has made a permanent change, but the radical decades are over, and the city's character and function are again being shaped by its circumstances, location, and vital role it can perform for the new China. As this happens, Shanghai will continue to push China to some extent away from the Maoist vision, and in so doing will perhaps revive old fears and resentments. But crowding, problems with housing and services, and galloping pollution seriously threaten the city's ability to function. Further growth must depend on careful planning and heavy new investment if Shanghai is not simply to stifle. Shanghai is indeed a mix of problem and promise, both to an extreme degree, and from whatever ideological perspective. Successive regimes have found it to be both—and have been unable to control its growth, probably Shanghai's biggest problem of all.

## Notes

1. Statistics from or about China present special problems, more than in many developing countries, partly because of China's immense size and population, partly for political or ideological factors which have greatly worsened the understandable shortage of competent statisticians or record keepers and the generally unsophisticated comprehension, collection, and use of statistics. Experts of all kinds were a target from the beginning of the revolution in 1949, and especially during the turbulent years of the Great Leap Forward (1957-58) and the Cultural Revolution (1966-76). Little was salvaged from the inadequate statistical services of earlier years; few or no statisticians were trained while those who survived were frequently hounded, often ignored. The situation has improved since 1978 and the official 1982 census appears to have been competently done with international assistance and observation. But caution is still in order in the use of most other statistical data, especially those before about 1980. There has also been a general tendency not to report, or not to pass on to foreigners, anything which could be seen as critical or negative. Shanghai has always offered better and more reliable statistical information than most of the rest of China, as the biggest and most modern city; it has also been much studied by Chinese and foreigners. In August 1983 the Shanghai Academy of Social Sciences published a large volume (in Chinese), *Shanghai Ching Chi 1949-1982* (The Economy of Shanghai 1949-1982), containing 1,296 pages of text and statistics compiled and written by a large team of research workers during the preceding year. Data presented are the result of serious professional work and certainly the best so far available. Statistical data given in this chapter for the period since 1980 are drawn largely from that volume and can be taken as in the main reliable.

2. Seybolt (1977) is primarily a translation of a 1973 compendium of articles appearing in the Chinese press at the time of an investigation of the program and the adoption of some remedial measures plus other Chinese articles on selection and adjustment of urban youth in rural areas published between 1973 and 1976. See also Chen (1972) and Singer (1977).

3. The apparent precision of these figures gives a misleading impression of accuracy. As Howe acknowledges, the data sources are incomplete and often conflicting.

4. Personal interview.

5. Interview with members of the Huangpu Conservancy Board, 1981.

# Bibliography

Bernstein, Thomas P. 1977. *Up to the Mountains and Down to the Villages: The Transfer of Youth from Urban to Rural China.* New Haven, CT: Yale University Press.

Chao, Kang. 1966. "Industrialization and Urban Housing in Communist China." *Journal of Asian Studies* 25:381-96.

Chen, Pi-chao. 1972. "Over-urbanization, Rustication of Urban-Educated Youths, and Politics of Rural Transformation." *Comparative Politics* 4:361-86.

*China Daily.* 1984a. "Shanghai Draws Up 5-Year Plan." (3 Sept.):2.

―――. 1984b. "Trade Swamps Shanghai." (22 Sept.):3.

―――. 1984c. "Shanghai Bids to Solve Traffic Chaos." (15 Oct.):3.

Fan, Zhen. 1984. "Shanghai Adopts Policies to Tempt More Investment." *China Daily* (11 Dec.):1.

Howe, Christopher. 1969. "The Supply and Administration of Urban Housing in Mainland China: The Case of Shanghai." *China Quarterly* no. 33, 73-97.

―――. 1970. *Employment and Economic Growth in Communist China, 1949-1957.* Cambridge, England: Cambridge University Press.

―――. 1973. *Wage Patterns and Wage Policy in Modern China, 1919-1972.* Cambridge, England: Cambridge University Press.

―――. 1978. *China's Economy: A Basic Guide.* New York: Basic Books.

―――. 1981a. "Industrialization under Conditions of Long-Run Population Stability: Shanghai's Achievement and Prospect." In *Shanghai. See* Howe 1981b.

―――, ed. 1981b. *Shanghai: Revolution and Development in an Asian Metropolis.* Contemporary China Institute Publications. Cambridge, England: Cambridge University Press.

Lardy, Nicholas R. 1975a. "Centralization and Decentralization in China's Fiscal Management." *China Quarterly* no. 61, 25-60.

―――. 1975b. "Economic Planning in the People's Republic of China: Central-Provincial Fiscal Relations." In *China: A Reassessment of the Economy. A Compendium of Papers Submitted to the Joint Economic Committee, Congress of the United States.* 94th Cong., 1st Sess. Washington, DC: U.S. Government Printing Office.

Lee, R. W. 1966. "The Hsia-Fang System: Marxism and Modernization. *China Quarterly* no. 28, 40-62.

Liu, Pei'en. 1980. "Some Suggestions on Environmental Protection Work." *Guangming Ribao* (Guangming Daily) (6 Mar.):2.

Ma, Lawrence J.C. 1979. "The Chinese Approach to City Planning: Policy, Administration, and Action." *Asian Survey* 19:383-55.

Ma, Lawrence J. C., and Edward W. Hanten, eds. 1981. *Urban Development in Modern China.* Boulder, CO: Westview Press.

Mast, Hermann, III. 1971. "Tai Chi-t'ao, Sunism, and Marxism during the May Fourth Movement in Shanghai." *Modern Asian Studies* 5:227-49.

Murphey, Rhoads. 1953. *Shanghai: Key to Modern China.* Cambridge, MA: Harvard University Press.

―――. 1977. *The Outsiders: The Western Experience in India and China.* Michigan Studies on China. Ann Arbor: University of Michigan Press.

―――. 1980. *The Fading of the Maoist Vision: City and Country in China's Development.* New York: Methuen.

Murphey, Rhoads, and Ellen Murphey. 1984. "The Japanese Experience with Pollution and Controls." *Environmental Review* 8:284-94.

Ningjun, Wang. 1984. "Mapping Out Shanghai's Role in World Economy." *China Daily* (29 May):4.

Roll, Charles Robert, and Kung-Chia Yeh. 1975. "Balance in Coastal and Inland Industrial Development." In *China: A Reassessment. See* Lardy 1975b.

Salter, C. L. 1976. "Chinese Experiments in Urban Space." *Habitat* 4:19-35.

Seybolt, Peter J., ed. 1977. *The Rustication of Urban Youth in China: A Social Experiment*. The China Book Project. White Plains, NY: Sharpe.

Shanghai. Municipal Council. 1939. *Annual Report*. Shanghai.

Shanghai Academy of Social Sciences. 1983. *Shang-hai Ching Chi 1949-1982* (The Economy of Shanghai 1949-1982). Shanghai.

Singang, Hu. 1984. "Shanghai to Use Foreign Trade for Development." *China Daily* (3 Aug.):1.

Singer, Martin. 1977. "The Revolutionization of Youth in the People's Republic of China." 2 vols. Ph.D. dissertation, University of Michigan.

Smil, Vaclav. 1984. *The Bad Earth: Environmental Degradation in China*. Armonk, NY: Sharpe.

White, Lynn T., III. 1978. *Careers in Shanghai: The Social Guidance of Personal Energies in a Developing Chinese City, 1949-1966*. Publications-Center for Chinese Studies. Berkeley: University of California Press.

———. 1981. "Shanghai-Suburb Relations, 1949-1966. In *Shanghai. See* Howe 1981b.

Whyte, Martin King, and William L. Parish. 1984. *Urban Life in Contemporary China*. Chicago: University of Chicago Press.

Zhen, Su. 1985. "Shanghai Begins to Clean Up Rivers." *China Daily* (3 Jan.):3.

Zhen, Xia, and Jian Chuan. 1982. "Shanghai Leads in Modernization March." *Peking Review* 25 (4 Jan.):19-27.

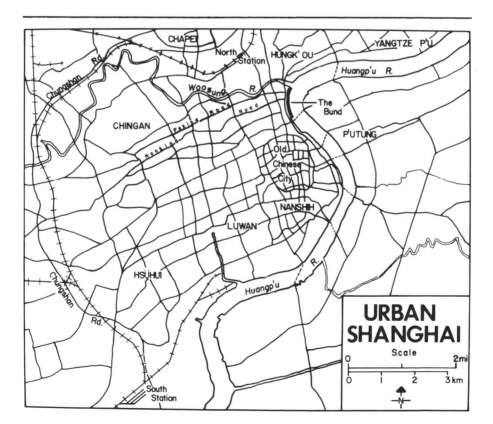

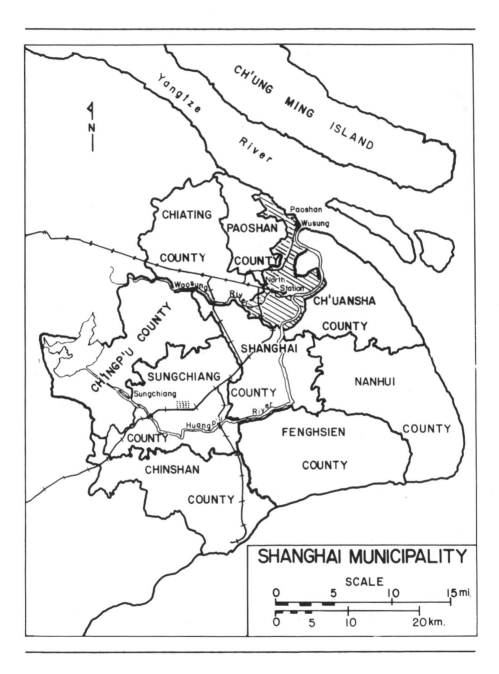

CH'UNG MING ISLAND

Yangtze River

N

CHIATING COUNTY

PAOSHAN COUNTY

Paoshan
Wusung

North Station

Woosung River

CH'UANSHA COUNTY

CH'INGP'U COUNTY

SHANGHAI

SUNGCHIANG COUNTY

Sungchiang

NANHUI COUNTY

COUNTY

Huangp'u River

FENGHSIEN COUNTY

CHINSHAN COUNTY

SHANGHAI MUNICIPALITY

SCALE

0    5    10    15 mi.

0    5    10    20 km.

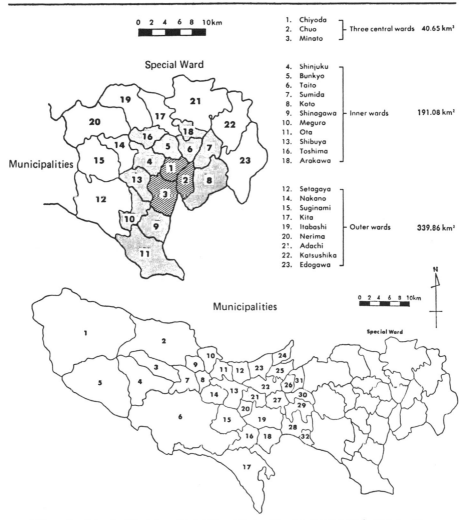

Adapted from fig. 1-1 in *City Planning of Tokyo*, TMG Municipal Library no. 13, Tokyo Metropolitan Government, 1978.

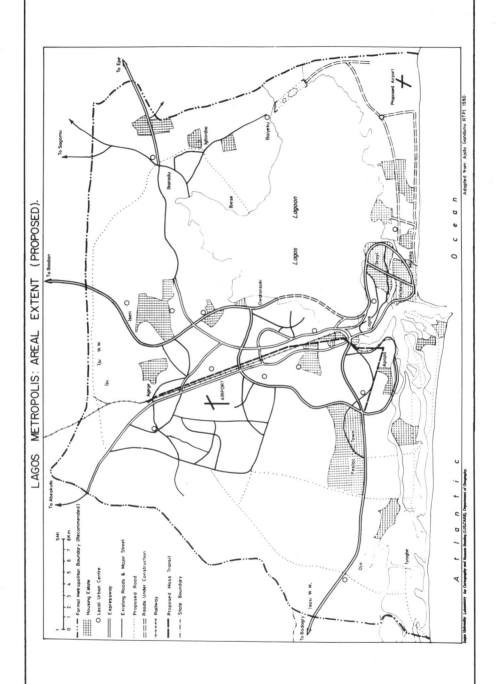

LAGOS METROPOLIS: AREAL EXTENT (PROPOSED).

Formal Metropolitan Boundary (Recommended)
Housing Estate
Local Urban Centre
Expressway
Existing Roads & Major Street
Proposed Road
Roads Under Construction
Railway
Proposed Mass Transit
State Boundary

Adapted from Ajato Gandonu RTPI 1980

Lagos University Laboratory for Cartography and Remote Sensing (LULCARS), Department of Geography.

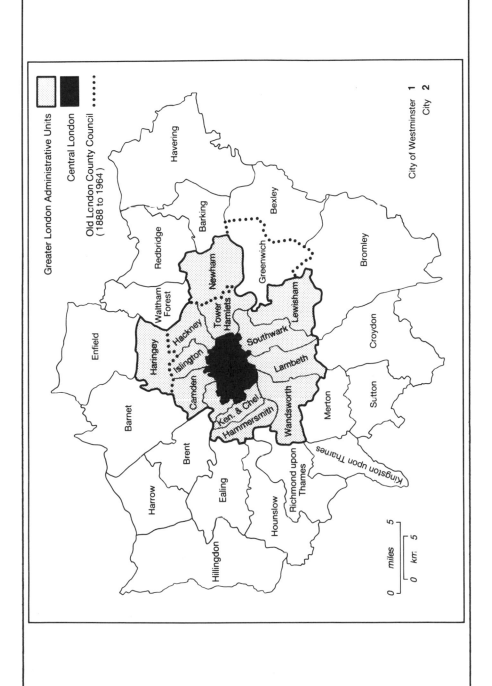

Greater London Administrative Units

Central London

Old London County Council
(1888 to 1964)

City of Westminster 1
City 2

Havering

Barking

Bexley

Redbridge

Greenwich

Newham

Bromley

Waltham
Forest

Tower
Hamlets

Lewisham

Haringey

Hackney

Southwark

Enfield

Islington

Lambeth

Camden

Croydon

Ken. & Chel

Wandsworth

Hammersmith

Merton

Barnet

Sutton

Brent

Kingston upon Thames

Harrow

Ealing

Hounslow

Richmond upon
Thames

Hillingdon

0        miles        5

0      km.      5

# 6

# Delhi

Hans Nagpaul

In 1981, Delhi had a population of more than six million. The city consists of two parts, Old Delhi and New Delhi. Old Delhi has been the capital of many mighty Hindu and Moslem empires. The only surviving link with past civilization in Old Delhi is the walled city of Shahjahanabad built by Moghul Emperor Shahjahan during the 17th century (Dayal, 1982). The British who displaced the Moghuls around 1803 had originally established their seat of government in Calcutta but transferred it to Delhi in 1911. Many old ruins, monuments, mosques, and forts are found in Old Delhi. Most of the city is crowded with narrow roads and streets, especially inside the walled city, and most business buildings and dwellings lack planned amenities. Throughout the city, unauthorized construction has been the order of the day for the past 30 years which has led to mixed land-use patterns of industrial, commercial, and residential activities.

New Delhi presents a contrast with its broad roads and open areas. Created as the capital of India and completed in 1931, New Delhi was originally planned as the seat of British government with government buildings to house its employees. The city has grown rapidly into a sprawling metropolis with several skyscrapers, deluxe hotels, museums, stadiums, cultural centers, and almost one hundred embassies and high commissions of foreign countries. Delhi has become a major travel gateway for national and international visitors. Here the British made a final proclamation 15 August 1947 to free India and handed over the country with its new city of spectacular architectural uniqueness and aesthetic planning (Delhi Administration, 1976).

184

# Major Demographic Characteristics

The Census of India divides the Union Territory of Delhi into urban and rural parts. According to the 1981 census, the urban part had a population of 5,768,200 (93 percent) while the rural had 452,206 (7 percent). This distinction seems meaningless in the context of the dominant framework of urban activities and communications in the territory. The census also divides Delhi into 30 towns and 231 villages. The towns are subdivided into five classes ranging from 100 thousand people for the first class to 5 thousand for the last.

In addition, the concept of Delhi Urban Agglomeration is used which includes 25 contiguous towns and excludes 5 which are not contiguous. Class 1 towns contain more than 90 percent of the population which shows a high degree of concentration (India, Registrar General, 1984). Since 1915, marginal modifications have been made in Delhi's boundaries and the total area has remained around 15 hundred square meters.

The city has always shown an increase in population (table 6.1). The birthrate, reported to be 28 per thousand in 1984, had been about 35-40 in 1941. The present birth and death rates are leading to a natural increase of more than 2 percent, Delhi's death rate of 7 seems underreported because the overall death rate in India is still about 15 per thousand.

Population growth between 1941 and 1951 was indeed spectacular mainly because of the influx of Pakistani refugees, displaced after partition of the country in 1947. Subsequent increases have been the result of internal growth and immigration from other parts of the country, mainly the neighboring states of Haryana, Punjab, and Uttar Pradesh and also from Rajasthan. By 1981, about one-third of Delhi's population reported being born outside of Delhi. This large migration flow has been the result of some dominant push factors operating in the countryside and many pull factors of the city attracting almost 100 thousand new migrants every year (Bose, 1980; India, Registrar General, 1984).

As in other major cities, Delhi's population is predominantly male. The disproportionate sex ratio is explained by the large number of men continuously moving to Delhi in search of employment, leaving their families in rural areas. The age composition in the 1981 census shows about 35 percent below 14 years, 10 percent between 15 and 19, 35 percent between 20 and 39, 15 percent between 40 and 59, and 5 percent 60 years and over. As in other developing economies, 45 percent of the population is below age 19 and requiring proper nutrition, shelter, medical care, education, and transport facilities as an investment for healthy growth. Five percent of the population is old and needing better amenities of life, medical care, and transport. Delhi's population lives in 1,211,784 households. See table 6.2.

Most people in Delhi speak Hindi; another major group speaks a modified version of Hindi, called Hindustani; Punjabi is a third major language, brought mainly by refugees from Pakistan in 1947. A small percentage of people working in government offices and national business organizations speak Bengali, Gujrati, Malayalam, Marathi, Telugu, Tamil, and Sindhi.

TABLE 6.1
Variations in Population—Delhi, 1901-1981
(in thousands)

| Year | Number | Percentage Decade Change | Population Density /km² | Sex Ratio (Females/1000 Males) |
|------|--------|--------------------------|-------------------------|--------------------------------|
| 1901 | 406 | — | 271 | 862 |
| 1911 | 414 | 2 | 245 | 793 |
| 1921 | 488 | 18 | 326 | 733 |
| 1931 | 636 | 30 | 429 | 722 |
| 1941 | 918 | 44 | 613 | 715 |
| 1951 | 1744 | 90 | 1174 | 768 |
| 1961 | 2659 | 52 | 1791 | 785 |
| 1971 | 4066 | 52 | 2738 | 800 |
| 1981 | 6220 | 53 | 4194 | 801 |

SOURCE: Census of India (1981).

TABLE 6.2
Households and Population by Sex and Religion—Delhi, 1981
(in thousands)

| Religion | Households | Population | Males | Females |
|----------|------------|------------|-------|---------|
| Hinduism | 1027 | 5200 | 2884 | 2316 |
| Islam | 85 | 482 | 272 | 210 |
| Sikhism | 73 | 394 | 209 | 185 |
| Christianity | 13 | 62 | 32 | 29 |
| Jainism | 12 | 74 | 39 | 35 |
| Buddhism | 1 | 7 | 4 | 3 |
| Total | 1212 | 6220 | 3440 | 2780 |

SOURCE: Census of India (1981).
NOTE: A household is generally considered to be a group of persons living together and taking meals from a common kitchen.

About 15 percent of college graduates age 20 and over speak English which continues to be the dominant language in schools, colleges, universities, courts, and government offices though propagation of Hindi has been going on for the past 20 years.

India is known for its castes and tribes. The census does not enumerate castes of the total population. However, the Indian Constitution prescribes some protections and safeguards for scheduled castes and tribes. According to the 1981 census, 23.5 percent of the country's population belonged to these groups. About 18 percent of Delhi's population was enumerated under scheduled castes; no one was identified as belonging to tribes. A vast majority of scheduled caste members in Delhi work in menial jobs of street sweeping and garbage collection.

Literacy for the Delhi population in 1981 was about 62 percent versus 36 percent for India. Male percentages were 68 in Delhi versus 47 in the country; female percentages were 53 and 24 percent. See table 6.3 for a comparison of 1981 Delhi educational levels.

TABLE 6.3
Population by Education—Delhi, 1981

| Educational Level | Males | Females |
|---|---|---|
| Primary | 429,894 | 314,067 |
| Middle school | 341,989 | 202,943 |
| High school | 603,198 | 324,016 |
| College other than technical | 295,489 | 175,390 |
| Technical diploma | 22,514 | 10,087 |
| Technical engineering degree | 22,973 | 488 |
| Medical degree | 8,358 | 3,637 |
| Agriculture and dairying degree | 649 | 25 |
| Veterinary science degree | 34 | 2 |
| Teacher training | 6,445 | 16,216 |

SOURCE: Census of India (1981).

## Dominant Ecological Patterns

The Union Territory of Delhi, a narrow strip of the Indo-Gangetic plain, lies between 700 and 1,000 feet above sea level. The territory is west of the river Yamuna, flanked by ridges to its west and south forming an alluvial triangle which contains about 70 percent of the total population and includes the old walled city, civil lines, New Delhi, and newly developed colonies west and south of the Red Fort area. Since 1950 the city has expanded and changed so much that it has emerged as one of the giant metropolises of the world.

Some Indian scholars, especially geographers, have attempted to explain the dominant ecological patterns of metropolitan cities including Delhi by applying American ecological models of zones and sector theories. One writer claims that in Old Delhi there is concentric growth from Chandni Chowk and much of the residential structure has taken place along radial roads in a wave-like fashion originating at Chandni Chowk. As for New Delhi, Yadav (1979) feels that the Connaught Place center does not provide a logical arrangement of concentric urban development and the sectoral form has much relevance in explaining the ecological pattern around Connaught Place. The relevance and validity of the sectoral explanation have become limited since 1960 with the establishment of many new shopping centers or because of the expansion of existing limited shopping centers into bustling shopping markets. In South Delhi, Lajpat Nagar, New Delhi South Extension, INA, Sarojini Nagar, Sunder Nagar Markets, and Nehru Place have emerged as the most flourishing markets, influencing the ecological pattern of adjoining areas during the past ten years. The old markets—Ajmal Khan, Kahn, Central, Bengali, and Kamla Nagar Markets—also continue to expand, are becoming busy centers, and have affected urban development considerably. Hence, American ecological models of multiple nuclei development seem more relevant today (Cousins and Nagpaul, 1979).

Delhi's initial extensive expansion began in 1948 to 1950 when an avalanche of refugees estimated to be about half a million arrived in Delhi.

The Indian Ministry of Rehabilitation established three main refugee camps—Kingsway, the largest; and others at Tibbia College in Karolbagh and at Shahdara across the river Yamuna. Soon after, more than two dozen refugee colonies were constructed in the south, west, and outside the walled city. Among the most well-known colonies which became focuses of further urban growth from 1960 to 1980 are Nizamuddin, Lajpat Nagar, Kalkaji, and Malaviya Nagar in the south; Rajinder Nagar, Patel Nagars, Moti Nagar, Ramesh Nagar, and Tilak Nagar in the west; and Gandhi Nagar in Shahdara in the east.

Another important feature of the landscape was the creation of several new colonies to provide residences for thousands of government workers especially during the 1960-70 decade. The Ministry of Works and Housing and the government's National Building Organization have been primarily responsible for constructing these colonies. Multilevel buildings are arranged in parallel rows and blocks. At least five categories of housing are made available to government employees based on their official status and income. One Indian scholar has called these colonies salaried apartheid (Mitra, 1970). Thus segregation on socioeconomic variables is widespread throughout Delhi. The most important new colonies housing a large majority of government employees are those of Kaka Nagar, Vinay Nagar, Laxmibai Nagar, Moti Bagh, Ramakrishna Puram, and Sadiq Nagar. Several old colonies like Gole Market where single-story buildings existed for almost two decades or so have been replaced by multistory apartments in recent years. The construction material used is reported to be poor; maintenance is infrequent; and the overall layout is shabby in many of these colonies.

Another important characteristic relates to land use by public and semipublic agencies and foreign embassies. The Rashtrapati Bhawan (former Viceroy's House) is one of the most impressive buildings just behind the Central Secretariat. The Indian Parliament House is another building unique in architectural style. In the postindependence era, many new buildings have been built, among them the Supreme Court, National Archives, Vigyan Bhawan, National Museum, National Gallery of Modern Art, International Dolls Museum, and Sapru House. Several theaters and auditoriums have also been completed; six massive stadiums have been constructed; and more than two dozen luxury hotels in the public and private sectors have opened. In South Delhi, a unique complex of buildings, the Asian village, was completed in connection with the Asian Olympics held in 1982. The huge skyscraper of the New Delhi Municipal Committee is yet another addition in the early 1980s. Many more multilevel public and private buildings are also under construction.

It is debatable whether a poor society where elementary needs of most residents cannot be met can afford the luxury of palatial hotels, stadiums, and skyscrapers which are high-capital intensive with low benefits to native residents. Moreover, questions have been raised whether such structures suit the climate and cultural ethos of the city.

Apart from new structures, Delhi is full of ancient and medieval monuments reported to be in excess of one thousand, scattered across the city's landscape. Among the most notable are Red Fort, Purana Quila; centuries-old tombs, mosques, and temples. In addition, the Qutab Minar, Iron Pillar, and ruins of Tughlakabad attract thousands of tourists every day. Older gardens still remain lovely places of interest as are the Jantar Mantar, India Gate Memorial, Rajghat, Shanti Van, and Nehru Memorial Museum (Dayal, 1982; Delhi Administration, 1976).

Though small-scale and cottage industries are scattered over the city, industrial land use is mainly concentrated along the Najafgarh Road and on Kalkaji Road where a large planned industrial estate in Okhla has been established. Industrial areas have also been created in Naraina, Vazirpur, and certain areas on Rohtak and Grand Trunk-Karnal Roads. Under the new National Capital Regional Plan, some new industries have been established and are still being encouraged in the cities of Bullabgarh, Faridabad, and Sonepat in the state of Haryana, and in Ghaziabad and Modinagar in Uttar Pradesh.

Several attempts have been made to relocate some of the old industries functioning in congested parts of Old Delhi. But even today large industrial units like the Delhi Cloth Mills, Birla Cotton and Weaving Mills, and Ganesh Flour Mills continue to function in the most thickly populated residential areas.

Perhaps Delhi's most prominent ecological characteristic relates to mixed use of urban land. On one hand are planned residential localities, either government or private; then there are unplanned localities which may be authorized or unauthorized. Within these two broad categories, Yadav (1979) has further identified almost a dozen subcategories such as old or newly planned localities, planned refugee localities, old authorized unplanned localities, old unplanned localities inside the walled city or outside, regularized unauthorized unplanned localities, and temporary unplanned unauthorized localities. Along with these distinctions a striking feature of Delhi is that even planned localities, posh or middle-class, contain a large number of temporary structures established haphazardly where domestic and lower-class workers live. Most residents lack kitchens, bathrooms, latrines, or even electricity and water. The need to provide shelters with proper amenities for this group must be recognized.

## Slums, Squatter Settlements, and Unauthorized Construction

According to the 1981 census, Delhi had about 1.3 million housing units. Almost 78 percent were residences; other uses include 7 percent as shops; 4 percent as factories, workshops, and worksheds; 2.8 percent as shop- or workshop-cum residences, including household industries; 1.2 percent as business houses and offices; .7 percent as restaurants or other eating places; .2

percent as places of worship; .1 percent as hotels and tourist homes; .07 percent as places of entertainment and community catering; the remaining 5.3 percent were used for other activities. The census treated a building or part of it as a census house when the structure had a separate main entrance from road, common courtyard, or stairway, or was recognized as a separate unit.

Ribeiro (1982) referred to housing units as shelters when he looked at housing stock from 1951 to 1981. In 1951, 86 percent of these shelters were classified as traditional and located in the walled city or in urban villages; by 1981 their share had declined to about 32 percent. The urban poor had occupied about 5 percent in 1951 while their share increased to 25 percent by 1981. Unauthorized dwellings consisting of 5 percent of the shelters in 1951 rose to 17 percent in 1981. Employer-provided shelters showed a slight increase from about 5 percent in 1951 to 6 percent in 1981. Multifamily housing developed by the Delhi Development Authority accounted for about 6 percent of Delhi's shelters in 1981; there were none in 1951. Individual housing on newly developed lots increased from zero to about 13 percent in 1981. Slum rehousing occupied about 1.5 percent in 1981 and cooperative group housing was the lowest at about 0.1 percent.

Since publication of the Bharat Sewak Samaj report on Old Delhi slums (1958), problems of slums have received considerable attention in Delhi and other metropolitan cities. This report highlighted that slum conditions prevailed in all parts of the city and 1,787 units were found unfit for human habitation because of congestion, dilapidation, lack of amenities, and unsuitable location. Later studies revealed deplorable conditions and grossly inadequate facilities. They found extensive unauthorized construction and squatter settlements in Old Delhi and New Delhi (Bijlani, 1977; Majumdar, 1983; Singh and De Souza, 1980).

India's first five-year plan had admitted in 1950 that most cities had a large number of substandard houses, unsanitary mud huts, and makeshift structures. Demolishing slums and other temporary structures was advocated in this context and steps were taken to formulate and enact suitable laws for town and country planning and establish administrative structures for their implementation (India, Planning Commission, 1983).

In subsequent five-year plans, preparation of master plans for metropolitan cities was undertaken and the demolition strategy was somewhat deemphasized. Instead some measures were adopted to introduce marginal improvements within existing structures and neighborhoods, and to prevent growth of new slums and unauthorized settlements. In 1962, the Indian government approved the final master plan for Delhi, completed with the assistance of many foreign consultants. The plan intended to stop haphazard growth, improve existing conditions, and promote planned and balanced urban development for a population of 4.6 million up to 1981. Under the plan, large-scale acquisition development and land disposal took place, and extensive programs for new construction were launched. At the same time, under a resettlement program started in 1960 by the government of India, about 200 thousand squatter families with about a million persons were provided sites of 21 square meters each with common services in 44 resettlement colonies. They were expected to build their own structures according to their economic abilities over a period of time (Ribeiro, 1982). Where shifting squatters was not feasible or was unsuccessful, some improvements were introduced in their living conditions. By 1981, 16 thousand units were also constructed to rehouse slum dwellers.

Problems of unauthorized housing and colonies continue. Yadav (1979) reported 130 colonies in 1960-62 but by 1977 more than 600 new colonies had sprung up and by 1981 about 200 thousand households were classified as being in such colonies. From time to time most of the unauthorized construction as well as unauthorized colonies have been regularized on grounds varying from humanitarian to political patronage (Yadav, 1979). Even in posh localities such as Sunder Nagar, Jorbagh, Vasant Vihar, Defence Colony, or Greater Kailash, squatter huts are found everywhere. Urban villages, rural enclaves, and the deteriorating colonies originally constructed for the Pakistani refugees are a common feature of the Delhi landscape.

The middle- and upper-class tradition of keeping servants attracts the urban poor who set up their living quarters in garages, on vacant lots, in streets, even in hallways (De Souza, 1978). The prevalence of the informal urban sector is yet another cause for overcrowded and haphazard urban development. The original master plan is now under revision and it is believed that the new plan will take into consideration the experience gained during the past 20 years.

The housing problem has become acute during the last ten years. Land prices are skyrocketing, rents have become very high (two rooms for about 100 U.S. dollars per month), unauthorized colonies and squatters have increased, traffic jams are neverending. Oriental type markets are springing up every-where and living conditions in the walled city are decaying day by day. The sixth five-year plan reported that nearly one-fifth of the total Indian urban population is estimated to be slum population (India, Planning Commission, 1981b). For Delhi, the estimate could range from one-third to half the population living under appalling conditions and lacking basic amenities.

The same plan reiterates that providing shelter is a basic need which must be met. In 1984, the National Planning Commission admitted once again that housing has emerged as one of India's greatest needs, perhaps next only to food. Delhi is no exception. But in the seventh five-year plan, announced in September 1985, housing and urban development have not been given the highest priority. This means that the urban poor are not likely to live better in the years ahead. The lower middle and the middle classes are also likely to suffer unless crash programs for constructing new dwellings at affordable prices are launched.

## Economy and Development

Delhi indeed has a mixed economy. In 1983, almost 574 thousand persons were employed in the public sector while about 200 thousand worked in the private organized sector of about 5 thousand establishments (Delhi Administration, 1985c). Though Delhi has many large industries, small-scale and cottage industries play an important role. Among the important industries are textiles, machine tools, metal and engineering goods, transport equipment, rubber and chemical products, and electrical machinery.

Small-scale and cottage industries, which originated mainly in the postwar period, continue to dominate the market in handloomed and embroidered products; costume jewelry; pottery, brass, plastic, leather, and ivory goods.

According to recent statistics, the relative share of the primary sector (agriculture, animal husbandry, forestry, fishery, mining, and quarrying) in the total product at constant prices has gradually declined from about 7 percent in 1970-71 to about 4 percent in 1982-83; the share of the secondary sector (manufacturing, construction, electricity, and water supply) has also declined from about 26 percent in 1970-71 to about 21 percent; and the relative contribution of the tertiary sector (all other activities, especially services) has increased from about 67 percent to about 76 percent (Delhi Administration, 1984). These statistics clearly establish the importance of trade and commerce, hotels, transport services, communications, storage facilities, banking and insurance, and public administration activities. See table 6.4.

According to the latest census, one-third of the population is in the formal economy, including about 52 percent of the males and about 7 percent of the females. In addition, the economy's informal sector may be employing about a million persons of all ages (India, Town and Country Planning Organisation, 1981).

In the context of the general economy, the role of banks, insurance companies, and other commercial houses is considerable. With nearly a thousand branches of various, mainly nationalized banks, Delhi is a national and international center of banking, trade, and commerce. Old Delhi continues to be a predominant wholesale distributing center. Within the general framework of Delhi employment, it is interesting to find almost 400 thousand applicants registered with different branches of the employment exchange at the end of 1983, distributed as follows:

| | |
|---|---:|
| Professional, technical, and related | 44,716 |
| Administrative, executive, and managerial | 78 |
| Clerical and related | 42,379 |
| Sales | 274 |
| Farmers, fishermen, hunters, loggers, and related | 959 |
| Production and related, transport equipment operators | 24,707 |
| Service | 1,186 |
| Unskilled | 14,185 |
| Unclassified by occupation | 261,301 |

More startling are the data on the educational level of these applicants: below middle school, 56,376; middle school, 29,107; high school and intermediate, 230,066; college degrees, 64,681; postgraduate degrees, 9,555. Among the college and postgraduate degree holders, the number of bachelor of arts

TABLE 6.4

Nonagricultural Establishments and Employment—Delhi, 1980

(in thousands and percentages)

| Economic Sector | Establishments | | Persons Usually Working | |
|---|---|---|---|---|
| | N | % | N | % |
| Manufacture and repair services | 47.1 | 38.7 | 408.0 | 33.0 |
| Wholesale and retail trade | 35.2 | 29.0 | 148.7 | 12.0 |
| Community, social, and personal services | 16.4 | 13.5 | 415.6 | 33.7 |
| Finance, insurance, real estate and business services | 8.5 | 7.0 | 107.3 | 8.7 |
| Restaurants and hotels | 8.0 | 6.5 | 45.3 | 3.7 |
| Transport | 2.1 | 1.8 | 39.2 | 3.2 |
| Storage and warehousing | 2.1 | 1.8 | 13.0 | 1.0 |
| Utilities | .7 | 0.6 | 20.0 | 1.7 |
| Communications | .5 | 0.5 | 20.6 | 1.7 |
| Mining and quarrying | .4 | 0.3 | 1.2 | 1.0 |
| Construction | .4 | 0.3 | 4.0 | 0.3 |
| Total | 121.7 | 100 | 1233.8 | 100 |

SOURCE: Delhi Administration (1985c).

graduates was highest at 23 thousand followed by education (22.4 thousand), commerce (14 thousand), science (85 hundred), engineering (14 hundred), medicine (about 2 thousand), and agriculture (about a thousand). While these data cannot uncover the total magnitude of Delhi's unemployment problem, they reveal some dominant trends. Widespread unemployment among the educated seems to reflect the economy's slow growth and failure of the education system. Massive unemployment among the unskilled presents other complex problems which have remained unsolved despite decades of economic planning (Delhi Administration, 1985c; India Planning Commission, 1979).

# Health, Education, and Welfare

## Health

The most common diseases for which patients seek clinic or hospital service are respiratory ailments, fevers, gastrointestinal infections, diarrhea, and dysentery. During 1981-82 within the territory covered by the Municipal Corporation of Delhi, 21,219 persons died from miscellaneous causes; the next major cause was tuberculosis (1,902) followed by injuries—including suicides (1,502), fevers (921), other respiratory diseases (708), and pneumonia (592). The figures were substantially higher for the total union territory. Delhi's overall death rate per 1,000 has been declining since 1941 from 41 to 7 in 1984; infant mortality has also fallen from 160 in 1941 to 45 in 1984.

In 1980-81, 84 maternal deaths were recorded and almost 45 percent of Delhi's deliveries were attended by untrained midwives. The birthrate of 28 per thousand is still high though extensive family planning services have functioned for the past 20 years. A recent study found 58 percent of eligible couples had received some type of family planning advice, 20 percent were sterilized, and 28 percent used other types of family planning methods (Delhi Administration, 1984, 1985b).

Delhi is served by hospitals and clinics run by the central government and local authorities of the municipal corporation, New Delhi Municipal Committee, the Cantonment Board, and many volunteer organizations. Central government employees and their families are covered by the Central Government Health Scheme initiated in 1954 with a number of clinics in different parts of the city (India, Ministry of Information and Broadcasting, 1984). Delhi has 38 general hospitals; 27 special hospitals for such problems as infectious diseases, tuberculosis, and mental illness; 7 hospitals of indigenous medicine; and 1 for homeopathic medicine. The total number of beds in all types of hospitals was 14,656 in 1984. In addition, there are 126 clinics with 1,121 beds, and 547 dispensaries to render medical service for minor ailments. The availability of government hospitals and clinics is about 0.09 in Delhi compared to 0.03 per thousand nationally; the number of beds is about 212 in Delhi compared to 74 per 100 thousand in India (Delhi Administration, 1985b). Despite Delhi's being India's biggest medical center with an extensive network of facilities, its hospitals, clinics, and dispensaries are perpetually crowded and medical personnel can hardly afford to spend more than a few minutes with each patient.

At the other end of the scale, is the All India Institute of Medical Sciences established in 1956 as an autonomous center for medical research facilities for various specialties of medicine and surgery. Delhi has also four medical colleges (Delhi Administration, 1976; Civic Affairs, Editors of, 1984). Until 1975, most graduates from these colleges migrated to foreign countries, particularly the United States and the United Kingdom. In recent years, a large number of graduates have gone to African and Middle Eastern countries. Thus the society which spends several hundred thousands of dollars on medical education remains deprived of the services of its own trained medical personnel.

## Education

Delhi, one of the largest educational centers in all of South Asia, has two major universities, the University of Delhi and Jawaharlal Nehru University. The former is both teaching-research and affiliating, granting degrees to students in all colleges in the Delhi area, while the second is primarily a teaching and research institution. In addition Delhi has about a dozen national institutes representing the fields of urban affairs, public adminis-tration, public cooperation, public policy, international trade, labor man-agement, international studies, town and country planning, Oriental studies, criminology, policy sciences, child development, and the environment.

In 1983, 52 general education colleges (34 for males, 17 for females), 13 professional technical colleges (11 male, 2 female), and 26 undergraduate intermediate colleges (19 male, 7 female) were functioning with a total of 74,922 in male colleges and 47,330 in female colleges. While female colleges admit only females, young women may attend male colleges. Delhi had 475 male higher secondary schools and 267 for females; 241 boys' middle schools and 95 for girls; 1,007 boys' primary schools, and 737 for girls, plus 50 prepri-mary schools, 3 vocational and technical schools, and 1,672 schools for special and other education. Total enrollment for all these institutions was 759,800 for males, 690,412 for females (Delhi Administration, 1985b). Although these figures seem large, the problem of high dropout after primary and middle schooling is beyond imagination as the 1981 census reports show.

Several problems are recognized in the field of education:

(1) Many schools function in tents and makeshift structures.
(2) Private schools existing under the label "public schools" admit children from the upper classes and create differential systems of high school education.
(3) Demand continues for new colleges as hundreds of young residents cannot gain admission to existing colleges.
(4) Education standards are reported to have deteriorated.

These problems are complex and no easy solutions are available. A national commission has once again examined the educational system and recom-mended a new set of priorities (India, 1985). One of the most pervasive problems confronting higher education relates to the need for promoting indigenous study material and making the system more socially relevant to conditions and requirements of Indian society (Nagpaul, 1980).

## Welfare

Under the Delhi Administration, the Department of Family Welfare, established in 1966, is mainly responsible for promoting family planning programs while the Directorate of Social Welfare, established in 1959, is primarily concerned with welfare of the weaker sections of the population through a network of services and institutions. Family welfare services such as family planning and immunizations are provided through 90 centers. During 1984-85, 26,217 sterilizations were performed.

Many types of social welfare services are given to destitutes, the old and infirm, widows, tuberculosis patients, and impoverished students. The depart-ment provides financial aid, counseling, day and foster care home services, and residential institutions for beggars, delinquents, and the disabled. Special centers render services to deaf and mute children and the blind. More than one thousand lepers receive residential accommodation and medical treatment. Since 1971, a crash-feeding program to combat malnutrition has been in existence. Special welfare services are also provided in the local prison. Public welfare services are supplemented by private services provided by more than a hundred voluntary welfare agencies.

The overall availability of welfare services remains inadequate and their quality poor in most cases (Delhi Administration, 1980). Delhi's graduate school of social work, one of the most reputable institutions in India, usually provides trained personnel for administrative and supervisory positions, though most of the workers lack professional training (Nagpaul, 1980).

## Transportation and Traffic Jams

Delhi presents a bewildering scene of crowds waiting daily at all hours near bus stops and other transportation terminals. Buses handle most of the traffic with the ring railway providing only limited service to specified railway stations. The number of all types of power-driven vehicles has increased substantially since 1979 (table 6.5).

In 1984-85, Delhi Transport Corporation buses had tripled in number over their 1970-71 level. Combined with even larger increases in private buses under corporation operation, 5,157 vehicles were available—a total of about 3.3 times the number in 1970-71. But the number of commuters had swelled from about a million to 3.85 million in the 14-year period, so problems of mobility remained acute.

Some relief has been provided to government employees by starting buses from specified points in the morning and evening to lessen commuting inconveniences. But for the vast majority of residents and tourists, Delhi's transport facilities are totally inadequate.

In Old Delhi, slow moving horse-drawn carriages, cycle rickshas, bicycles, and bullock carts are still common. Even in New Delhi, bicycles continue to be the commoner's vehicle (Civic Affairs, Editors of, 1984). Middle and upper-middle classes use three-wheel scooters, while taxis, being expensive, are primarily restricted to rich people and foreigners. Occasionally the Delhi Transport Corporation, supervised by the central government, has initiated proposals and partially implemented them to provide an efficient fleet of vehicles to ensure regular, punctual, and adequate transport facilities. But unfortunately nothing seems to have worked to meet the requirements of Delhi's growing population and increasing volume of traffic in a constantly expanding city.

Delhi is also an important terminal for interstate buses and railways. Several years ago the interstate bus terminal was set up outside the Kashmiri Gate not far from Delhi Railway Station. Most buses originating here provide transport to many cities in neighboring Haryana, Punjab, Uttar Pradesh, and Rajasthan. Old Delhi and New Delhi Railway Stations handle several hundred thousand passengers daily.

TABLE 6.5
Transport Data—Delhi, 1970-1980
(in thousands)

| Vehicles | 1970 | 1980 | Percentage Increase |
|---|---|---|---|
| Motorcycles and scooters | 89 | 561 | 530 |
| Cars and jeeps | 54 | 152 | 181 |
| Goods vehicles | 13 | 50 | 285 |
| Auto rickshas | 10 | 29 | 190 |
| Taxis | 4 | 9 | 125 |
| Buses | 3 | 14 | 367 |
| Total registered | 173 | 815 | 371 |

SOURCE: Delhi Administration (1985b).

Air travel has increased considerably in Delhi. Palam Airport, built during World War II, has emerged as one of today's busiest air terminals. More than 40 airlines offer service here beside the government-owned Air India and Indian Air Lines, handling international and internal traffic respectively. The airport is constantly being modernized to cope with ever-increasing traffic. Ground facilities, though much improved during the past five years, remain inadequate and somewhat unsatisfactory.

## Social Disorganization: Problems of Law and Order

Traditional family structure and family life have been changing in Delhi and other metropolitan cities. The old authoritarian structure is loosening up considerably. Arranged marriage has been undergoing modifications. Matrimonial alliances are increasingly being made through advertisements in local newspapers. Young persons attending higher secondary schools and colleges are increasingly asserting their independence to choose their own spouses, though with parental approval. Divorce is no longer uncommon.

Problems of wife abuse and social evils of the dowry system are increasingly being discussed, debated, and even given some attention by social welfare agencies and judicial bodies. The Directorate of Social Welfare, Delhi Administration, has formulated and implemented several measures to provide needy women counseling and protection (1984). Judicial bodies have begun intervening by imposing moderately severe punishments on abusive husbands and their parents. However, the due process of law and order safeguards guaranteed under the Constitution of India have often delayed timely action by the courts.

Because of the changing family system and values and the mass media, traditional processes of socialization and social control have undergone considerable changes in Delhi and other parts of urban India. Varying forms of social disorganization are becoming visible, reflected through crime, deviant behavior, juvenile delinquency, drug addiction, divorce, suicide, and traffic accidents. Until recently crime rates in Delhi and other metropolitan cities remained surprisingly low (Nagpaul, 1976). But in 1984, Delhi had 317 murders, 275 attempted murders, 264 robberies, 1,580 burglaries, 13,940 thefts, and 125 arson cases (Delhi Administration, 1985b). Data on suicide are distressing. During 1981, more than 40 thousand suicides were committed in India; Delhi, with 346, ranked third after the cities of Bangalore and Madras with almost 700 each (India, Ministry of Home Affairs, 1984). A recent survey of crimes against women reported that rape, molestation, kidnapping, abetment to suicide, and indecent acts have increased considerably during the past decade. Fear of crime is so widespread that travel by public transport or taxis or going to movie houses at night is discouraged (Misra and Arora, 1983). In 1984, more than seven thousand accidents killed 1,240 persons and injured 5,751 in Delhi streets and roads. Delhi fire departments responded to almost seven thousand calls for help (Delhi Administration, 1985c).

Coping with rising crime and other forms of deviant behavior is presenting serious problems. More and more police stations and posts have been established, up from 72 in 1970 to 113 in 1984. Police strength has increased from 16,687 to 31,010. Some modernization is taking place by providing the police more automobiles and vans equipped with wireless communications so that they could become more responsive to people's needs.

The judicial system, traditionally cumbersome and slow, is also being reformed. On 25 September 1985, the Chief Justice of the Supreme Court of India, P. N. Bhagwati while inaugurating a seminar in Delhi on "Judicial Reforms—Role of the Bar," said:

After all the legal system has to work to the satisfaction of its consumers, i.e., the public. The endeavor of the Bench and the Bar should be to create a sense of confidence among the public so that the system can grow and evolve effectively (*Hindustan Times*, 1985).

The frightening murders, rapes, and robberies, coupled with the widespread incidence of pickpocketing by lower-middle- and middle-class juvenile gangs, require special attention from police and courts to prevent disrupting patterns of life and save thousands of innocent lives. Being a capital city, various agitations and demonstrations—an integral part of democratic societies—frequently take place, bringing additional pressure on the machinery of law and order. Moreover, because Delhi has more than one hundred embassies and consulates, the police perform yet another important task in protecting foreign diplomats and tourists and a considerable amount of resources are diverted in this direction.

## Quality of Life

Quality of life is a normative concept and cannot be dealt with in absolute terms. Various reports of the five-year plans and other government publications on food, shelter, employment, poverty, transport, environment, and human services, even in the 1980s, admit that living conditions for most metropolitan residents continue to be poor (India, Planning Commission, 1981a, 1981b, 1983, 1985). Many Indian scholars writing about life in Delhi during the 1970s noted that about 80 percent of the households had a monthly income less than 30 U.S. dollars; about 60 percent lived in only one room and almost 50 percent had no structural amenities such as kitchen, bathroom, or latrine (Bhattacharya, 1976; De Souza, 1978; P. S. Majumdar and Majumdar, 1978; Yadav, 1979). Conditions inside the old walled city were reported to be still more deplorable because of widespread filth, congestion, and water scarcity. Even in planned housing projects in planned localities, scarcity of water and erratic electricity supply were the order of the day.

Have these conditions changed in the 1980s? According to recent statistics the per capita income in Delhi at current prices was about 350 U.S. dollars for the year 1982-83 (Delhi Administration, 1984). Even doubling this amount to arrive at an average household income, the low level remains obvious. If incomes have remained low, the living standard of most households continues to be characterized by subsistence conditions notwithstanding the inflationary spiral of the last decade which imposed additional hardships.

Actually half or more of the households may still have an annual income of 350 U.S. dollars or below. The recreational and cultural life of these low-income people is limited; most of their free time is spent on commuting to work and back home. Most people have to spend a considerable amount of time just to procure the daily necessities of life, so many stay home after their work. Most people visit places of worship periodically, enjoy celebrating religious festivals, and frequently participate in folk activities. People also go regularly to movies—often the single most important activity of their lives.

But Delhi is a city with a split personality. Though more than three-fourths of its residents live under poor conditions, at least a fourth live in a somewhat better urban environment, and about 5 percent enjoy apartments and houses in planned localities where civic amenities are reasonably well provided. Another small group with household heads occupying higher administrative, executive, and managerial positions in government lives in comfortable shelters. Diplomats and other personnel of foreign embassies, consulates, foundations, and multinational corporations constitute another small group living in palatial houses and enjoying lives of luxury with teams of servants at their disposal.

Whether a family is rich or lower-middle class, it is likely to have some domestic help. The upper classes have full-time servants; the lower-middle is satisfied with part-time help. In all localities, regardless of the status of residents, numerous children from the poor classes are seen at all hours engaged in such petty activities as shoe shining, cleaning, or collecting garbage. Though beggary is considerably less than ten years ago and mainly concentrated in Old Delhi, scattered beggars do wander around all over the city. The living styles, consumption habits, and leisure time activities of the wealthier population groups tend to overshadow the true picture of deprivation and misery of the millions struggling to eke out subsistence living in an environment where housing, water, and transport facilities are woefully inadequate.

Individual privacy is limited. Most people live in one- or two-room dwellings and cannot even think of privacy. Frequently the entire family— parents, sons, and daughters—share the same sleeping room. Consequences of this type of living on growth of the human personality are generally unknown though there are many studies of the urban poor living under crowded conditions (T. K. Majumdar, 1983; Mehta, 1977; Trivedi, 1980; Wiebe, 1975).

In households with four or more rooms, individuals do get separate rooms to themselves but this privilege is available to only a very small percentage of the population. These people are also likely to be members of prestigious clubs with indoor swimming pools and facilities for other sports activities.

More than a decade and a half ago, Mitra (1970), a scholar on urbanization and a top level administrator for 20 years, described Delhi's problem as a capital city as one of integration. Consumption levels needed to be narrowed and the conspicuous expenditure of the minuscule privileged group in New Delhi had to be curtailed. Basic amenities of life needed a more equitable distribution and a leveling among all groups. Mitra felt that such equalization would bring close ties of a common identity to the people involved. Unfortunately his suggestion has been accepted more in proclamations and pronouncements by the five-year plans and in public relations statements of the Delhi Development Authority than in practice.

## Mass Communications and the Revolution of Rising Expectations

Delhi has become a metropolitan center. Thousands of Indians from other parts of the country arrive daily as do hundreds of international passengers. Foreign radios, televisions, stereos, and jeans are flooding the markets of Delhi. Through a network of satellite, submarine cable, high-frequency radio, and special communication facilities, modern telephone, telex, and telegraph services are provided to almost all parts of the world. The introduction of television transmitting national and international programs is bringing considerable changes in the life-styles, work habits, and leisure activities of Diliwalas. Surprisingly in 1984, 686,000 licensed televisions were reported, more prevalent than Delhi's 462,000 radios (Delhi Administration, 1985b). Of course thousands of radios exist without licenses. The average price of a good TV set is about 300 U.S. dollars, fairly high in the context of Indian income levels; small sets are available for about 150 U.S. dollars.

Educated Indians are avid readers of newspapers and periodicals. In 1982, 49 daily papers were published—18 in Hindi, 16 in English, 15 in other languages—and 2,419 periodicals were published—741 in Hindi, 1,046 in English, 632 in other languages. Besides the several hundred thousand subscribers to newspapers and periodicals, thousands also visit reading rooms maintained by municipal authorities and voluntary organizations in different parts of the city. Many foreign governments have libraries and reading rooms where their publications are available to Delhi residents (Delhi Administration, 1985c).

Another source of contact with the outside world is the well-established postal system. Diliwalas get their mail twice a day, a luxury not available in many other parts of the world. In 1983-84, Delhi had 587 post offices, 143 telegraph centers, and about 400 thousand telephones (Delhi Administration, 1985c). But telephone service is grossly inadequate and inefficient because of outdated equipment and work habits of the Indian personnel.

Thus through printed material, radio, television, and personal contact, not only the educated but all persons increasingly come to know about the affluence of Western societies, their music, dances, life-styles, food habits, and a host of other behavioral patterns. As a consequence of this cultural contact and diffusion of foreign ways of life, a revolution of rising expectations is taking place in Delhi and metropolitan cities. Initially stimulated by the high-sounding goals incorporated in the Indian Constitution, this revolution is being reinforced by cultural contact. Upper-class youths in particular and youths from the middle-class in general are increasingly attempting to acquire more material possessions even if they have to use illegitimate means. They are becoming more individualistic and are drifting away from the authoritarian rules of the traditional family system. The philosophy of social Darwinism emphasizing survival of the fittest is rapidly spreading in urban India replacing traditional conceptions of communal patterns of living based on caste and kinship. Hedonism is increasingly gripping the minds of the young in a world where the cost of living has soared, the economy is characterized by scarcity, differential rewards based on class and caste are widespread, and individual hopes, ideals, and dreams are being shattered even for the college educated.

People realize that there is a better way to live, but social and economic constraints continue to force them to live under the most unhealthful and deplorable conditions. People's rising expectations generate more violence, crime, mental illness, drug addiction, and suicide. Fulfillment of some of these expectations can only be achieved through more employment opportunities, decent housing, and a better urban environment. Such provisions will reduce some of the basic undesirable social problems now rampant in metropolitan cities.

## Urban Mental Health
## Through Religious Commitment

If the Delhi environment is congested, crowded, and polluted for the vast majority of its population living in structures lacking basic amenities of life, what keeps people going without collapsing mentally? Of course, crime is increasing; more delinquents roam the streets every night; the divorce rate has gone up; and other forms of deviance are becoming more visible. But religion continues to play an important role in the life and activities of ordinary people. Religion not only protects their mental health but also provides many other types of psychological and social satisfactions.

Whether people are Hindu, Moslem, Sikh, or Jain, they believe in God and His supernatural powers. People try to comprehend God through a number of deities. According to the 1981 census, Delhi has more than 25 hundred places of worship and people visit them frequently. In almost every locality, some kind of religious activity goes on under tents or in open air several months of the year. Beliefs in destiny, evil spirits, and ritualism continue to be widespread. Whenever people are confronted with failure or undergo frustration, they are likely to seek help from their personal god to cope with the problem. Whether one visits a barber shop, grocery, or modern department store, one is likely to find a number of pictures or statues on the walls representing varied images of God to whom owners and employees alike pay their respects or even worship daily and thereby seek success in their trade or profession and peace of mind. The same rituals are followed in dwellings as well. Thus we find the growth of urbanization has not modified religious beliefs and practices of the common people to any large extent and they continue to derive support from religion just as urbanites in Western societies seek help from psychiatrists, psychologists, and therapists.

Ironically, this general commitment to religious beliefs and values does not seem to present major conflict among individuals who are believed to use illegitimate and dubious means to attain their goals. Almost everywhere one hears about the prevalence of widespread corruption in Delhi and other parts of India. It is alleged that one could become a member of a housing cooperative society illegally and thereby obtain a lot to construct one's house or one might choose to sell it in the underground economy. Should one start construction, beginning with the initial phase of having plans approved by the local municipality to the last phase of obtaining a completion certificate and all through the intermediate phases of arranging for building materials and public utilities, the process will require payment of money to the officials and contractors concerned. Otherwise one is confronted with endless bureaucratic procedures and neverending delays.

Recently about a hundred employees of the Delhi Development Authority were found guilty of dishonesty in their relationships with the public (Delhi Development Authority, 1985). Corruption is so widespread in all walks of life that it has become a way of life. Whether one wants a telephone or permit to pour cement, one is forced to get it through the underground economy. In the context of urban development in Delhi, corruption is believed to be rampant today in the private and public sectors of the economy. One can only hope the religious beliefs and values will eventually prevail and bring people back within the legitimate and legal framework of society.

## Delhi's Administration Structure

Since 1958, there are essentially three organs of Delhi metropolitan government: (1) Municipal Corporation of Delhi (MCD), (2) New Delhi Municipal Committee, and (3) the Cantonment Board. Only the MCD is an elected body; the committee is nominated by the central government, and the board consists of partially elected and partially nominated ex-officio members (Khosla and Datta, 1972). There are also the Delhi Transport Corporation and Delhi Milk Supply Scheme, other local government organs.

The MCD is charged with municipal government of the entire area of the union territory, excluding the areas assigned to the New Delhi Municipal Committee and the Delhi Cantonment Board. The respective spheres of activity between the Delhi Administration and the MCD are different, though some overlapping exists. The corporation has obligatory and discretionary functions. The former include all those generally associated with municipal government and the latter relate primarily to educational, cultural, and welfare activities for the poor (Delhi Administration, 1976). The MCD has a number of statutory, special, ad hoc, and standing committees, besides the two responsible for administering Delhi's water and sewage program and its electrical power. In 1981-82, 29 separate MCD departments carried out municipal functions. Slum clearance has been one of its major activities since 1959. MCD also maintains several hospitals, schools, community centers, and family planning centers. Its revenues are primarily derived from taxation and government grants and loans (Delhi Administration, 1976; Municipal Corporation of Delhi, 1982). In recent years, there have been attempts to decentralize activities of the corporation by dividing the city into ten zones and delegating municipal powers to zonal committees which include local councillors and one or more aldermen (Civic Affairs, Editors of, 1984).

The New Delhi Municipal Committee, created in 1933, has jurisdiction over about 43 square kilometers, including government buildings, the diplomatic enclave, parts of Connaught Circus, and some adjacent areas. Financial resources come mainly from property taxes, other types of taxes, and profits from the distribution of electricity. Since 1931-32 when the committee had a budget of about 8,000 U.S. dollars, the scope of its activities has expanded so vastly that it spent more than 64 million U.S. dollars during 1982-83.

The cantonment is administered under provisions of the Cantonment Act, 1924, and extends over an area of about 43 square kilometers. The Cantonment Board, in charge of municipal functions and administration, essentially functions under supervision of the Ministry of Defense (Delhi Administration, 1976; Khosla and Datta, 1972).

The central government plays the most dominant role in the administration of Delhi. The lieutenant governor of the Delhi Administration links the MDC and central government with certain supervisory and controlling powers. The *Delhi Gazetteer* points out that the by-laws and regulations framed by the MDC are approved by the central government; certain specific administrative matters such as city improvement schemes, classification of urban areas, and levy of certain taxes also require approval.

The central government has a reserve power to interfere in any matter and enforce its views. In the case of the New Delhi Municipal Committee and the Cantonment Board, central government powers are still more extensive and sweeping.

Khosla and Datta (1972) discuss some of the anomalies which continue in Delhi's present administrative structure. The metropolitan council and other elected bodies have no real power in the formulation and implementation of some of the most vital programs. The overall administrative structure does not seem responsive to people's basic needs. Moreover, the management of organizations like the Delhi Development Authority, Delhi Transport Corporation, Delhi Milk Scheme, or the Mother Dairy is least accountable to Delhi's elected representatives.

## Planning and Intervention in Urban Development

Soon after independence in August 1947, the central government had to develop several new colonies with thousands of structures to accommodate about half a million refugees. In 1956, the Delhi Slum Areas (Improvement and Clearance) Act was passed to provide a legal framework to slum improvement and clearance efforts, including demolition and redevelopment or replacement of unfit housing. The government was empowered to acquire such areas compulsorily for redevelopment purposes, emphasizing rehabilitation of slum dwellers with minimum dislocation of work place and residence.

In 1972 this program shifted toward environmental improvement of slums through provision of minimum facilities like water taps, storm drains, sewers, latrines, paved lanes, street lighting, and community play areas. Intervention also came in relocating several hundred thousand squatters at predetermined sites while some unauthorized colonies were regularized and a limited number of civic services were provided.

Delhi's master plan (1961-81), prepared by the Town Planning Organisation, came into effect in September 1962. The Delhi Development Authority, created in 1957, was also involved in completing the plan and had overall primary responsibility for its effective and successful implementation. Other local bodies, mainly the Delhi Administration, MCD, New Delhi Municipal Committee, Central Public Works Department, and Ministry of Works and Housing, had responsibility for the plan's detailed implementation.

Under the old 1894 Land Acquisition Act and the 1976 Urban Land Ceiling Act, about 45.5 thousand acres were acquired through February 1983—13,900 for constructing dwellings; 7,200 for resettling Delhi squatters in the late 1970s; 11,200 for industrial, commercial, and institutional purposes; 7,100 for parks, forests, and recreational grounds; 2,300 acres were reported to be under unauthorized use and 3,800 remained vacant (Bijlani and Balachandran, 1980; India, Planning Commission, 1983).

The major effort to construct new dwellings was launched in 1968-69 when 1,200 units were completed and in subsequent years extensive construction was undertaken. During the past 25 years, about 200 thousand units have been built, of which more than half were sold by the end of 1984 and the remaining are in the process of being completed. Besides, 250 thousand lots were developed to resettle squatters and slum dwellers and 30 thousand developed lots were allotted to cooperative societies. About 4 percent of the lots were auctioned for commercial purposes, some bringing as high as 500 U.S. dollars per square meter. In addition, many massive structures were completed especially in connection with the Asian Olympics held in Delhi in 1982. Finally, a program to improve the environment was introduced and more than 80 thousand trees were planted, along with development of several hundred parks, forests, and rock gardens (Kumar, 1984).

Currently, Rohini Residential Project is under way to develop about 100 thousand lots in some 6 thousand acres divided into 19 sectors (Kumar, 1984). This project will also include about five thousand dwellings of different types.

The first master plan is over and Delhi Development Authority is now developing another master plan for the next 20 years to cover a larger area, the National Capital Region, which includes Delhi and several cities from neighboring states (see figure 6-1). This plan will involve development of 18 regional towns, rural growth centers, and some centrally placed villages to divert population concentration and selected economic and governmental activities from Delhi. Population of the region is projected to be about 30 million by 2001 in an area of about 30 thousand square kilometers of which Delhi Metropolitan Area will occupy about 31 hundred with a population of some 20 million. The present city of Delhi is projected to have over 14 million people (Bhargava, 1984; India, Town and Country Planning Organisation, 1975; Ribeiro, 1985).

These figures are, however, tentative projections based on prevailing birth, death, and migration rates in the region. The task of providing even basic services would be colossal. What will happen to land prices, housing supply, and living conditions are questions which have not been adequately investigated and answered as yet. But if existing policies continue and if the past is any indicator of future urban development and planning, living conditions may further deteriorate. Elementary problems concerning water shortages, open sewage, and lack of proper garbage disposal methods continue to plague all corners of the city.

It is true that the Delhi Development Authority (DDA) has changed the entire landscape of Delhi in a period of 20 years, has built several thousand new dwellings, resettled about a million lower-class persons, and has completed massive concrete structures of national and international fame. The authority has spent the equivalent of millions of U.S. dollars and continues to do so. It combines in itself functions of planning, land-use

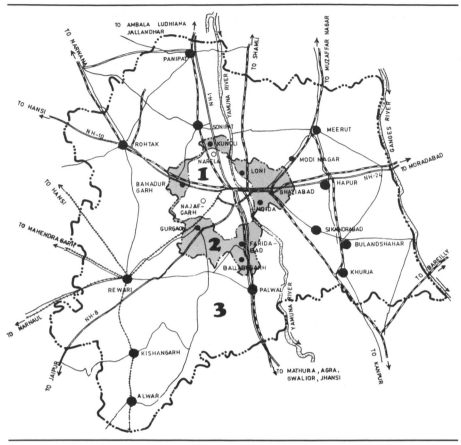

**Figure 6.1: National Capital Region, 2001**

SOURCE: *Vikas Varta* 2, no. 3 (1985: cover).
NOTE: 1 = existing Delhi area; 2 = proposed Delhi Metropolitan Area; 3 = proposed National Capital Region.

control, municipal services, horticultural development, establishment of social facilities and city centers, and house construction. We are told by the authority that it has received laurels and acclamations from all over the world (Kumar, 1984). I pose the following questions to the authority:

(1) Has DDA solved the scarcity of housing in Delhi?
(2) Has DDA reduced the problem of unauthorized colonies of squatter settlements?
(3) Has DDA improved living conditions in old slums or reduced growth of new ones?
(4) Has DDA provided adequate housing to the urban poor?
(5) Has DDA achieved balanced and integrated development?
(6) Has DDA equalized services and amenities of life among different parts of the city?
(7) Has DDA enlisted any degree of public cooperation and participation in its activities?

Currently one of the most popular activities in Delhi is surmising on what Delhi will be like in the year 2001. During the last two decades, Indian academicians, administrators, journalists, and even politicians have written hundreds of articles and published dozens of books on the problem of metropolitan India in general and Delhi in particular. Several professional journals such as *Nagarlok*, published by the Center for Urban Studies of the Indian Institute of Public Administration, and the *Journal of the Institute of Town Planners* and some popular journals such as *Civic Affairs, Seminar*, and *Delhi Vikas Varta* have brought out special issues on Delhi's problems in the context of the Delhi Master Plan (Nagpaul, 1979). More than one hundred seminars and conferences have been held to discuss and evaluate the progress of the Delhi experiment in urban planning and intervention.

Asok Mitra (1970) wrote about general deficiencies in Delhi living conditions. In 1973 Prakasa Rao and Sundaram presented a dismal picture of public utilities. In 1976 Bhattacharya stressed the wide gap between major policy pronouncements and actual achievements of urban planning, and that urban planning had remained predominantly physical land-use planning with little integration with socioeconomic planning. In 1980, Sundaram highlighted the inadequacy of physical structure and infrastructure and problems of coordinating and administering the Delhi plan. In 1984, Datta and Jha brought out forcefully some of the major lapses and failures in implementing urban development in Delhi. The Town and Country Planning Organisation itself had reviewed the master plan in 1973 and the Indian Ministry of Works and Housing produced an evaluative report in 1978. Since then planners have written and spoken about major shortcomings in the Delhi experiment (Ribeiro, 1983; Shafi, 1981). Thus the questions raised above can only be answered negatively, notwithstanding the main achievements gained in more housing to accommodate millions of residents and more massive structures to enhance Delhi's international image.

The Delhi Development Authority combines many conflicting functions. There are numerous anomalies in relation to other governmental bodies responsible for urban development and housing construction; DDA is too much engrossed in physical and spatial planning; and it seems obsessed with its international reputation.

The authority has grown into a huge bureaucratic organization with about 35 thousand employees. Most of the planners, social scientists, and administrators are trained in foreign countries or have received their professional education through foreign teachers and materials. There is a desperate need to divide the authority into several independent units and to incorporate indigenous elements in the training of planners, administrators, and social scientists. Problems of coordination with local bodies such as the Delhi Municipal Corporation, New Delhi Municipal Committee, Delhi Transport Corporation, state governments, and other branches of the central government have to be tackled smoothly to achieve successful implementation of various programs in the future.

Finally, two most important areas which need serious attention and action relate to (1) public accountability to the community and development of organizations for wider participation of community groups; and (2) exploring ways to bring down the artificially inflated prices of land and property so that suitable legislation can be developed to reduce them. If speculation for nationalized land in the hands of the authority cannot be curbed, one of its primary objectives, then it must admit failure to serve the common good of the total population. In this context the need to improve living conditions in rural areas and develop selected regions with a view to reducing the continuous flow of migration to Delhi and other metropolitan areas become imperative (India, Town and Country Planning Organisation, 1982).

## Misplaced Priorities and Misallocated Resources

For the past three decades Indian elites representing science, technology, economics, public administration, architecture, city planning, and other fields have been busy planning for the generation of economic growth and development and the equitable distribution of wealth and other economic benefits among different population groups. They have held hundreds of conferences and seminars to discuss and debate issues, problems, and solutions and continue to do so frequently. They have published dozens of volumes of recommendations on almost every facet of life reiterating reduction of poverty and unemployment and introduction of the socialistic pattern of society.

In this context urban planning and development have been given special attention for two decades, especially in Delhi. It is true that the Delhi Development Authority (DDA) and some other public organizations have improved the quality of life in Delhi. But let us examine some other aspects of DDA projects.

For example, in 1982, DDA created the Asian Village to hold the Asian Olympics with dozens of well-furnished, heated, and air-conditioned houses, several dining halls, a tower restaurant, and a viewing gallery surrounded by flowering plants, fountains, and an undulating lake. The main auditorium, fitted with all types of modern lighting and sound equipment, seats 2,500 and is suitable for ballets, operas, musical recitals, and dramatic performances. The village has several administrative buildings, a disco center, coffee shop, and hospital. But since completion of the olympics, this ultramodern complex is serving little purpose in the life of Delhi and has not become the hub of sports, cultural, and recreational activities its planners had originally forecast; perhaps it will never be.

The Delhi Development Authority has created another sports complex, Indraprastha, modeled on the Houston Astrodome and New Orleans Super-dome. DDA brochures describe the new stadium as India's first and among the world's finest—it shimmers in the bright winter sun, a bowl-like structure, girdled with eight massive pylons supporting a domed roof that has won admiration for its sheer architectural ingenuity, Two thousand metric tons of structural steel support the pylons. In contrast, Delhi has hardly any public swimming pools for ordinary citizens and facilities of other sports are woefully inadequate.

Delhi offers yet another example of the waste of scarce national resources in search of recognition and prestige in the world community. About a dozen Western type hotels with modern facilities like air-conditioning have been built under public and private auspices. Some charge between 80 and 100 U.S. dollars for a single room per day. Meals are expensive; drinks are sold at exorbitant prices. True, India is interested in providing comfortable places for tourists and to accumulate foreign exchange, but most of these islands of prosperity seem out of place. Comfort can be achieved through less ostentatious alternatives. The high consumption of water, electricity, and other municipal services by these islands is unbelievable and likely to create more pressures on utilities provided to other Delhi residents.

Consequences of the collaboration of wealthy foreigners and natives for such enterprises seem not to have been fully examined. The recent trend of erecting massive high-rise structures in the tropical climate of Delhi, imitating Western cities mainly located in temperate zones, may prove to be one of the most disastrous blunders of Indian planners in recent history. Perhaps intervention from the Delhi Urban Art Commission is appropriate.

Finally, the decision to produce small cars instead of developing mass transit and improving the existing local transport system reflects wholesale an American orientation in coping with internal circulation and mobility problems. Existing roads are narrow, inadequate, and already choked with human traffic and all sorts of vehicles. The need to modernize the ring railway around the city and its hinterland should receive a higher priority than a strategy of producing more new cars. No doubt, as the latter strategy is pursued, problems of air pollution will worsen an already bad environmental situation.

## Bibliography

Bharat Sewak Samaj. 1958. *Slums of Old Delhi*. Delhi: Atma Ram.

Bhargava, Gopal. 1983. *Socio-Economic and Legal Implications of Urban Land Ceiling and Regulation Act 1976*. New Delhi: Abhinav.

———. 1984. "The National Capital Region." *Delhi Vikas Varta* (July-Oct.).

Bhattacharya, Mohit. 1976. "Urbanization and Policy Intervention in India." In *Asia Urbanizing. Population Growth and Concentration—and the Problems Thereof. A Comprehensive Symposium by Asian and Western Experts in Search of Wise Approaches*, edited by Social Science Research Institute, International Christian University. Tokyo: Simul Press.

Bijlani, H. U. 1977. *Urban Problems*. New Delhi: Indian Institute of Public Administration.

Bijlani, H. U., and M. K. Balachandran, eds. 1980. *Law and Urban Land*. New Delhi: Indian Institute of Public Administration.

Bose, Ashish B. 1980. *India's Urbanisation, 1901-2001*. 2d ed. New Delhi: Tata McGraw Hill.

Civic Affairs, Editors of. 1984. "Delhi on the Move." *Civic Affairs. Monthly Journal of City Government in India* 31(12):1-174.

Cousins, Albert N., and Hans Nagpaul. 1979. *Urban Life: The Sociology of Cities and Urban Society*. New York: Wiley.

Datta, Abhijit, and Gangadhar Jha. 1984. "Delhi: Two Decades of Plan Implementation." In *Urban Innovation Abroad. Problem Cities in Search of Solutions*, edited by Thomas L. Blair. New York: Plenum Press.

Dayal, Maheshwari. 1982. *Rediscovering Delhi, the Story of Shahjahanabad.* New Delhi: S. Chand.
Delhi Administration. 1976. *Delhi Gazetteer.* New Delhi.
———. 1980. *Delhi Social Welfare.* New Delhi.
———. 1984. *Estimates of State Income of Delhi.* New Delhi.
———. 1985a. *Annual Administration Report.* New Delhi.
———. 1985b. *Delhi Economy in Figures.* New Delhi.
———. 1985c. *Delhi Statistical Handbook.* New Delhi.
Delhi Development Authority. 1985. "Delhi 2001." *Delhi Vikas Varta* 2(2).
De Souza, Alfred, ed. 1978. *The Indian City. Poverty, Ecology and Urban Development.* New Delhi: Manohar.
India. 1985. *Challenge of Education, a Policy Perspective.* New Delhi.
India. Ministry of Home Affairs. Bureau of Police Research and Development. 1984. *Crime in India, 1979.* New Delhi.
India. Ministry of Information and Broadcasting, Research and Reference Division. 1984. *India 1984. A Reference Annual.* New Delhi.
India. Planning Commission. 1979. *Study of Special Employment Programs for the Educated Unemployed.* New Delhi.
———. 1981a. *Report of the Expert Group on Programs for Alleviation of Poverty.* New Delhi.
———. 1981b. *Sixth Five Year Plan 1980-85.* New Delhi.
———. 1983. *Task Forces on Housing and Urban Development. Planning of Urban Development.* New Delhi.
———. 1984. *The Approach to the Seventh Five Year Plan, 1985-90.* New Delhi.
———. 1985. *The Seventh Five Year Plan.* New Delhi.
India. Registrar General. 1984. *Census of India, 1981.* Series 28, Delhi, General Economic Tables and Social and Cultural Tables. Delhi: Directorate of Census Operation.
India. Town and Country Planning Organisation. 1973. *Review of Master Plan for Delhi.* New Delhi.
———. 1975. *National Capital Region, Regional Plan.* New Delhi.
———. 1981. *Informal Sector in a Metropolis.* New Delhi.
———. 1982. *Regional Planning Efforts in India.* New Delhi.
Journal of the Institute of Town Planners, Editors of. 1977/1978. "Towards a Development Plan for Delhi." *Journal of the Institute of Town Planners* 96/97 (Oct.-Jan.).
Khosla, J. N., and Abhijit Datta. 1972. "Delhi." In *Great Cities of the World. Their Government, Politics and Planning*, vol. 1, edited by William Alexander Robson and D. E. Regan. 3d ed. London: Allen & Unwin.
Kumar, Tej. 1984. "The City of Delhi, from Hamlet to Megalopolis." *Delhi Vikas Varta* (July-Oct.).
Majumdar, P. S., and Ila Majumdar. 1978. *Rural Migrants in an Urban Setting: A Study of Two Shanty Colonies in the Capital City of India.* New Delhi: Hindustan.
Majumdar, Tapan K. 1983. *Urbanising Poor in Delhi: A Sociological Study of Low-Income Migrant Communities in the Metropolitan City of Delhi.* New Delhi: Lancers.
Mehta, J. 1977. "Life Styles of Low Income Rural Migrants in Metropolitan Delhi, a Micro Case Study." *Nagarlok.*
Misra, S., and J. C. Arora. 1983. "A Study of Crimes against Women." *Police Research and Development.*
Mitra, Asok. 1970. *Delhi—Capital City.* New Delhi: Thomson Press.
Municipal Corporation of Delhi. 1982. *Annual Administrative Report for 1981-82.* Delhi.
*Nagarlok, Urban Affairs Quarterly.* Selected issues.
Nagpaul, Hans. 1976. "The Ecology of Crime in India." *Indian Anthropologist* 6(2):1-16.
———. 1979. "Approaches and Strategies for the Improvement of Slums and Squatter Settlements in Metropolitan India: A Bibliographical Essay." *Indian Journal of Social Research* 20:105-18.
———. 1980. *Culture, Education and Social Welfare: Need for Indigenous Foundations.* New Delhi: S. Chand.

New Delhi Municipal Committee. 1984. *Fifty Golden Years of Dedicated Service (1933-1983)*. New Delhi.

Prakhasa Rao, V.L.S., and Kavasseri Vanchi Sundaram. 1973. "Delhi." In *Encyclopedia Brittanica*.

Ribeiro, E.F.N. 1982. "Shelter for Urban Delhi by 2001." Unpublished manuscript.

———. 1983. "Planning for the Capital." In *The Environmental Concern*, edited by S. C. Bhatia. Delhi: University of Delhi.

———. 1985. "National Capital Region, Framework for Integrated Growth." *Delhi Vikas Varta* (Jan.-Mar.).

Shafi, Sayed S. 1981. "Planning the Indian Metropolis: Some Reconsiderations." In *Urban Problems and Policy Perspectives*, edited by Gopal Bhargava. New Delhi: Abhinav.

Singh, Andréa Manafee, and Alfred De Souza. 1980. *The Urban Poor. Slums and Pavement Dwellers in the Major Cities in India*. New Delhi: Manohar.

Sundaram, Kavasseri Vanchi. 1980. "Delhi, the National Capital." In *Million Cities of India*, edited by Rameshwar Prasad Misra. New Delhi: Vikas.

Trivedi, Harshad R. 1980. *Housing and Community in Old Delhi: The Katra Form of Urban Settlements*. New Delhi: Atma Ram.

Wiebe, Paul D. 1975. *Social Life in an Indian Slum*. New Delhi: Vikas.

Yadav, D. S. 1979. *Land Use in Big Cities: A Study of Delhi*. New Delhi: Inter-India Publications.

# 7

# Lagos

Michael L. McNulty and Isaac Ayinde Adalemo

Lagos is fast gaining an international reputation as one of the worst cities in the world. Newspaper headlines like these: "Lagos, A City Struggling with Itself" (*Christian Science Monitor*, 29 February 1984), "Nigeria's Grimy Capital: A Bustling City of Aggressive, Rude Wheeler-Dealers" (*Des Moines Register*, 1 March 1983), and others in the *New York Times*, the *Wall Street Journal*, and *Le Monde*, as well as television documentaries, present a picture of a metropolis out of control. Lagos teems with inadequate services, uncollected garbage, unmoving traffic, inefficient institutions, and unbridled corruption in the public and private sector. Are these symptoms of serious malaise, from which Lagos may never recover—or simply manifestations of rapid growth which can be expected to abate?

In this chapter, we examine the growth of Lagos in terms of primary functions the city has performed from precolonial days to the present. This review of the changing functions of Lagos is not intended as historical, but rather as a basis for analysis of current structural and institutional problems. We view the city as a physical manifestation of the social, economic, and political institutions that create its various physical parts. The problems of Lagos are not of recent origin, although the petroleum-fueled expansion of the economy has seriously exacerbated them. The different functions Lagos has performed through its early period as a commercial center, later as a colonial capital, and more recently as the federal capital and center of economic and political life in the country each gave rise to physical structures and institutions, which have not disappeared as growth has taken place. Many have survived and frequently operate at cross-purpose to one another.

Lagos is not one city but many. There is the Lagos of the international businessman; the Lagos (Eko) of the indigenous population whose families have lived there for generations; the Lagos of the migrant laborer from Cross River State; the Lagos of the market women in Ajegunle and the taxi-driver from Ikeja. Each of these and many more people and institutions have created a city which is often in conflict with itself. These conflicts continue to occur because the different functions and institutions of Lagos have not given way to each other, but have come together to occupy the same physical space, while operating in vastly different social, economic, and political space. Although this situation is true of many major cities, it is particularly evident in Nigerian cities. We analyze the different functions, physical and social patterns, and conflicts which result. Throughout the development of Lagos, efforts have been made to direct and control its growth. The approaches taken, institutions involved, and instruments of planning and control will be critically examined. Is Lagos a monument to ineffective and inefficient planning? To what extent could present-day problems have been anticipated and avoided? Has planning failed Lagos, or is Lagos simply incapable of being planned?

## The Nigerian Urban System

Lagos is part of Africa's most extensive urban system. Unlike the rest of the continent, Nigeria has a long tradition of urbanization. The contemporary urban system of Nigeria reflects precolonial patterns of urbanization in the Yoruba towns of the southwest and Hausa-Fulani areas of the north, and the impact of colonial administration and economy on growth of towns along major transport lines and sources of natural resources and agricultural products.

In analyzing the growth of metropolitan Lagos, it is important to recognize this national setting. The structure and functions of Lagos reflect its role as national administrative and commercial center. Many problems seen as metropolitan problems in fact reflect conditions well beyond the boundaries of Lagos. Nigeria's sociopolitical and economic problems are most visible and most concentrated in the great cities. Nonetheless, origins of some of those problems (and perhaps their solutions) lie in the rural areas of Nigeria. Lagos reflects not only the result of urban, industrial, and administrative economies of scale, but also the effects of serious decline in the rural economy. Growth of the Nigerian urban system was greatly affected by the oil economy which boomed after the civil war and contributed to the relative neglect of rural areas and agriculture. Rural migrants moved to the cities not only because of opportunities present there, but because of the lack of alternatives. The problems of Lagos are national problems.

Despite the promise which the cities, and Lagos in particular, seemed to hold for rural migrants, employment opportunities failed to be generated in anywhere near the required numbers. Failure of the urban economy to provide adequate employment is at the heart of Nigerian urban problems. Mabogunje

(1978) saw Nigerian urban centers failing in employment, liveability, manageability, and serviceability.

That so many Nigerian cities remained administrative and service centers rather than becoming industrial, manufacturing, and agro-processing centers meant that many urban inhabitants were unemployed or underemployed. Low incomes of the population, coupled with the relatively meager fiscal resources of the urban centers made it impossible to provide adequate service. Inadequacy of urban services and deterioration of those few that were available further contributed to poor living conditions. The capacity of local authorities to manage continued urban growth proved inadequate, and their ability to meet new demand for housing and essential services was greatly exceeded. The results are cities with seemingly intractable urban problems. Nowhere in Nigeria are these problems so evident as in Lagos. Here problems of rapid urbanization were exacerbated by a very difficult physical site for development, growth of initially separate communities, and absence of an effective administrative structure.

## The Metropolitan Area

Historically, the city of Lagos has grown from the initial settlement of Lagos Island to cover areas which now include a group of islands and adjacent mainland areas. At the center is the lagoon after which the Portuguese named the city of Lagos. There are creeks to the south with other islands which together may serve recreation needs of the metropolitan population. Lagos has grown in areal extent from a settlement of 1.5 square miles (3.8 square kilometers) in 1881 to a metropolis of about 102 square miles (265 square kilometers) by 1981, not counting the area covered by water. (See table 7.1.) The Master Plan for Metropolitan Lagos anticipates that the metropolis will expand ultimately to about 15 hundred square miles (3,885 square kilometers), or about ten times its 1981 areal extent, as parts of Badagry and Ikorodu within Lagos State and parts of Ogun State adjacent to Lagos effectively become integral parts of the rapidly expanding metropolitan area (Lagos State, Ministry of Economic Planning, n.d., vol. 1, p. 7). Current and anticipated growth of Metropolitan Lagos has implications for providing facilities and administering a rapidly urbanizing area.

Lagos has grown in a variety of ways. Apart from northward expansion along the firm and better drained land areas, there has been considerable reclaiming of portions of the wetlands which constitute some 15 percent of the land area of Lagos City (island and mainland). In addition to this expansion of the built-up area, the city has grown by encroaching on and absorbing essentially nonurban settlements located at its ever expanding fringes (Adalemo, 1981).

The land areas which now constitute Metropolitan Lagos have never been under the same political jurisdiction. Until 1967 when Lagos State was created as part of the 12-state structure of Nigeria, the city of Lagos (Lagos

TABLE 7.1
Areal Extent and Population, Metropolitan Lagos,
1881-1981

| | | Population (in thousands) | | |
|---|---|---|---|---|
| Year | Areas in Square Miles | City | Surrounding Area | Total |
| 1818 | 1.5 | 37 | NA | 37 |
| 1891 | 1.6 | 33 | NA | 33 |
| 1901 | | 42 | NA | 42 |
| 1911 | 18.0 | 74 | NA | 74 |
| 1921 | 20.2 | 100 | NA | 100 |
| 1931 | 25.6 | 126 | NA | 126 |
| 1952 | 27.2 | 272 | 74 | 346 |
| 1963 | 32.0 | 665 | 470 | 1135 |
| Creation of Lagos State | | | | |
| 1976[a] | 65.0 | 1021 | 2479 | 3500 |
| 1981[a] | 102.0 | | | 4710 |

SOURCE: Mabogunje (1968); Onakanmi (1979); National Census (1952, 1963); Lagos Metropolitan Master Plan Unit estimates.
a. Estimates. The report of United Nations experts in regard to the Master Plan for Metropolitan Lagos (p. 119) expects the Lagos population to be approaching 13 million by the turn of the century, making it one of the world's most populous metropolitan areas.

Island and Lagos mainland) was administered as Lagos Federal Territory while other areas now part of the metropolis were under the jurisdiction of the then Western Region government. Even now, except for the role of the Lagos State government in administration of the area, there is still no single identifiable administrative setup for Metropolitan Lagos nor any formally delimited metropolitan area.

In the absence of a formally delimited area, it becomes necessary to indicate the boundaries and components of what we refer to as Metropolitan Lagos. The 15 hundred square miles referred to earlier can only be meaningful in the context of a metropolitan region. The metropolis proper must be much more limited in areal extent. For purposes of this chapter, Metropolitan Lagos refers to the contiguous built-up area and immediate urban fringe. See table 7.2 and figure 7.1. The older parts of the city are fairly well built-up (over 80 percent) and there is not much room for further growth in them except through further increasing density. Much future growth is likely (and has started) to occur in the fringe areas such as Ketu (23 percent built up), Oregun (28 percent), Amuwo Odofin/Festac (41.5 percent), and Shomolu/Bariga/Oworonsoki (48 percent). Amuwo Odofin/Festac area was opened for development by the Badagry Expressway which runs westward from the core area of the city. The other areas are adjacent to the Lagos-Ibadan Expressway northeast of the city. Construction of the Apapa-Oworonsoki Expressway connecting Badagry Expressway to the northeast has opened up large tracts of land for development in the western sector bringing within the limits of the growing midwestern edge of the metropolis such settlements as Isolo. Growth

TABLE 7.2
Areal Extent and Residential Densities for Component Units—
Metropolitan Lagos, 1977-78 Estimates

| Unit | Areal Extent in Hectares | | | Estimated Number of Inhabitants (thousands) | Gross Residential Density |
| | Total Area | Built-Up Area | Gross Residential Area | | |
|---|---|---|---|---|---|
| Lagos Island | 524 | 439 | 134 | 253 | 1885.1 |
| Victoria Island | 505 | 358 | 194 | 14 | 74.2 |
| Ikoyi | 1129 | 908 | 638 | 85 | 133.4 |
| Maroko | 514 | 234 | 222 | 85 | 384.7 |
| Apapa Ajegunle | 1763 | 705 | 453 | 324 | 716.1 |
| Apapa Wharf | 922 | 893 | 303 | 65 | 213.2 |
| Ebute Metta/Ido/ Iganmu | 1757 | 1193 | 815 | 382 | 468.2 |
| Yaba | 1001 | 647 | 295 | 140 | 474.6 |
| Surulere | 751 | 663 | 490 | 174 | 354.3 |
| Mushin | 2680 | 1954 | 1267 | 999 | 788.8 |
| Shomolu/Bariga | 1837 | 881 | 697 | 481 | 689.4 |
| Ilupeju/Maryland | 785 | 497 | 264 | 47 | 177.7 |
| Oshodi-Sogunle | 750 | 450 | 362 | 131 | 361.3 |
| Ikeja (including airport) | 5931 | 3610 | 891 | 236 | 265.3 |
| Agege | 2601 | 1177 | 934 | 394 | 422.2 |
| Amuwo Odofin/ Festac | 3670 | 1522 | 528 | 6 | 12.5 |

SOURCE: Adapted from Lagos State Ministry of Works and Housing/United Nations Office of Technical Cooperation Technical Report, Lagos (February 1977).

of the metropolitan area is accelerated as such new areas are opened up for residential and industrial development.

## Physical Characteristics of the Site

Four landform types are identifiable in the Lagos metropolitan region: sand ridges and depressions, swamps, lagoons, and creeks. The sand ridges run parallel to each other and follow the same east-west trend as do the creeks. No streams drain them but because of their three-to-five-meter elevation and relative slopes, water manages either to drain away or percolate into the soil which is underlain by unconsolidated sands. The sand ridges offer the most optimal site for settlement.

The scattered villages are on the sand ridges, as are the early planned estates such as on Lagos, Ikoyi, and Victoria Islands and in Ebute-Metta, Yaba, and parts of Surulere. Even today, where it is not possible to complete reclamation schemes, the built-up areas follow this trend of sand ridges as exemplified by the Iganmu and Yaba east areas. The sand ridges contributed to the scattering of villages over the region, instead of the more familiar agglomerated Yoruba settlements characteristic of this part of Nigeria.

The depressions have their bottom layers of ground waterlogged and

**Figure 7.1: Growing Edges—Metropolitan Lagos**

SOURCE: Adalemo (1981). Reprinted by permission.

flooded during the wet season because drainage into the lagoons is very slow and the underground water level is high with water always near the surface. The poor drainage, pools of water, and brackish water conditions discouraged the first settlers who soon began to use the depressions as dumping grounds. But in 1931 antimalaria drainage canals were begun, which helped improve drainage. However, with the great demand for land, private developers encroached on and blocked the drainage canals by erecting buildings across them, worsening the already poor drainage. Blocking the drainage channels increased flooding in the suburban areas, notably Surulere and Idi-Araba, and parts of Yaba, Lagos Island, and Ebute-Metta since flood water from these areas could no longer flow freely to the lagoon.

The swamps form a flattish terrain extending to the waterfronts. The swamps are characterized by a high level of underground water and scattered pools with salinity as high as in the lagoons. Generally water-logged and covered largely by aquatic vegetation, the swamps are inaccessible and difficult to exploit for human needs. The soils are highly compressible when the water has been drained away, and hence are unstable. Buildings erected on them have suffered cracks, and collapsed. The swamps are also a breeding ground for mosquitoes.

Before swamps can be used, they need to be properly reclaimed. The first attempts at swamp reclamation were on the Okesuna (Lagos Island) and Ikoyi swamps. Work was extended to Apapa, Surulere, Ebute-Metta, and Yaba, and recently to Iganmu (Lily Pond now partly occupied by the National Theatre) and the FESTAC Village. Among problems associated with reclamation schemes have been neglect of initial surveys of the depth of clay and silt in the swamps to be refilled, estimation of the amount of materials to be used, and the type of materials to be used to reduce shrinkage. In particular, private developers rushed into constructing buildings, roads, and so on, in the newly reclaimed areas, thus contributing to the problem of the land's settling when heavy structures were loaded on relatively unstable materials. Some schemes use household wastes for reclamation; this method is crude, slow, unhygienic, and gives very poor results, and unfortunately is common in Yaba East, Makoko, Iwaya, west of Surulere, Ajegunle, Isolo, and Mushin.

Finally, lagoons constitute 40 percent of the study area. It has now become possible to reclaim deeper parts of the lagoon as the reclamation technology (hydraulic sand-filling) has greatly improved. Lagoons are being exploited in this way for use as alternative transport routes (construction of the ring roads) for Lagos Metropolis. Their presence has been an attraction for fishing settlements such as Oluwa, Igbon Ijin, Tomaro, Ikuata, Agala, Oka Ogbe, and Ogogoro.

## Growing Edges of the Metropolitan Area

As Lagos has grown, land for expansion has been at a premium. Through land reclamation and further inland extension of the urban fringe, planned and unplanned growth has extended the limits of the urbanized area. Two types of growing edges are identified: (1) the edge lining planned growth zones which is more regular and predictable and a product of deliberate planning; (2) the edge lining areas of unplanned growth which is dynamic and has no easily predictable extent and direction—it relates to unplanned and haphazard growth.

Examples of the first type are few. They can be found in association with growth into the lagoons (the ring roads) and the new planned areas along Badagry Road and also the outer boundaries of the airport. Along Badagry Road the next trend of growth should be northward if not overtaken by the dynamic westward growth trend of Isolo, Mushin, Surulere, and Ikeja. The second type best describes the rest of the growing edges of the Lagos Metropolis. Different zones of this edge are outlined below:

(1) *Maroko and Ilado*. Expansion is south to the Atlantic Ocean and eastward into swamps and marshes. Nearby villages threatened include Moba, Ogoyo, and Alaguntan.

(2) *Ebute-Metta and Yaba*. Iwaya and Makoko villages are expanding toward each other through the swamp. Yaba and Ebute-Metta residential areas are expanding through crude reclamation and filling between Iwaya and Makoko on one side and Yaba and Ebute-Metta on the other.

(3) *Bariga-Oworonsoki-Ogudu*. Filling in depressions and swamps, Oworonsoki and Somolu are expanding in all directions. Oworonsoki is fast approaching Ogudu. The new Lagos-Ibadan expressway passes through Ogudu and is attracting new activities which will change Ogudu's status and facilitate its absorption into the Lagos metropolis. Numerous unnamed villages are springing up. The street pattern follows the footpaths in Somolu, predetermining the future street pattern when the area becomes filled in and part of the metropolis.

(4) *Ojota and Ketu*. These settlements emerged mainly as squatter settlements for travelers along the Ikorodu road. Formerly serving mainly as overnight areas, they have now become residential areas for Lagos daily commuters. This zone is expanding southward to meet the fast-growing edge of Maryland. The new Lagos-Ibadan expressway will speed the settlement process, and the new zone will also spread to Ogudu.

(5) *Northern extremity*. Oregun, Agidingbi, and Alausa are fast expanding. Omole and Aguda have merged and now perform a suburban function and house urban commuters. The Lagos State Secretariat has moved from Ikeja to Alausa.

(6) *Along Agege motor road and railway line*. Located here are the fast-growing edges of Agege and Ikeja, rapidly growing northward toward Agege and east to meet the growing villages of Oregun, Alausa, and Agidingbi. Ikeja's westward growth is halted by the international airport, but evidence shows the trend for growth north of it, suggesting that growth will spill over to its western side. Agege has shown a tendency to grow along major transport routes. Westward it has elongated to absorb a number of villages. The new Agege bypass road has facilitated opening up the area between Agege and Isolo for development. Still further north the Iju junction and water works are about to merge and are growing toward Agege and Isheri.

(7) *International airport area*. The airport area has expanded westward to accommodate the new international airport, temporarily arresting westward growth of Oshodi which has already encroached on Mafoluku village. Like Ikeja, it seems Oshodi will grow southward until it surrounds the airport.

(8) *Isolo and Idi-Oro.* Idi-Oro has long been absorbed by the metropolis. It still has sharp edges, now smoothed by the ring road, toward which there has been fantastic sprawling growth. New growth is taking place on the western side of the Apapa-Oshodi-Oworonsoki Expressway. Isolo and Isagatedo have merged and expanded to join Idi-Oro. These areas are still expanding north to meet Oshodi and west and south to merge with the planned satellite town. This area features extensive residential development, government and privately financed.

(9) *Surulere and Iganmu areas.* With the villages which have expanded and merged with those absorbed, plus opening of the ring road, has come an urban sprawl to fill in the area east of the road involving the planned estates and the area occupied by private developers, mainly from the old villages of Aguda and Abebe. Iganmu follows a planned growth, aligned to the sand ridges, except along its northern and western contact with Surulere and the coastal swamps where growth is haphazard.

(10) *Southern lagoons and creeks.* Expansion thus far has been arrested by lagoons and creeks so that many villages still retain their identity. Because of transport problems to the rest of the metropolis, this area has not attracted commuters to the extent of the areas discussed above.

# *Dynamic and Complex Land-Use Patterns on Lagos*

The core area of Lagos developed as a commercial and administrative center. Consequently, land uses predominating on Lagos Island are commercial and institutional. There is, however, a high-density residential zone to the north of the island. Lagos harbor, once the main seaport of Nigeria, was the focus around which these land uses developed. While the Lagos quays have now been moved, with Apapa and Tin Can Island facilities providing the port services formerly provided in Lagos at a much enlarged level, the pattern to which it gave rise has persisted.

Development of the Central Lagos Business District (CBD) has been further accentuated by the Central Lagos Slum Clearance Scheme which the Lagos Executive Development Board (LEDB) embarked on in 1955. Some 70 acres of densely populated land in the northern part of Lagos Island was to be cleared and replanned in such a way as to prevent the reoccurrence of slum conditions. About 30 thousand residents of the area were to be temporarily accommodated in a new mainland residential district, known as New Lagos (now Surulere), until the replanning was completed and they could return to their former residential area. The scheme did not work out as planned. New Lagos became a permanent development and the slum clearance scheme area was allocated partly for residential and partly for commercial land-use development. Trading activities, which had developed along the main road connecting the harbor with the mainland as well as diffused through the old residential district, could now expand into the newly available land. Retail shops sprang up all over the area. The western half of Lagos Island is almost entirely devoted to commercial use while the eastern half is occupied by government offices and residential development.

Apart from the CBD, commercial land uses can be found along roadways and side streets and particularly in the traditional market sites (Sada, et al., 1978). According to a 1978 survey, these markets occupied a total area of about 100 hectares distributed in close proximity to population centers in different parts of the metropolis. Trading activities are so widespread that one observer, when asked how many markets there were in Lagos, commented that "all of Lagos is a market!" (Chief Dare, Head of Market-Women's Association; personal communication, April, 1971). The CBD on Lagos Island, however, continues to be the main focus of commercial activities for Metropolitan Lagos. (See table 7.3.)

The earliest industrial developments were located in close proximity to the Apapa Port Complex. Subsequently, however, the government's desire to encourage industrial development led to establishment of estates in different parts of the metropolis. These industrial estates were carefully laid out and provided with facilities. Because of this planned development, industrial land use is distributed much more evenly throughout the metropolis and in a less haphazard manner than residential land use.

The largest industrial estate (180 hectares) is located in Ikeja and was developed in 1959 just outside the city limits when the area was still under jurisdiction of the former Western Regional government. With the creation of Lagos State, Ikeja, Mushin, Badagry, and Ikorodu districts came under jurisdiction of the new state government. The Lagos State Development and Property Corporation (LSDPC), the agent of state government charged with responsibility for the development of industrial and residential land use, took advantage of the additional land in these districts to provide for future expansion of the metropolis.

Residential land use covers slightly more than 50 percent of the total built-up area of Metropolitan Lagos and varying proportions of each district. Quality ranges from the high-class residential districts (with population density of less than 100 persons per hectare) of Victoria Island, Southwest Ikoyi, Ilupeju, and parts of Apapa to the poor and near slum conditions (over a thousand persons per hectare) in Mushin, Agege, Shomolu-Bariga, and Ajegunle. Between the two extremes are the residential districts of Yaba and Surulere on the Lagos mainland. There is a definite contrast between the quality of residential facilities in the former Federal Territory of Lagos and the settlements not originally administered as part of that territory. The 1976 Household Survey of Metropolitan Lagos shows that the rooming type of housing, in which several families share a single building, predominates in those areas located outside the former federal territory and city limits. Living conditions are generally poor in such housing and basic urban facilities are often not available to residents. Bringing services to these areas of the metropolis is one of the most intractable problems facing those charged with

TABLE 7.3

Land Use in Standard Urban Fringe—Metropolitan Lagos,

1976-1979

| Land Use | 1976 | | 1979 | | Change 1976-79 | |
|---|---|---|---|---|---|---|
| | Total Area in Hectares | Hectares/ 1000 Persons | Total Area in Hectares | Hectares/ 1000 Persons | Total Area | Hectares/ 1000 Persons[a] |
| Residential | 2841.3 | 2.24 | 5147.2 | 2.45 | 2305.9 | 0.21 |
| Commercial | 346.0 | 0.27 | 434.7 | 0.21 | 97.7 | −0.06 |
| Industrial | 301.3 | 0.24 | 825.1 | 0.39 | 523.9 | 0.15 |
| Institutional | 597.0 | 0.47 | 2571.3 | 1.23 | 1974.3 | 0.76 |
| Transport | 489.3 | 0.39 | 801.3 | 0.38 | 312.0 | −0.01 |
| Open space | 16.0 | 0.01 | 28.0 | 0.01 | 12.0 | 0.00 |

SOURCE: Master Plan for Metropolitan Lagos (n.d.: 110, vol. 1).
a. Negative values indicate intensification of use.

the responsibility for planning and managing development of the metropolitan area.

The nature of the terrain on which the original city developed is an important factor in determining its eventual form while development of transport routes (rail and road) has done much to bring about changes in the structure of the metropolitan area. The motor road linking the port of Lagos with the rest of the country ran south to north along the least problematic ecological zone. The railroad was constructed parallel to the motor road. Residential and other developments followed the pattern of the transport routes thus giving Lagos its initial linear form.

The complex system of lagoons, creeks, islands, coastal ridges, and depressions ensured that the city grew as isolated communities linked only through the one main axial south-north road. The discontinuous transportation system is still a peculiar feature of Lagos even though government agencies through reclamation efforts and bridge construction, have gone to great lengths to provide links between pairs of communities in recent times.

For a long time, Lagos Island was linked to other areas of the metropolis and the rest of the country only by Carter Bridge. In the late 1960s Eko Bridge was constructed to link the island and mainland west of the aging Carter Bridge. The Third Mainland Bridge was constructed east of Carter Bridge in the latter half of the 1970s and linked with Eko Bridge via a new road built over the reclaimed part of the lagoon formerly occupied by the Lagos quays. The Third Mainland Bridge was also designed to be linked with the Lagos-Ibadan Expressway further north through an elevated roadway built on the Lagos lagoon between Ebute Metta in the south and Oworonsoki in the north. This in turn would be linked with the Apapa-Oworonsoki Expressway in the west to form an integrated system of major urban transport routes designed to relieve pressure on the original main axial road running through the center of the city. Unfortunately, the extension to Third Mainland Bridge dropped in rating and was not constructed during the 1980-85 plan period. Since all such road projects are federal projects, the Lagos State government can do little to change the priority rating especially as there are complaints that the federal government is committing a disproportionate level of investment or development funds to Lagos relative to the rest of the country.

Federal government investment in the Lagos metropolitan area during the Second National Plan period (1970-75) was greater than the total investment for the entire country during the first plan period (1962-68). The need to make substantial funds available for development of a new federal capital at Abuja also means that less and less money will be available for Lagos.

Because the federal government is concerned solely with linking Lagos with the rest of the country, practically all of the road projects mentioned above run south to north. The only major east-west road, Badagry Expressway, was initiated by the Lagos State government. Apart from a portion of the Apapa-Oworonsoki Expressway, there is no comparable east-west route in the city. The railroad which runs south-north through the center of the city continues to be a major barrier to the easy flow of east-west traffic in the expanding metropolis. In many instances, long detours have to be taken to cross from east to west or vice versa.

The overall effect of these developments is to continue to foster the growth of isolated communities linked to one another only through tenuous routes to the main arterial routes. The narrow spine-like form of the initial city has been broadened considerably as developments follow construction of the arterial roads. Construction of these roads has rendered many parts of the metropolis relatively more accessible for development. It has also facilitated developments to the east and west of the former thrust.

## Primacy: A Debatable Concept

In the majority of developing or underdeveloped countries, a characteristic feature of their spatial structure is the existence of a primate city with a disproportionate share of total urban population. Statistics available for Lagos clearly demonstrate that it could be very misleading to define primacy in demographic terms only. In spite of evidence that the Metropolitan Lagos population is increasing at a rapid rate and that the rate of spatial expansion is also phenomenal, Lagos could not assume the title of Nigeria's primate city on the basis of population figures alone. Its population of almost four million (by the most recent estimates) is probably twice as large as the next largest city; but it represents only about 4 percent of the total population of Nigeria and, at best, only about 20 percent of the total urban population.

The dominance of Lagos becomes evident however when one considers the proportion of the nation's economic activities located within the metropolitan area (see table 7.4). Nigeria's largest port complex is located within the metropolitan area. The dearth of suitable sites along other sections of Nigeria's coast for development of port facilities has conferred tremendous locational advantage on Lagos. The Lagos port complex is also better linked with the rest of the country than any other port facility. Consequently the

TABLE 7.4
Absolute Share[a] of Industrial Activity by Major Urban
Centers—Nigeria, 1965-1969

| Major Center | 1965 | 1969 |
|---|---|---|
| Metropolitan Lagos | 37.8 | 50.0 |
| Kano | 9.8 | 10.6 |
| Kaduna | 8.1 | 13.9 |
| Ibadan | 6.1 | 6.4 |
| Sapele Area | 6.5 | 4.5 |
| Zaria | 2.5 | 2.9 |
| Jos | 2.2 | 2.2 |
| Ewekoro | 1.7 | 1.6 |
| Total | 74.7 | 90.1 |

SOURCE: Schatzl (1973: 232).
a. Defined as the arithmetic mean of the percentage shares of the number of people employed, their wages and salaries, gross output, and gross value added.

share of Lagos in Nigeria's foreign trade is about 70 percent. It rose to 90 percent during Nigeria's civil war when the eastern ports (Port Harcourt and Calabar) were out of commission. The share of Lagos in Nigeria's import trade is about 80 percent.

Almost all industrial and commercial enterprises in Nigeria have their headquarters in Lagos (see table 7.5). Their activity ensures that the city accounts for not less than half of the total value added by manufacturing and a higher proportion of capital investment in the industrial sector. Unfortunately, because of the capital-intensive nature of these industrial establishments, too few employment opportunities are generated by these investments. Nevertheless, together with the public administration sector of the economy, they have brought about a high concentration of skilled manpower within the Lagos metropolis. Fifty-six percent of value added in utilities, 48 percent in services other than transportation, communication, and distribution, and 30 percent in public administration are produced within the metropolis. The trend has until recently been in the direction of greater and greater concentration of economic activities in the Lagos Metropolitan Area. Its share of gross national product rose from 10.5 percent in 1958 to over 20 percent in 1975 and the annual growth rate of the concentration ratio was 2.7 percent for 1958-62, 4 percent for 1966, 5.6 percent for 1973, and 9.6 percent for 1975 (Lagos State, Ministry of Economic Planning, n.d., vol. 1, p. 72).

Concentration of industrial investment in the Lagos metropolitan area has also been encouraged by a number of policy decisions on the part of the federal and state governments. Among these are promotion of the so-called import substitution industries, establishment of industrial estates in various parts of Lagos, and the directive given to government agencies which made the economic performance criterion (profitability) the sole basis for giving loans to or participating in industrial enterprises. Since import substitution industries rely almost entirely on imported raw materials, the temptation to locate as close as possible to the port is great. The establishment of industrial

TABLE 7.5
Headquarter Locations of Leading Commercial and Industrial
Enterprises—Nigeria, 1960

| State | Location | N | % | |
|-------|----------|---|---|---|
| Anambra | Enugu | 3 | 1.8 | |
| Cross River | Calabar | 2 | 1.2 | |
| Imo | Umuahia | 1 | 0.6 | |
| Kaduna | Kaduna | 2 | 1.2 | |
| | Funtua | 1 | 0.6 | |
| Kano | Kano | 5 | 3.0 | |
| Lagos | Lagos Islands | 56 | 33.0 | ⎫ |
| | Lagos Mainland | 43 | 25.0 | ⎬ 87% |
| | Ikeja | 28 | 17.0 | ⎪ |
| | Mushin | 21 | 12.0 | ⎭ |
| Oyo | Ibadan | 3 | 1.8 | |
| Plateau | Jos | 1 | 0.6 | |
| Rivers | Port Harcourt | 1 | 0.6 | |
| Total | | 170 | | |

SOURCE: Adapted from Adegbola (1983: 22-25, Tables 6 and 7).

estates (see table 7.4) which included the provision of infrastructural facilities was extremely attractive to industrialists. It ensured litigation-free land and other facilities at subsidized rates. For quite some time, location within Metropolitan Lagos was the surest way of satisfying the profitability criterion.

The need for a change of policy has been recognized. The Third National Development Plan document is replete with policy statements indicating a shift toward decentralization of economic activities. More favorable consideration is likely to be given to financing industrial enterprises willing to locate in other parts of the country and use local raw materials. With respect to location within the metropolitan area itself, there is evidence of locational shift over time from the central core (Lagos Island) to the periphery. The pattern of growth of manufacturing gives a clear indication of this shift. Growth of manufacturing had started to decline in the inner core during 1956-62. Capital investment intensified during that period even though the total number of establishments declined. In subsequent periods, the area suffered a net decline in growth of industrial establishments and manufacturing investment as cheaper land and better facilities became available at the periphery.

The same pattern is evident in the Ebute Metta/Yaba/Surulere/Mushin/Igbobi zone except that there was a slight degree of capital intensification in the 1966-70 period. Additional manufacturing establishments continued to move into the Apapa/Iganmu/Iddo/Ijora zone until 1970 and particularly during 1962-66. The only zone which continues to enjoy positive growth in number of establishments and magnitude of industrial investment is the Agege/Ikeja/Ilupeju zone. More recently evidence of new growth can be found in the Oregun-Ojota, Oshodi, Isolo, Matori, Gbagada, and Ogba areas not covered by data in table 7.3, but which have attracted industries largely as a result of the creation of industrial estates. High land values at the core areas have forced industries requiring extensive land area to move to the periphery where cheaper land is available.

## Labor Force and Employment Structure

The phenomenal areal expansion of Lagos is a response to the needs of the rapidly expanding urban population. According to the technical reports of the Master Plan Unit, the total population growth rate for Metropolitan Lagos is much higher than the national average. Between 1953 and 1980 the population of Lagos grew at the rate of 9.4 percent per annum whereas the national average was 2.5 percent. A breakdown of the total population growth rate for Lagos shows natural increase to be 4.0 percent leaving a net immigration rate of 5.4 percent. Natural increase and net immigration are each higher than the national average. Net migration is more than double national average population growth. The inference is that Lagos attracts a disproportionate number of people from other parts of Nigeria.

Recent studies show that the main motivation for migration to Lagos is economic. Income levels are relatively higher in Metropolitan Lagos than in other regions of Nigeria. A rather high proportion of jobs is concentrated in Lagos. About 70 percent of national industrial investment is in Metropolitan Lagos. Thus people tend to gravitate to Lagos from all parts of Nigeria in search of job opportunities which they perceive to be more readily available.

In the absence of reliable census and demographic data, it has not been possible to establish changes which may have occurred since the studies referred to above. From what we have observed recently, however, it is safe to infer that if there has been any decline in the rate of migration to Lagos, it is hardly noticeable. The Lagos Metropolitan Master Plan Report expects a slight decline in net immigration during the 1980-90 period. This expectation is based on a number of assumptions which, in the light of current developments in Nigeria, may not be entirely realistic. In any event, the absolute population of Metropolitan Lagos continues to increase through immigration and absorption of populations of fringe areas, placing considerable pressure on existing facilities.

Most migrants come to Metropolitan Lagos in search of employment. The labor force is estimated to have grown from 1.6 million in 1978 to about 2.8 million in 1985. This rate of growth is at least twice the average annual growth rate of projected employment. The indication is that, at the current rate of population increase in Metropolitan Lagos, an increasing proportion of the labor force will be out of work or underemployed.

The distribution sector of the economy provides the greatest number of jobs (more than a third), followed by other services, manufacturing and crafts, and public administration. But in terms of the productive nature of the labor force, manufacturing and crafts accounts for the highest value added (a third), followed by public administration, building and construction, distribution, and other services. Distribution activities concentrated in the economy's informal sector provide the largest number of jobs but at the lowest wages. Much of the labor force involved in retail trade is underemployed making the sector's productivity level as measured by value added low compared to other sectors.

The informal sector is the entry point into the labor force for a large number of migrants who came to the city poorly educated and with inadequate skills. Those who can improve their educational and skill levels may eventually join the formal sector where recent studies show that about 20 percent of skilled jobs remain unfilled. Quite a number remain, however, in the informal sector. Some acquire skills through the apprenticeship system while others continue to eke out a living in the petty trade sector. An army of retailers dealing in a few items at a time line the merging points on expressways or road intersections inside the metropolis. This petty trade along routes of slow-moving traffic is a special phenomenon known as the "go slow market" (Aradeon, 1979:12). An even larger number of under-employed persons can be found at the traditional markets in different parts of the metropolis. Practically every building in all but the best residential areas has a shop or shops in front from which groceries and other items are sold to members of the immediate community.

## Problems of Metropolitan Growth

In preceding sections of this chapter, we have tried to outline the spatial, demographic, and economic characteristics of Lagos. We have also tried to indicate the pattern of growth so far as available data have permitted us to map a trend. The nature and pattern of growth bring with them consequential problems begging for solutions. Many problems have physical manifestations such as transport route congestion, high room occupancy rates, and flooding. Generally, the problems are complex and do not have simple causal explanations. In this section we attempt to identify the problems of rapid urban growth as they manifest themselves in Lagos as well as the steps taken in trying to solve these problems. We examine the effects of current government policies and make suggestions about the direction future policy should take. Finally, we hope to make projections which will enable us to get a fair picture of what conditions in Metropolitan Lagos are likely to be at the turn of the century.

### The Physical Terrain Around Lagos

One problem confronting Lagos throughout its developmental history is the difficult nature of the physical terrain which has created impediments to continuous and systematic development of the metropolis. Obviously the areas being added to Metropolitan Lagos from Badagry and Ikorodu areas have worsened rather than improved the wetland/dryland proportions even though the absolute area of dry land will have almost doubled. The area covered by water and the wetlands has more than trebled (table 7.6).

The flat unconsolidated coastal sand is susceptible to flooding and its load-carrying capacity is quite low. Development takes place along the sand ridges while the depressions occupied by swamps are much more difficult to develop, leading to discontinuous, nonsystematic development in Lagos. Transport routes follow the crests of the sand ridges with hardly any connections. Communities occupying the sand ridges are virtually isolated. Some crude reclamation in the past allowed some measure of intercommunity linkage. Technological developments in hydraulic sand filling and foundation piling have made it possible for these areas to be developed but at extremely high costs.

## The Land-Tenure System and Land-Occupation Patterns

Family ownership is the most common pattern of land tenure in Lagos. Dispute over ownership leads to nondevelopment of some parcels of land. Since vacant land is not subject to taxation, land parcels remain undeveloped for long periods of time, constituting environmental hazards within the community. Except in areas designated as Government Residential Areas (GRA), land development was largely unregulated during the colonial period. Even now, the generally weak enforcement of planning regulations has led to what Aradeon (1979) describes as discordance between plan and plan implementation. In spite of zoning plans, multiple land uses characterize most parts of Lagos. Virtually every house in the high-density residential area has facilities for retail trading activity. Noncompatible activities are often found together in the same area.

There is evidence of high occupancy rates and overcrowding within buildings in most parts of Lagos (table 7.7). Because of the rapidly increasing population and consequent increasing demand for housing, the situation has not changed appreciably despite efforts of the federal and state governments to provide a new stock of houses. The rooming house type of housing still provides low-cost accommodation for about 50 to 60 percent of the population in the metropolis. Many of these have no urban services because the rate of providing them has not kept pace with the rapid rate of growth.

## Rapid Population and Labor Force Growth

The high rate of natural increase has been documented. Concentration of commercial and industrial investment in the Lagos metropolitan area has brought about a differential between wages and expected income in Lagos and other parts of Nigeria in spite of efforts to standardize income and wages in the

TABLE 7.6
Land, Water, and Wetland Areas—Metropolitan Lagos
(in km²)

| Area | Water | | Wetland | | Land | | Total |
|---|---|---|---|---|---|---|---|
| Lagos Island | 38 | | 28 | | 144 | | 210 |
| Lagos Mainland | 54 | | 2 | | 32 | | 88 |
| Ikeja | — | | 45 | | 247 | | 292 |
| Shomolu-Bariga | — | | 3 | | 34 | | 37 |
| Mushin | — | | 19 | | 52 | | 71 |
| Subtotal | 92 | (13%) | 97 | (14%) | 509 | (73%) | 698 |
| Part of Badagry | 65 | | 201 | | 214 | | 480 |
| Part of Ikorodu | 152 | | 29 | | 280 | | 461 |
| Total | 309 | (19%) | 327 | (20%) | 1003 | (61%) | 1639 |

SOURCE: Adapted from Lagos Metropolitan Master Plan Unit/United Nations Development Programme Project staff/Wilber Smith and Associates.

public sector. This wage inequity has led to large-scale migration into the metropolitan area in search of employment. The concentration ratio increased from 2.7 percent per annum in the 1958-62 period to 5.6 percent, evidence that the employment opportunities which this level of concentration is expected to produce are not being realized. The growth of productive urban employment is inadequate because most manufacturing establishments invest in capital-intensive methods, thus dampening the demand for labor. A high proportion of the labor force is in the informal sector where productivity is low. Consequently, a high level of underemployment exists and many resort to crime to make a living. Meanwhile, the stream of immigrants keeps increasing with consequent pressure on urban services.

The rate of population growth far outstrips the rate at which facilities and services can be provided to meet minimum needs. Unhealthy living conditions and long journeys to work keep workers' productivity low.

## Administrative and Financial Aspects

The multiplicity of administrative authorities within the metropolitan area has meant that developmental efforts are poorly coordinated. Inadequate human resources and management have further complicated the problem. Because of inadequate financial resources resulting largely from poor tax collection and financial management, the various administrations are incapable of providing the necessary level of services and facilities. Recent estimates suggest that for local governments within Lagos State, including those which administer parts of the metropolitan area, internal sources are a declining percentage of their total revenues, accounting for approximately 60 percent in 1980, 50 percent in 1981, and only 23 percent in 1982. This decline is attributed to inadequate revenue collection efforts. There is as yet no metropolitan government to administer the metropolitan area. Various

TABLE 7.7
Occupancy Rates—Metropolitan Lagos, 1976

| Zone | Number of Houses | Inhabitants per House | Inhabitants per Room |
|------|------|------|------|
| Lagos Island | 5,487 | 44.3 | 5.4 |
| Agege | 13,404 | 29.4 | 5.0 |
| Mushin Central | 4,301 | 67.4 | 4.5 |
| Iganmu, Ajegunle-Apapa | 9,318 | 55.7 | 4.2 |
| Yaba, Ebute Metta | 5,864 | 41.4 | 4.1 |
| Ikoyi West | 1,247 | 27.1 | 4.1 |
| Ijeshatedo, Itire | 5,943 | 43.1 | 4.0 |
| Ikeja | 6,040 | 23.9 | 3.8 |
| Airport-Shogunle | 4,006 | 32.9 | 3.7 |
| Maroko | 3,117 | 27.4 | 3.7 |
| Mushin West, Surulere North | 4,627 | 64.6 | 3.5 |
| Oregun, Ketu | 2,513 | 54.9 | 3.2 |
| Mushin East, Bariga | 9,431 | 50.9 | 3.1 |
| Mushin Northwest | 1,115 | 27.4 | 2.5 |
| Amuwo, Festac Town | 721 | 16.9 | 1.6 |

SOURCE: Adapted from Lagos Stage Ministry of Works and Planning (1977).

agencies of the state government provide services for the metropolitan area and the Master Plan Unit of the state's Ministry of Works is charged with responsibility for planning but not plan implementation. Implementation is the responsibility of the various ministries and agencies which may have their own priorities or operate under differing terms of reference and with varying resource endowments.

The size of the metropolis, the large population involved, and the enormity of the problems would confound any administrative or institutional organization. Steps have been taken by various institutions to provide relief and solutions to the problems identified.

## Toward Solutions to Problems in Lagos

We now review some of the actions taken and plans made to grapple with the problems of Metropolitan Lagos.

To alleviate the perennial flooding brought about by the nature of the physical terrain and human interaction with the natural environment, steps have been taken to remove obstructions to drainage channels by demolishing structures (including residential buildings) built across them. The Ministry of Environment has made clearing drainage channels a priority. Comprehensive action plans have been designed to dredge natural channels to improve their capacity to drain their natural basins. Work is currently proceeding on dredging projects and some beneficial effect is already being felt in the reduction in flooding of areas drained by these stream channels. Major artificial drains are being provided to complement the natural channels.

Additional land for residential and industrial development is being made available within the metropolis through hydraulic sand filling of wetland swamps and even shallow parts of the lagoons. This filling makes for a more continuous development of land and allows for better transport linkages as streets can be planned in the reclaimed areas to maximize circulation rather than being restricted by the physical conditions of the land. Hydraulic sand filling of the lagoon has made possible (albeit at great expense) the Lagos inner ring road which now routes traffic away from the central part of the inland, thus reducing congestion.

Regular ferry service now links the southwestern part of the city with the CBD and more is being planned to utilize the extensive lagoon in the eastern part of the metropolis. One hopes emphasis will be placed on providing standard rather than luxury ferry boats. An effective and well-coordinated metropolitan bus service is essential to the successful operation of a ferry service.

## Comprehensive Plan and Impact of Government Policies

Through cooperative efforts of indigenous and foreign consultants and United Nations experts, a comprehensive physical plan has been drawn up for Metropolitan Lagos. This plan, completed in 1981, provides short-range (1981-85) and long-range (up to the year 2000) guidelines for the city's future growth. The Lagos State government has started to follow these guidelines as they are periodically reviewed. The period of uncoordinated development seems to have come to an end even though planning regulations are still not being enforced as they should.

The Land Use Decree promulgated by the federal military government in 1978 vests all land in the state government, so that new development can only take place with the approval of the governor through a Certificate of Occupancy or formal consent to deal in land. This development in the administration of land in Nigeria holds forth the promise that if properly applied, unplanned and speculative activities which pose serious problems for urban planning could be curbed.

The Third National Development Plan and recent actions of the federal government indicate the desire to decentralize investment away from Lagos. Creation of 19 states in Nigeria solved some political problems, but it has resulted in the creation of more centers of administration and economic investment. Local government reform—a continuing exercise—has as its principal objective establishing and strengthening the third tier of government and creation of a role for the small- to medium-sized urban centers which serve as the headquarters of local government areas. The most ambitious effort at reducing pressures on Lagos is the decision to move the federal capital to Abuja near the country's geographic center. It is hoped that as the administrative center moves, the focus for industrial investment will also shift. While the decision to move the federal capital cannot by itself bring about a shift of focus, the recent policy decision to encourage industries which use local raw materials is likely to bring about a more decentralized industrial development. As long as emphasis was on import substitution industries which relied on imported raw material or semifinished inputs, the importance of Lagos as a port continued to confer locational advantage.

These efforts at decentralization are not likely to affect the rapid growth of population and labor force immediately. The Master Plan Study envisages no reduction in the rate of migration within the next decade. As more centers of investment become established and grow, however, their beneficial impact should assist in relieving pressure on Lagos. Ironically, efforts of the state government to improve environmental conditions and the provision of infrastructural and social facilities may produce the opposite effect of encouraging more people to migrate to the metropolis.

The Lagos State government has moved to improve living conditions in Lagos despite this possible effect of encouraging migration. Housing stock is being increased by government-built low-cost houses. Miniwater works (boreholes) are being built in various locations in the metropolis to supplement the existing main water supply which is grossly inadequate (supplying only about 40 percent of the minimum daily requirement). Many areas of the metropolis, especially the high-density residential areas, rely on water from private boreholes and hand-dug wells. Most industrial establishments still depend on on-site groundwater withdrawal and electrical power generation which adds tremendously to production costs for which the consumer ultimately pays.

Perhaps the most ambitious project to date is the attempt at improving transportation by building an elevated Metroline or rapid rail mass transit (RRMT) system. Because of the increase in traffic generated within the metropolis without a commensurate increase in the physical infrastructure, congestion has become noticeable especially since 1975. Worker productivity is seriously affected by long journeys to work, usually about two hours. Massive road construction and improvement of existing roads do not seem to have brought about any appreciable change. The Lagos State government has, after considering feasibility studies, decided to build the first phase of an RRMT system. A 0.75 billion naira (a naira equals about $1.35) contract was signed with a French consortium for construction and financing of a 28-kilometer elevated line from Lagos Island north to Agege. A recent review of the project ordered by the military government allowed for better coordination between state and federal government agencies. When the cost of relocating services to make way for the Metroline is added to the construction cost, the total cost is likely to approach 1.25 billion nairas. The Metroline is expected to carry some 45 thousand passengers during each peak hour trip. Planners are fully aware that this level of performance can only be realized if the RRMT system is fully integrated with a coordinated and improved road transport system.

The need to reduce rapid population growth is still a great concern to the government. While not being directly involved, both the federal and Lagos State governments have encouraged the activities of private organizations involved in family planning and population control. The impact of these efforts is, however, negligible and population increase continues unabated.

The Lagos State government requested the World Bank to advise it on how administrative and financial performance (especially revenue generation and investment priorities) could be improved. The World Bank (1983) confidential report and recommendations are already available to the government.

## Future Prospects

The problems of Nigerian cities are many, but they arise from individual and collective human action (or inaction). Their solutions lie in those same individual and collective human actions. Solving the problems of Lagos will require much greater effort at exercising control over urban growth, serious efforts to strengthen the total revenue base of the city, and maintenance of at least minimal service delivery capability. The real challenge to Nigerians will be to muster the political and social will to address these issues.

Ursula Hicks (1974), after a comprehensive study of large cities in various countries, concluded that no country has really succeeded in coming to grips with the socioeconomic or even the physical problems of catching up with the needs of modern large cities. Such a conclusion holds forth little hope for appreciable success in the attempts by developing countries in the Third World to solve the problems which now confront their large cities, especially when limited resources and lack of competent administrative personnel further limit the prospects of development. The future need not be that gloomy, however, if we can escape the temptation of comparing the situation in these cities with those in the developed countries. We should not expect Third World cities to develop in exactly the same manner as cities of technologically advanced countries.

## Bibliography

Adalemo, Isaac A. 1981. "The Physical Growth of Metropolitan Lagos and Associated Planning Problems." In *Spatial Expansion and Concommitant Problems in the Lagos Metropolitan Area*, edited by D. A. Oyeleye. Occasional Paper no. 1. Lagos: Geography Department, University of Lagos.

Adegbola, Olukunle A. 1983. "Towards a More Enduring Settlement System in Nigeria." Paper presented at the Second Annual Conference of the Population Association of Nigeria, University of Ife, Ile-Ife, 23-25 November.

Akintola, Joseph Olafioye. 1974. "The Pattern of Growth in Manufacturing in South-Western Nigeria, 1956-71 and the Role of Direct Public Policy in That Growth." Ph.D. dissertation, Boston University.

Aradeon, David. 1979. "Metropolitan Lagos: Conflicts between Design Intentions and the Realities of Spatial Use." In *Draft Final Report on Planning the Metropolis of Lagos—A*

*Series of Planning Studies*. Vol. 1. *Background Papers and Research Reports*, edited by A. Gandonu. Report submitted to the Lagos Metropolitan Master Plan Unit, Ministry of Economic Planning and Land Matters.

Ayeni, Bola. 1981. "Lagos." In *Problems and Planning in Third World Cities*, edited by Michael Pacione. New York: St. Martin's Press.

Hicks, Ursula K. 1974. *The Large City: A World Problem*. New York: Wiley.

Lagos State (Nigeria). Ministry of Economic Planning and Land Matters. n.d. *Master Plan for Metropolitan Lagos*. Prepared by United Nations subcontractor Wilber Smith and Associates in collaboration with United Nations Development Programme Project staff and the Lagos State Government.

Lagos State (Nigeria). Ministry of Works and Planning. 1977. *Residential Densities in Metropolitan Lagos*. Master Plan Bulletin no. 3. Lagos.

Mabogunje, Akin L. 1968. *Urbanization in Nigeria*. London: University of London Press.

Onakanmi, O. 1979. "Land, Property Values and Valuation Rates in Metropolitan Lagos." In *Draft Final Report. See* Aradeon 1979.

Sada, P. O., and Andrew G. Onokerhoraye. 1984. "The Emergence of Secondary Cities and National Development in Nigeria." Paper presented at the Ninth Conference on Housing in Africa, Dakar, April.

Sada, P. O., Michael McNulty, and Isaac A. Adalemo. 1978. "Periodic Markets in a Metropolitan Environment: The Example of Lagos, Nigeria." In *Market-Place Trade—Periodic Markets, Hawkers, and Traders in Africa, Asia, and Latin America*, edited by Robert H. T. Smith. Vancouver: Centre for Transportation Studies, University of British Columbia.

Schatzl, Ludwig H. 1973. *Industrialization in Nigeria: A Spatial Analysis*. Munich: Weltforum Verlag.

Williams, Babatunde Abraham, and Annmarie Hauck Walsh. 1968. *Urban Government for Metropolitan Lagos*. Praeger Special Studies in International Politics and Public Affairs. International Urban Studies no. 2. New York: Praeger in cooperation with the Institute of Public Administration.

World Bank. 1983. *Nigeria: Lagos Urban Sector Review*. Report no. 4479-UNI.

# 8

# Cairo

## Ahmed M. Khalifa and Mohamed M. Mohieddin

Egypt is a highly urbanized society. It has always maintained a large urban population and is rapidly approaching the point where half its population will be living in urban areas. In 1976, 44 percent of the total population lived in urban centers, up from 37 percent in 1960 and only 19 percent in 1907. Egypt had three cities greater than one million: Cairo, 5.1 million; Alexandria, 2.3 million; and Giza, 1.2 million in 1976. The growing trend toward megalopolitan concentration is magnified if the ratio is drawn between the population of these three big cities and the urban population alone. Together the three cities exceed the rest of Egypt's total urban population by 2.6 times. No transitional large cities with a population of 500,000 to a million stand between Cairo, Alexandria, and Giza and smaller middle-sized cities—a basic imbalance in the pattern of Egyptian urbanization. If current rates of population growth and urbanization continue, Egypt would have ten cities equal to the size of present-day Alexandria and four equal to today's Cairo by the end of this century (El-Khorazaty, 1984:5). In 1907 there were only two large urban centers (100,000 inhabitants plus); today's Egypt possesses 20 such cities. Together with the big three, the 20 accounted for 75.3 percent of the total urban population in 1976 and 33 percent of the total population (CAPMAS, 1981:17).

The Egyptian urban scene is also characterized by the dominance of a large primate city, Cairo. Cairo alone accounted for 13.8 percent of Egypt's population and 31.6 percent of the total urban population in 1976, with Alexandria running a distant second with 6 percent of the Egyptian population and absorbing 14.4 percent of the urbanites. See table 8.1. The gap between Cairo and Alexandria has been constantly increasing during this

TABLE 8.1
Comparative Growth Rates—Cairo and Egypt, 1897-1976
(population in thousands)

| Year | Cairo Population | Annual Growth Rate (%) | Egypt Population | Annual Growth Rate (%) |
|------|------|------|------|------|
| 1897 | 590 | — | 9,717 | — |
| 1907 | 678 | 1.4 | 11,183 | 1.4 |
| 1917 | 791 | 1.6 | 12,670 | 1.3 |
| 1927 | 1,071 | 3.1 | 14,083 | 1.1 |
| 1937 | 1,312 | 2.1 | 15,811 | 1.2 |
| 1947 | 2,091 | 4.8 | 18,806 | 1.7 |
| 1960 | 3,349 | 3.7 | 25,771 | 2.5 |
| 1966 | 4,220 | 3.9 | 29,724 | 2.4 |
| 1976 | 5,074 | 1.9 | 36,656 | 2.1 |

SOURCE: Waterbury (1982). Calculated from official census results.

century. In 1907 Alexandria was a little over half the size of Cairo. By 1947, the percentage had declined to .43, and when the next census was taken in 1960 it was .39. By 1976, Alexandria was only one-third of Cairo's size (see Morcos, 1980). If one were to consider Greater Cairo, with its 22 percent share of the total population, the primacy ratio would be even sharper.

A city may manifest its primacy in realms other than population concentration, and Cairo does. While the city accounted for about 14 percent of the nation's population, it has 66 percent of all television sets, 52 percent of all telephones, 33 percent of all pharmacies and medical doctors, 27 percent of all hospital beds, and 62 percent of all university graduates. Of all degrees at all educational levels in Egypt, 23 percent are awarded to Cairo's students. Cairo consumes 27 percent of the nation's entire electrical output, and 40 percent of its vegetables, fruits, and meat. Over half the nation's skilled tradesmen reside in Cairo and 36 percent of its production workers. In transformative industries alone, Cairo houses 42 thousand establishments (30 percent of the nation's total) and has 52 percent of national employment in this sector (Waterbury, 1978; see also Al-Neklawi, 1973).

The heavy concentration of population in Cairo is attributable to a variety of factors including: (1) accumulation of foreign ministries and national and regional governmental units; (2) rapid population growth and consequent increase in consumption which further results in the increase of commercial activities; (3) relatively higher wages; (4) its dominating role in national financial, commercial, and professional activities; and (5) its being the all-time Mecca of all other Arab countries because of its pioneering role in education, modernization, culture, and arts.

This chapter focuses on post-1960 changes in Cairo's population size and distribution, economic activities, and chronic problems of housing, transportation, and sanitation. We shall also discuss anticipated future conditions of Cairo in the year 2000.

Such an enterprise is faced with several hurdles. First, there is more than one Cairo to talk about—Cairo city proper; metropolitan Cairo (Cairo City, its adjacent city of Giza, and the northern industrial city of Shubra al-Khaymah); and greater Cairo, a vast region of about 29 thousand square kilometers that includes in addition to metropolitan Cairo, the provinces of Giza, Imbaba, Al-Badrashin, al Saf from Giza Governorate, plus the provinces of al-Qanatir, al-Khanka, Shibin al-Qanatir, and Qalyub. The focus here will be on Cairo city proper, and reference will be made to the two other cities when appropriate.

Second, Cairo has been growing by leaps and bounds and as it grew, the city has been annexing areas not previously under its jurisdiction, and has been increasingly breaking down its districts for administrative purposes. For example, the district of Shubra in 1960 was composed of eight census tracts with a total population of 296 thousand. In 1976, the district was broken down into two independent districts, Shubra (129 thousand inhabitants) and al Shurabiyah (443 thousand) (CAPMAS, 1978). Thus, a study of intracity population distribution would constantly have to adjust for district boundaries to avoid faulty conclusions about spatial redistribution of the city's population. To the best of our knowledge, the professionally innovative study of Janet Abu-Lughod, *Cairo: 1001 Years of the City Victorious*, is the only attempt to adjust systematically for the changing boundaries of Cairo city.[1] Unfortunately, her data stop at 1960.

Administratively, Cairo was divided into 26 districts in 1976. Each district is divided into smaller sections called *shyakhat*, which serve as census-taking units as well as for other purposes. Accordingly, the city was divided into five major regions or subcities. The northern city is by far the largest with nine districts starting from Bulaq on the eastern shore of the Nile, and including the districts of Rod al-Farrag, al-Sahil, Shubra, al-Wayli, al-Matariyah, al-Zaytun, Misr al-Jadidah, and Madinat Nasr.[2] The eastern city includes al-Jamaliyah, al-Darb al-Ahmar, and al-Khalifah. The central buffering zone encompasses the districts of al Zahir, Bab al-Shariyah, al-Muski, Abdin, and al-Sayidah Zaynab. This subcity separates the eastern medieval city from the modern western city consisting of the districts of al-Azbakiah and Qasr al-Nil. Finally, the southern city is composed of four districts, Misr al-Qadimah, al-Ma'adi, Helwan, and al-Tebeen.[3]

Attempting to predict Cairo's future is hazardous. Data on important variables for planners and policymakers, such as return migration, recurrent migration, sectoral distribution of migrants, size and role of the informal sector in the city's economy—let alone income distribution—are hard to come by. While our ignorance of the future is understandable, our ignorance of the present state of affairs is definitely less understandable.

## Population Growth and
## Its Changing Distribution

By the end of the 19th century, Cairo consisted of two major subcities, more or less independent, plus an appendage to the south. By the middle of the 20th century, a third new city had been added to the original complex north of the existing area of settlement.[4] As the city expanded, its center has continuously moved west, leaving behind the decaying eastern city. First the center moved to the district of al-Muski, then to al-Azbakiah, and, finally, to Qasr al-Nil.

Cairo's spatial expansion to the north, west, and south is, of course, a function of population growth. During the 20th century, Cairo's population increased from 678,011 in 1907, to 791,000 in 1917—an annual growth rate of 1.6 percent. By the 1927 census, Cairo exceeded one million inhabitants for the first time. Because of worldwide economic crises, Cairo's growth slowed to 2.1 percent per annum between 1927 and 1937. The census of 1947 revealed that Cairo had almost doubled its population over the previous 20 years and the two million hurdle collapsed under the phenomenal peak of a 4.8 percent annual growth rate. Cairo's annual population growth slowed to 3.7 percent between 1947 and 1960. Still the city pushed the three million figure. During the intercensal period of 1960-66, the city's growth was 3.9 percent annually because of the active industrialization movement. When the 1976 census was taken, the city was toying with the five million figure, despite the slowdown in population growth to 1.9 percent per annum, its lowest since 1917.[5]

The two major components of Cairo's growth are derived from the rates of natural increase of its resident population and net migration into the city. From 1947 to 1966, the city's total growth hovered around 4 percent with migration and natural increase each accounting for about half the total growth. It is also significant that throughout the period 1938-63, Cairo's crude birthrate was generally higher than Egypt's as a whole. Only in the last decade, and especially since 1967, has the urban environment produced its depressing effects on Cairo's birthrate. Today migration probably accounts for more of Cairo's growth than does natural increase (Waterbury, 1978:127-29).

Several studies of Cairo have noted that most migrants to Cairo as well as to other cities have had rural origins. Since 1917, only 3 percent on the average of Cairo's population has come from other cities (see Abu-Lughod, 1971:174; Hijazi, 1971:16-30; Ouda, 1971). The bulk of these migrants come from the Delta villages. Although the Delta is numerically the most important source of migrants to Cairo, it is Suhaj governorate in Upper Egypt that as a proportion of its population has the highest rate of outmigration. Suhaj also has the highest rural density level and lowest literacy rate of any Egyptian governorate (Waterbury, 1982). Migrants from Delta villages tend to settle in the North, while those who come from Upper Egypt reside in the South around the area of Sayidah Zaynab and Old Cairo.

Despite the massive migration, Cairo does not show serious imbalance in the sex ratio. Delta village migrants move primarily in family groups, while those from Upper Egypt come more frequently as unaccompanied males who either remain single or have left wives and children behind in their home villages (Abu-Lughod, 1961).

In terms of occupation, a sample survey in Cairo in the early 1970s revealed marked differences between migrants and nonmigrants; see table 8.2 (Ibrahim, 1982a). It is noteworthy that migrants are better represented in the uppermost and lowest level of occupations, but underrepresented in middle-level occupations. Significant occupational differences exist between Upper and Lower Egyptian migrants. Upper Egyptian migrants go mostly into domestic and other personal services and have greater concentration in unskilled occupations, especially construction, while Lower Egyptian migrants follow varied occupations, have a slightly higher edge in the upper-level occupations, and are less likely to include housing as a part of wages (see, e.g., Abu-Lughod, 1961; Ibrahim, 1982a:19-20).

Of more importance to us than the mere growth of Cairo is the changing population distribution within the city itself. We have already divided the city into five subcities: the northern, eastern, central or transitional, western, and southern cities.

Growth is never a smooth or an even process; to the contrary, it is always uneven, and Cairo's is no exception. Ever since the city started moving its business center westward leaving behind its historical center of population gravity, the eastern city has been decaying, and a new center of gravity has been established in the northern city. By 1947, 49 percent of Cairo's population was living in that city, by 1960, 52.6 percent lived there, and 59.0 in 1976. The phenomenal growth of the northern city may be attributed to its proximity to the Delta villages which supplied the larger share of migrants to Cairo (about 60 percent), and its housing of some of Egypt's most advanced industries.

At the same time, the southern city, which until 1947 housed a mere 6 percent of Cairo's population, has been systematically increasing its share of inhabitants. In 1960, the city accounted for 11.6 percent of Cairo's population and ranked fourth in its share of Cairo inhabitants. By 1976, this city was experiencing a 2.1-fold increase in population, rising from 389,618 inhabitants in 1960 to 819,663 (16.2 percent of all Cairenes), and second only to the northern city. This city is the legitimate daughter of Naserrite Egypt; its motto is socialism. See table 8.3.

Growth of the northern and southern cities came at the expense of the eastern, central, and western subcities. The eastern city increased in absolute size from 1960-76 which could be accounted for by excess births over deaths, but its proportional share has been declining, from 16.0 percent in 1947, to 13.5 in 1960, and 9.8 in 1976. The city has apparently reached saturation level and is unable to absorb any more migrants. Any addition to its population would have to be crammed into the existing space. Furthermore, the new generation knows jobs exist outside the city in the modern sector up north and down south, and not in the traditional guild system on which the city was organized. Finally, the city is physically decaying.

The same story holds for the central buffering zone. This city with its five districts is losing population in absolute and proportional terms. Except for al-Zahir district with a 4,737-inhabitant gain during the 1960-76 period, the other four districts—Babal-Shariyah, Al-Muski, Abdin, and Sayidah

TABLE 8.2
Percentage Distribution of Recent Migrant and Nonmigrant
Males by Occupation—Cairo, 1973

| Occupation | Migrants | Nonmigrants |
|---|---|---|
| Professional | 6.5 | 5.5 |
| Technical | 4.9 | 3.8 |
| Managerial | 1.0 | 1.8 |
| Clerical work | 5.1 | 6.7 |
| Sales | 4.0 | 9.5 |
| Skilled work | 13.7 | 27.9 |
| Semiskilled work | 6.7 | 4.0 |
| Unskilled work | 42.6 | 15.6 |
| In school | 15.4 | 22.3 |
| Inactive | 0.1 | 3.1 |
| Total percentage | 100.0 | 100.0 |
| Sample size | 3,853 | 12,307 |

SOURCE: Ibrahim (1982a).

Zaynab—combined showed a loss of 31,688 inhabitants, 48 percent of the total loss of all districts that experienced population decline. Decay, oversaturation, and departure of the relatively better off are probably the most important reasons for population decline of these quarters.

The western city—though for different reasons—has also been losing population. In 1947, the districts of Azbakiyah and Qasr al-Nil housed 107,000 or 5.1 percent of the city's population.[6] By 1960, they accounted for 3.2 percent of the total residents of Cairo, and only 1.9 percent in 1976. Departure of foreigners after the 1952 revolution probably accounted for the population decline from 1947-60. Probably the open-door policy and increasing commercialism and transformation of downtown Cairo account for the decline from 1960-76. As foreign companies and banks poured into Egypt along with a surge of private investment—with the downtown their primary location—they seduced residents of downtown apartments to relocate by offering them soaring sums of money which enabled them to pay the required deposits in other residential quarters.

Thus, one can sum up the general features of Cairo for the last 30 years as follows: (1) continuous decay in the eastern historic and central subcities, although both have shown until now a remarkable ability for survival; (2) increasing commercialization of the downtown or western city; and, finally, (3) uninterrupted growth marked by continuous domination by the northern city, and (4) increased incorporation of the southern city into the mainstream of the city's growth. This elongated expansion coupled with westward movement of the business district, as well as expansion of residential quarters across the Nile in Dokki, Ajouza, Giza, and so on, is not without its implications for Cairo's current problems and future prospects, issues we address later.

TABLE 8.3

Population by Subcity, Districts, and Average Annual Rates
of Growth—Cairo, 1960-1976
(according to 1960 boundaries)

| Subcities and Districts | Population (thousands) | | Percentage Rate of Annual Growth |
|---|---|---|---|
| | 1960 | 1976 | |
| The Northern City | 1759 | 2918 | 4.11 |
| Bulag | 202 | 177 | −.86 |
| Rod-al-Farag | 265 | 272 | .16 |
| Al-Sahil | 303 | 438 | 2.77 |
| Shubra | 296 | 572 | 5.82 |
| Al-Wayli | 307 | 377 | 1.43 |
| Al-Matariyah[a] | 160 | 610 | 17.46 |
| Al-Zaytun | 100 | 241 | 8.78 |
| Misr al-Jadidah | 124 | 229 | 5.22 |
| The Eastern City | 452 | 499 | .65 |
| Al-Jamaliyah | 141 | 163 | .94 |
| Al-Darbal-Ahmar | 148 | 150 | .06 |
| Al-Khalifa | 161 | 186 | .94 |
| The Central City[b] | 639 | 612 | −.27 |
| The Western City[c] | 107 | 97 | −.50 |
| The Southern City | 389 | 819 | 6.89 |
| Misr al-Qadimah | 212 | 270 | 1.70 |
| Al-Maadi | 83 | 266 | 13.83 |
| Helwan | 94 | 282 | 12.46 |
| Total Cairo[d, e] | 3348 | 4948 | 2.98 |

SOURCE: CAPMAS, Detailed Census Results, Cairo Governorate (1960, 1976).
a.  Excludes Shyakhat Arab Abu-Tawila, Qalubiya Governorate, in 1960—population 27,337.
b.  Including districts of Al-Zahir, Al-Muski, Bab al-Sharyiah, Abdin, and Al-Sayidah Zaynab.
c.  Including districts of Al-Azbakiah and Qasr al-Nil.
d.  Excludes Madinat Nasr—population 64,892—and Al Tebeen—population 33,575—for total Cairo population according to 1976 boundaries of 5,074,016 and an average annual growth rate of 3.2 percent.
e.  Since 1960 the districts of al Wayli, Misr al-Jadidah, and Shubra have been divided into two, making the total number of districts in Cairo 26.

## Growth of Metropolitan Cairo

Despite the modest growth rate of Cairo city during the intercensal period 1966-76, other parts of metropolitan and greater Cairo continued to grow rapidly. Greater Cairo growth rates were 3.1 percent annually and Cairo's twin city of Giza grew at 8.0 percent per annum, a figure against which Cairo's peak growth rate of 1937-47 humbly bows. Even more spectacular is growth of the industrial suburb of Shubra al-Khaymah. A city just exceeding 100 thousand inhabitants in 1960, its population increased to 173 thousand in 1966, a 9.4 percent annual growth rate. Ten years later, with an annual growth rate hovering around 8.6 percent, its population reached 394 thousand (see CAPMAS, 1978; United Arab Republic, 1962).

It is interesting to note some of the differences between the pattern of migration in Cairo and Giza. In 1960, Giza City was composed of four districts encompassing 19 *shyakhas* or census tracts. By 1976, a transformed Giza City was divided into six districts (*aqsam*) with 42 shyakhas. While migrants to Cairo City poured into the city proper, migrants to Giza apparently migrated to its adjacent villages, gradually transforming them into large quasiurban population agglomerations, which, because of their proximity to Giza, are systematically annexed by it.

The former village of Bulaq al-Dakrur is a good example. Originally an agricultural village near Giza City and like a dozen or so villages within ten miles of central Cairo, it has simply been absorbed by the expanding city. The population increased 2.16-fold between 1960 and 1976. Bulaq al-Dakrur no longer has any significant agricultural activity but is instead mainly a dormitory for Cairo's poorest commuters. Like much of Cairo, it has not expanded physically; people have simply been crammed into existing space. Migrants made up 86 percent of Bulaq al-Dakrur's population, and nearly half came from 1967 to 1971. Significantly, 24 percent had migrated from Cairo and Giza—most probably salaried urbanites in search of lower rents. The population density reached 18,969 persons per square kilometer in 1976.[7]

While some of these villages were by 1960 well on their way toward urbanization, others were totally dominated by agriculture as the main source of income.[8] This stage of migration and the process of incorporating these villages into the city are not without their implications for the adaptation of migrants and native villagers, and, of course, for what type of urban settlements these villages will be.

It is interesting to note that Giza City, adjusted to its 1960 boundaries, did not grow as fast as its hinterland. Between 1960 and 1976, Giza's population rose from 491 thousand to 654 thousand, an annual growth rate of 3.9 percent. Most of this growth, however, took place in the most ruralized or quasiurban district of Imbabah; the other three districts grew at the very modest rate of 1.5 percent annually—even slower than Cairo City—which could be easily explained by natural increase.

## Changing Employment Structure in Cairo City

The employment structure of Cairo City is almost equally divided between the government—including its public sector—and the private sector, with the government employing 50.5 percent and the private sector accounting for 48 percent of all Cairenes age six and older. The remainder work for foreign and international agencies, such as embassies, the United Nations, Arab League, and so on (CAPMAS, 1978, Vol. 1, p. 18). Salaried employees dominate, representing 77.35 percent of all economically active persons in Cairo.

Employment distribution by economic sector shows that service, transformative industries, and commerce dominate the employment structure accounting for 81.2 percent in 1960 and 77.9 percent in 1976. The major decline within these sectors took place in the service sector, falling in relative importance from 43.4 percent in 1960 to 36.3 percent in 1976, despite an increase in absolute volume. In contrast, the transformative industry gained absolutely and proportionately, increasing its share of employment from 21.6 percent to 26.8 percent during the same period. Meanwhile, commerce showed a slight decline of 1.4 percent during the same period. The construction sector has almost doubled its relative share of employment from 4.7 percent in 1960 to 8.3 percent in 1976. The major contraction in the Cairo City economy appears to have been taking place between the industrial and service sectors.

Migration and population increase in urban areas, it is usually assumed, are products of greater employment opportunities and better wages in urban centers as compared to those in rural areas. Migration to urban areas in Egypt in general and to Cairo in particular has continued at a faster pace even though employment opportunities in the organized sector have failed to expand fast enough to keep pace with the growing number of job seekers. Thus, while unemployment hovered around 4.7 percent for the whole country in 1961, it was 7 percent in urban areas, and 7.5 percent in Cairo. By 1970, the situation was not much different. Overall unemployment was around 2 percent, while open urban unemployment was 4.2 percent. In Cairo and Alexandria open unemployment rates were 3.5 and 7.6 percent, respectively, and these two cities alone accounted for 59 percent of total urban unemployment (Mohie-el-Din, 1975).

There is little doubt that the urban informal sector employment plays an important role in sustaining those who cannot find employment elsewhere in the city's economy. The most important occupational categories in informal service employment are domestic servants, petty traders, traditional non-mechanized transport workers, waiters, potters, caretakers, and garbage collectors, to name a few. The breakdown of the informal sector by type of activity reveals an interesting feature: most of the urban-based activities are geared to providing basic consumer services (see Abdel-Fadil, 1980).

Despite its importance in offering employment to a substantial portion of migrants and poor urbanites, there is evidence that some informal sector occupational categories suffer from the open-door economic policy. For example, tailors are experiencing the crunch of ready-made suits and dresses and garbage collectors have been experiencing increasing difficulties selling the products they sift from the trash, as a result of increasing demand for foreign imported products because of the changing tastes of Cairo urbanites (see, e.g., Haynes and El Hakim, 1979).

## Occupational Structure
## and Intracity Inequality

Of more importance than the sectoral distribution of economic activities are changes in the occupational structure of employment in Cairo City between 1960 and 1976. Since the Egyptian census is a de facto one, it shows us how diverse population elements within the city tend to become segregated from one another in the degree to which their requirements and modes of life are incompatible with one another (see Wirth, 1938).

The general trend in Cairo City during 1960-76 shows that the proportional share and absolute volume of professionals, clerks, and production workers increased from 7.8 to 14.1 percent, 10.3 to 15.4 percent, and 33.3 to 36.7 percent, respectively. While the absolute volume of unclassified or odd jobs has increased by almost 270 percent, its proportional importance has risen by 59 percent, accounting for 4.3 percent (1960) and 7.3 percent (1976) of all employed persons age 15 and older. Meanwhile, other occupational categories, such as managers and administrators and sales personnel, have recorded a decline in their relative importance, despite an increase in their absolute share. Agriculture is the only occupational category that has been losing both ways, showing a 53 percent decline between 1960 and 1976, with the greatest decline taking place in the South.

Distribution of various occupations within the five subcities shows an increase across the board in the proportional share of professionals—the product of spreading educational opportunities to all citizens. At the same time, service occupations declined in all five subcities. The importance of production workers has increased only in the southern city. Coupled with the sectoral trends discussed earlier, and the increasing share of the southern city's population, it is clear that incorporation of the southern city into the city's economy was a major accomplishment of the 1960-76 period. See table 8.4.

Differences within the various subcities could not be revealed, however, at this aggregate level. As it stands, Cairo appears to be divided into internally homogeneous areas. The district of Misr al-Jadidah where an incipient upper-middle class and native elite is segregating itself is totally different from Bulaq, a predominantly working class residential area, or al-Wayli, or even al-Zaytun which falls across the major transportation axis linking Misr al-Jadidah to the city and containing a less exclusive and pretentious indigenous population (see Abu-Lughod, 1971). The occupational structure of the above-mentioned districts in 1960 and 1976 shows that professionals have been more than able to double their relative importance in all four districts. However, coupled with managers and administrators—the two higher categories among white-collar jobs—they explained 30.3 percent of all employees residing in Misr al-Jadidah in 1960, and had increased their share to 46.3 percent by 1976. At the same time the share of blue-collar workers, while experiencing a 6.5-fold increase, declined proportionately from 13.0 to 8.1 in the corresponding years.

In direct contrast, in the district of Bulaq, blue-collar workers accounted for 50.0 percent (1960) and 47.3 percent (1976) of all occupations held by district residents. However, while the district has lost 12.2 percent of its population between 1960 and 1976, it only lost 8.7 percent of those who held blue-collar jobs, which means that they were probably less prone to move than other wage earners despite the decline in their relative importance. Meanwhile, the top two categories of white-collar jobs accounted for 5 percent (1960) and 6.9 percent (1976). Proportionately, the only occupational category that proved to be gaining was clerical workers, increasing their relative share from 5.6

TABLE 8.4

Percentage Distribution of Working Population by Occupation—Cairo, 1960-1976

| Occupation | North | | East | | Center | | West | | South | | Cairo | |
|---|---|---|---|---|---|---|---|---|---|---|---|---|
| | 1960 | 1976 | 1960 | 1976 | 1960 | 1976 | 1960 | 1976 | 1960 | 1976 | 1960 | 1976 |
| Professionals | 8.3 | 18.0 | 4.4 | 8.1 | 8.0 | 15.4 | 13.8 | 23.1 | 7.4 | 12.4 | 7.8 | 14.1 |
| Managers, administrators | 4.0 | 3.3 | 1.6 | .9 | 4.3 | 2.8 | 8.7 | 7.7 | 3.3 | 1.5 | 3.7 | 2.5 |
| Clerical workers | 10.7 | 16.0 | 7.5 | 11.6 | 12.7 | 16.1 | 11.5 | 13.0 | 6.8 | 10.7 | 10.3 | 15.4 |
| Sales persons | 13.0 | 9.8 | 18.2 | 14.1 | 15.7 | 13.5 | 12.9 | 12.0 | 10.1 | 6.0 | 13.9 | 10.1 |
| Service workers | 18.9 | 13.5 | 15.4 | 11.5 | 20.4 | 11.4 | 29.4 | 17.5 | 16.7 | 11.2 | 18.4 | 12.9 |
| Agricultural workers | 2.5 | .3 | .4 | 1.1 | .5 | .6 | 1.0 | 1.7 | 8.1 | 1.3 | 2.1 | 1.0 |
| Production, transport, mining, and quarrying workers | 38.0 | 31.3 | 48.1 | 45.0 | 34.3 | 31.9 | 19.4 | 16.4 | 43.7 | 49.8 | 33.3 | 36.7 |
| Unclassified | 4.6 | 7.4 | 3.9 | 7.1 | 4.1 | 8.2 | 3.4 | 8.8 | 3.9 | 6.4 | 4.3 | 7.3 |

SOURCE: CAPMAS official census results (1960, 1976).
NOTE: Columns sum to 100 percent.

245

percent in 1960 to 9.5 percent in 1976, about a 59 percent increase. The districts of al-Zaytun and al-Wayli exhibit a striking similarity. Despite a decline in the proportional share of production workers in both districts, they still account for almost one-third of all positions held by residents of the two districts. Top white-collar jobs accounted for about one-fifth of all occupations in 1976, about a twofold increase from the proportional share they exhibited in 1960. The two districts also have shown an increase in clerical workers: in al-Zaytun they accounted for 11.3 percent in 1960 and 18.2 percent in 1976; in al-Wayli they increased by 60.6 percent in the same period.

Variations exist not only between districts, but also within them. The district of Ma'adi is a case in point. Inherited by the Egyptian elite and upper-middle class from the excolonial powers, the original suburb, divided into two shyakhas, is at odds with the rest of the district.[9]

In 1960, when average room density for the entire district was 2.2 persons, it was 1.1 person per room in the suburb of Ma'adi. At the same time, density per room in the shyakhat of Dar es Salam was 2.6, 2.5 in Turah al-Haggarah, and 2.75 in al-Basatin.

The change in occupational structure within the district from 1960 to 1976 reveals some interesting features which might help to clarify the picture. Like Misr al-Jadidah, in the suburb on Ma'adi, the two uppermost white-collar occupational categories accounted for 45.3 percent and 44.9 percent of all occupations held by their residents. By 1976, and despite the decline of their relative share of total white-collar upper categories employment in the district, the relative importance of these two uppermost categories—professionals and managers—within their respective shyakhas rose to 59.1 and 53.3 percent.

By contrast, production workers who represented 9.6 percent in 1960 of all gainfully employed residents in the two shyakhas, were only able to boost their share of residents to 11.3 percent in the suburb, registering an average annual growth rate of 1.7 percent. Their absolute numbers remained small—176 in 1960 and 224 in 1976. For all the mumbo-jumbo of revolutionary ideology in the 1960s, it appears that the Egyptian bourgeoisie has been successful in insulating its residential quarters from being penetrated by the working class. More boldly stated, what we have here is an economic apartheid with bourgeois enclaves scattered here and there, surrounded by a large slum. A similar trend has already been noted in Misr al Jadidah; it also exists in Qasr al Nil, al-Rodah Island as opposed to the rest of Misr al Qadimah district, and al-Monira as opposed to the district of Sayidah Zaynab, though to a lesser extent than the others. Across the Nile in Giza, one might refer to al Muhandiseen, Mena house in al-Ahram district, the Giza Nile coast, Dokki, and Ajouza as variations on the same theme.[10]

## Income Distribution Pattern in Cairo

Despite the absence of solid empirical data on income distribution in Cairo City, it is suspected that its distribution pattern is more highly skewed than in the rest of the country. Saad Ibrahim (1982b), capitalizing on his survey of 634 households on income distribution in Cairo in 1979, divided Cairo society into six stratification configurations, using data on income, education, occupations, and life-style. From top to bottom these categories are the rich (1 percent), the secure (15 percent), the upwardly mobile (36 percent), the borderline (27 percent), the poor, and the destitute. He estimates that 21 percent of Cairo's population falls into the lowest two categories. Ibrahim suggests that mobility tended to be smaller in the lower income categories and greater at the middle, and that in the lowest four income categories movement tended to be mostly one step in either direction. He also stresses that by the late 1960s class structure appeared to have been crystalizing again.[11]

An earlier survey in 1971 revealed that 75 percent of families in Sayidah Zaynab earned a monthly income less than LE 20 per month, and only 4 percent earned over LE 50. In direct contrast, a similar sample in Misr al-Jadidah revealed that 60 percent of families earned over LE 50 per month and 27 percent over LE 100. Together Misr al-Jadidah, Qasr al Nil, and some shyakhas of the suburb of Ma'adi stand at odds with the rest of Cairo in terms of income as Sayidah Zaynab stands in harmony (Al-Neklawi, 1973:235).

A 1972 survey in Bulaq Dakrur found that 90 percent of the households earned less than LE 400 per year; the average for Egypt's cities is 56 percent. Using Ibrahim's estimate of 21 percent living at or beneath the poverty line as an absolute minimum benchmark, Waterbury (1982:329-31) estimates that something like 40 percent of Greater Cairo's households can be considered in an income range similar to that of Bulaq Dakrur.

How do the poor get by in a city like Cairo? Unni Wikan (1980), who has done anthropological fieldwork in a Cairo slum, identified a host of mechanisms. The standard solutions for the fundamental discrepancy between income and expectation from life include: reducing outlay for food, postponing consumption and payments as long as possible, borrowing money, selling belongings, buying the cheapest kinds of goods, practicing hired purchase to a large extent, establishing saving clubs with neighbors, avoiding involvement in occasions calling for material expenditure, wives working outside the home, and husbands usually holding two or probably three jobs at the same time. Despite these heroic efforts, Wikan finds most of the families she studied ending with a negative balance at the end of each month, and that amounts allocated for food are hardly sufficient to keep hunger at bay.

## Dilemmas of a Metropolis

Cairo is presently caught in a multitude of urban problems, all interrelated in a vicious circle. Rapid population growth, inadequate employment opportunities, low incomes, and lack of adequate housing are linked together in a depressing but indivisible chain of causation (Abu-Lughod, 1971:163). We have already alluded, implicitly or explicitly, to increasing regional and socioeconomic inequalities in Cairo.

The city is increasingly suffering from the burden of supporting a population steadily growing at a rate faster than it can cope with at any level, either in terms of its structure or services it can provide. It is customary to hear that Cairo is a city designed for 1.5 million people—a figure surpassed in the early 1940s—but estimates of the current population of greater Cairo range from 9 to 12 million. It is roughly estimated that an already supersaturated Cairo must deal with some 800 additional residents each day (Conservation Foundation, 1983).

## Housing Problems

In this urban dilemma, housing looms as a major concern. In 1947, Cairo's 2,090,064 residents were housed at an average of 4.6 per family in 448,333 dwelling units with a total of 1,039,742 rooms—an average of 2 persons per room. By 1960, the population of the newly delimited metropolitan Cairo was 3,348,779, housed in 687,858 dwelling units, with 1,439,158 rooms, and a room density of 2.3 (Abu-Lughod, 1971:164). John Waterbury (1978) has estimated that average room densities rose from 2.3 per room to 3.1 between 1960 and 1972. He (1982:337) later suggested the figure of 3.32 for 1976. An average, however, is never true and because of severe inequalities in distribution of housing space in Cairo, the detailed picture is really grim. Almost half the occupied dwelling units in Cairo in 1960 consisted of only one room and were not necessarily occupied by small families. The median size of households occupying a one-room unit was in excess of four, and families with as many as ten persons or more were found in units of this size (Abu-Lughod, 1971). Bab al-Shary'ia occupancy in 1975 was six to a room, and often a dozen or so villagers shared a tiny cubicle and slept in shifts (*The Economist*, 1975).

Another question of great relevance to the housing crisis is whether new additions to the Cairo housing stock are being made at rates sufficient to replace units lost through demolition. A graphic answer to this question comes from Hamdi Ashur, ex-governor of Cairo, who stated that about 400 thousand Cairo dwellings are structurally unsound, and about 40 percent are on the verge of total collapse (Abdel-Fadil, 1980:128-29). Cairo Governorate estimates that the minimum the city needs is an additional 56 thousand housing units per year to cover population increase and migration (39 thousand units), emergency cases such as total collapse of old buildings (16 thousand units), and to alleviate density (a thousand units). To keep the crisis at its level and assuming an ultraconservative average cost of LE 1,000 to construct a single unit of low-cost housing, would mean LE 50 million per year at 1975 prices, or 750 million from now until the year 2000. This figure would be nearly the equivalent of 37.5 percent of all public and private investment in the high-priority industrial sector between 1952 and 1975. In Cairo the private sector has been building 10 thousand units a year, while the public sector provides only 3,375 units—well below the minimum required.

With such a severe housing shortage, where do all these people live? In addition to increasing room density, the most prominent building phenomenon in Cairo is informal housing which covers a wide range of types. Not only are there a big number of squatter settlements in different parts of Cairo, but informal housing also extends to include other forms of unlicensed construction. In fact, informal housing has been estimated as constituting 80 percent of the buildings in the last few years. Most notorious was a building in al Dokki, a middle-class residential district, built by Hassan Bayyumi, which collapsed a few days after it was finished in December 1974. Bayyumi had been authorized by city engineers to build a six-story structure; he had exceeded his authorization by five stories.

Bayyumi was typical of private sector building entrepreneurs. He violated the building permit and used no steel reinforcing at all. The incident provoked a flurry of public protest and official indignation. The leftist editor of *al-Tali'a*, Lutfi al-Kholi (1975), borrowed Bayyumi's name to characterize a whole group of speculators, exploiters, and profiteers who trafficked in basic needs. He dubbed them "al-Bayyumiyun" and had in mind more than private housing entrepreneurs (Waterbury, 1983:183-87).

Another form of informal housing for the poor is in Dar es Salam. The district was originally a rich, fertile land, ideal for agriculture, and as late as the 1960s was characterized by a pure rural life-style. Since then, it has undergone a significant change in ecological structure and socioeconomic profile.

The boom in urbanization of Dar es Salam started in the 1970s following the increase in labor migration to the Arab states. Since it provided a good target for labor migrants able to save money and come back to Cairo looking for residence, Dar es Salam served as an ideal location for investment in building new houses, being close to Cairo with cheaper land, compared to other districts inside the city. Consequently, unlicensed houses were built in the area and several squatter quarters also emerged. That low-cost housing has not been able to equal the rising demand in this respect has aggravated the problem of informal housing, especially on Cairo's fringes. Living and sanitation conditions are understandably very poor. Dwellings are constructed of whatever materials people find available; some are built for temporary residence, others constitute permanent dwelling units.

One squatter area with legal status is in Zainhom, which lies in proximity to public housing and shelters a big number of homeless families whose houses have collapsed. The dwelling units are built of different materials and meant to be temporary, yet some families have been living in the area for as long as 20 years (El-Safty, 1984).

Another form of clandestine housing is Cairo's so-called Second City of rooftop dwellers who build huts, shacks, poultry runs, and so forth on the roofs of buildings of Cairo's poor districts. If there were one such housing unit for half of Cairo's buildings—about 110 thousand units with five people to a family—well over half a million Cairenes might be living in the Second City (see Stewart, 1965:23).

Yet, Cairo's unique contribution to urban housing is the City of the Dead; Desmond Stewart (1965:25) provided this colorful description:

It is a lion-coloured, unlofty city, with streets like streets elsewhere, with the houses numbered, as though the postman might bring letters by the first delivery. But if he bangs the door and, no one opening, pushes in, he enters the parody of an ordinary house—two adjoining rooms, dust-carpeted, in each an oblong shape of stone or plaster. Under one floor lie the male members of the family; segregated in death in the adjoining cellar are the women.

This cemetery constitutes 2.5 percent of the total inhabited area of Cairo, while housing areas represent nearly 34.5 percent. No one knows exactly how many Cairenes are squatters there, but one authoritative source, Cairo's ex-governor Hamdi Ashur, has quoted a figure of one million (Abdel-Fadil, 1980:129).

Normal functioning communities are found in the City of the Dead. Services are even provided there: shops, mosques, and street vendors. Markets are held regularly. A network of renting and subletting operates to provide residence for those who seek it there. Legal status is at stake, but the inhabitants continue to live in peace in the City of the Dead. The government tolerates the technically illegal population because it has no other housing to offer. It has supplied the City of the Dead with running water and electricity but no sewage system (Copetas, 1981). A police station, medical clinic, and post office have been set up, along with four public schools that operate three shifts a day. One teenager reported there were 74 children in his class. Some people have been living among the dead for two or three generations (Wren, 1978).

The necropolis seems to include a highly heterogeneous population separating those who have outside jobs from the wretchedly poor. Some of the tombs contain television sets, stereo phonographs, and even video players. Attitudes differ sharply. One resident who works for the Ministry of Religious Endowments says, "You have two or three rooms with a garden, electricity, and water, and what's wrong with that? If you also have a dead man under you, so what?" Others voice their discontent: "The government builds great hotels for the tourists and offices for their business and forgets about us." They still, however, consider themselves lucky when they compare their condition with the lot of others.

Housing shortage is a problem at all income levels in urban Egypt, but is most felt among the poor. Since the poor cannot pay enough to make a lucrative return for private investment, the government has to help. Direct government intervention in this area appeared only after 1952. Before the 1952 revolution, only one publicly subsidized low-rent project was constructed in Cairo, the Workers City in Imbabah consisting of 1,100 dwelling units (Abu-Lughod, 1971:166). Since then, construction of "popular dwellings" (masakin) has become essential in government housing policy. The principle new cities of popular housing in Cairo are Zeinhum and Ain-el-Sirra, slum clearance projects in Bulaq, Rod el Farrag, Sahil, Mataryiah, and Helwan.

The Central Agency for Public Mobilization and Statistics (CAPMAS) has ascertained that the public sector constructed about 92 thousand low-cost housing units in urban areas over the 1960/61-1969/70 period (Abdel-Fadil, 1980:131). Yet, the gap between the urgent needs of urban lower classes and the state's ability to provide low-cost housing constitutes one of Egypt's major crises. Cairo suffers no more than any other city, but the magnitude of its housing deficit is growing with the city. The crisis is more acute in the two industrial suburbs of Helwan and Shubra al-Khaymah. In 1969, the director of the Greater Cairo Planning Commission estimated that Helwan would face a deficit of 130 thousand housing units by 1975 (Abdel-Fadil, 1980; Waterbury, 1978).

## Transportation Problems

Judging by the frequency it appears in the newspapers and on equal footing with the housing problem is Cairo's transportation problem. Cairo may be a quasiurbanized village, or ruralized city, but it manifests at least three characteristics of the urban environment: great physical mobility among its inhabitants, relatively vast movement of goods, and extremely high population densities in residential districts (Waterbury, 1978). At least half the city's working population has to cover substantial distances to get to work. In addition, about a million people move in and out of the city daily. Thus, public transportation is absolutely critical to Cairo's economy. The number of passengers carried by public transportation has been growing for over a decade by 10-15 percent each year while available means of public transport have scarcely grown (Waterbury, 1982:332).

In 1978, the system handled four million passengers per day. By 1990, the system will likely need to accommodate 13 million passengers per day (Waterbury, 1982:332). During rush hours (7 a.m. to 8 p.m.), Cairo buses are crammed beyond theoretical capacity, so passengers cling to the outside or sit on the roof or in windows. The city's traffic chaos reduces the speed of the crowded buses, so they never stop. Getting on and off is an art, a sport, and proof of one's spirit. Boarding is to take a running leap and clutch at whatever one can, window or door openings, bumpers, or fellow passengers, who often extend a helping hand. Getting off is the real challenge, requiring what one connoisseur calls the "flying dismount," which is to leap from the bus and run furiously to compensate for the loss of forward momentum, spinning, dodging, and darting around oncoming traffic and would-be passengers swarming to get onto the bus just escaped.[12]

With fares fixed and unchanged since 1953, greater Cairo Transportation Authority was operating in 1979 at a deficit of LE 26 million, and interest on its outstanding debt amounted to LE 6 million. Yet, Cairo's transport system receives more public investment annually than the rest of the country combined (Waterbury, 1982:332).

A fundamental question of equity reflects the conflicting interests of the employed urban mass and the better-off professional and administrative strata. While some people advocate placing maximum financial emphasis on developing sound public transportation at the expense of private convenience, the physical planning of Cairo's transportation grid is being substantially determined by the 150 thousand automobile owners. Financial resources are being diverted from public transport and the millions who use it daily. For example, a major expenditure involves a number of overpasses, elevated roadways, ring-roads, and Nile bridges to move, for the most part, private automobiles from middle-class suburbs in Helipolis, Giza, and al Ma'adi to the central business district and to one another. Cairo is being ripped apart to cater to its growing private vehicle fleet. Between 1972 and 1977, the number of private autos increased from 80,559 to 133,599, about an 11 percent annual growth rate. By 1990, it is expected to grow to 400 thousand (Waterbury, 1982:332-34). To handle this growth, Cairo would have to increase its road surface by 300 percent (El-Baradei, 1985). Meanwhile, Cairo's public bus fleet declined at an average annual rate of 6.3 percent from 2,190 in 1972 to 1,363 in 1977. See table 8.5.

The only benefit for public transportation of the new roads—in addition to using them—is in diverting private automobile traffic from congested arteries connecting Cairo's most densely populated districts. On the other hand— noting one study—these roads will strengthen the enclave existence of the well-to-do, joining their place of work and residence by private cars, and concealing the decaying popular quarters that cause much unease to the middle class (Waterbury, 1982:334). Even official government studies doubt the wisdom of such road construction. "There will be an exceedingly low economic return on these investments . . . It is presumed that they are being carried out for other than economic development purposes" (Egypt, Ministry of Transport, 1977).

One major development in favor of Cairo's masses is construction of the subway which has the advantage of carrying at least three times as many passengers per hour as any other means of transportation. For if Cairo grows to 14 million by 1990, it would be inconceivable to expand the roads to handle the projected load of 13 million passenger miles per day. Yet, intense criticisms were raised about the utility of a subway for Cairo by several sophisticated white-collar engineering professors.[13]

We referred earlier to Cairo's elongated growth along the north/south axis and the move of the city's center westward as a major ecological problem. All traffic going in either direction (east/west or north/south and vice versa) has to go through the downtown area, creating several points of choked traffic. Yet this elongated shape could prove positive. As of 1978 only 12 Nile ferries were operating in Cairo. Clearly the Nile surface in a city that extends 34 kilometers in length and at the most 7 kilometers in width is underused. The Nile could provide an uninterrupted path of transportation, with buses for the most part running horizontally as opposed to the current vertical planning of transportation routes, and moving people to ferry stops along the Nile shores. Another possible alternative is to encourage bicycles instead of private cars or buses, especially for short-distance commuting of less than three kilometers. These alternatives should prove useful by reducing air pollution in an already

TABLE 8.5
Motor Vehicle Fleet—Greater Cairo, 1972-1977
(in thousands)

| Year | Private Autos | Buses | | Taxis | Total |
| | | Private | Public | | |
|---|---|---|---|---|---|
| 1972 | 80.6 | 1.1 | 2.2 | 14.5 | 98.3 |
| 1973 | 83.0 | 1.2 | 2.0 | 15.6 | 101.8 |
| 1974 | 87.4 | 1.4 | 1.6 | 17.7 | 108.1 |
| 1975 | 94.6 | 1.7 | 1.5 | 28.3 | 126.0 |
| 1976 | 109.5 | 2.2 | 1.4 | 30.2 | 143.4 |
| 1977 | 133.6 | 2.8 | 1.3 | 29.4 | 167.2 |

SOURCE: Waterbury (1982); based on unpublished CAPMAS data from 1978.
NOTE: Excludes the large number of private vehicles operating on temporary, import, and customs plates.

overpolluted city, and by having a positive impact on the general health of the population.

## Health Problems

Three major factors influencing the health of human settlements are environmental hazards, communicable diseases, and inadequacy of health care services. Air pollution is a major concern. The prevailing winds in Cairo blow north or south. One brings toxic fumes from the lead and zinc smelters in Shubra al-Khaymah; when the winds shift, they bring poisonous pollutants from the steel and cement factories in the south in Helwan, notable for its myriad of dead trees. The main cause of Cairo's air pollution, though, is its traffic jams, caused not only by swarms of cars, but also by the 80 thousand animal carts that clog the narrow streets and create their own pollutants. Carbon monoxide levels in some areas are three or four times what is considered dangerous (Friedrich, 1984).

Cairo has no publicly organized domestic solid waste disposal system. Cairo's solid household waste is almost entirely digested within the city in a unique manner contrary to the usual one-way flow of materials and with little cost to the public. Three main types of waste—public area waste, bulk waste, and domestic waste—must be distinguished in Cairo. Only public areas of the city are serviced by municipal waste collection operations carried out by regular, salaried employees. Organic waste is taken to a city-owned compost factory that produces fertilizers. Bulky household waste, such as broken furniture or unwanted papers and magazines, is sold to collectors wandering Cairo's quarters who are called *rubabikya* dealers—a corrupted version of the Italian word for old clothes.

Household or domestic waste is collected by the Zabaline (see Haynes and El Hakim, 1979:102 for an outstanding description and analysis of the internal structure and organization of this craft). From the refuse they extract such items as tin, glass, paper, plastic, rags, and bones, on which hundreds of craft workshops and factories within the city depend for raw materials. Organic matter from waste food in the refuse is fed to the pigs the Zabaline breed for the meat and sausage market. What the pigs do not eat is turned into compost and sold for agricultural purposes (Haynes and El-Hakim, 1979).

This system employs 40 thousand households in Cairo. While the system is economically and ecologically sound, it is not without its human cost. Only 40 percent of all children born to the Zabaline survive the first year (Consevation Foundation, 1983:7). A new technology should not address the entire waste-handling system but only those aspects that are undesirable—the human cost, the temporary and inadequate shelter, and the poor living conditions. For example, one could start with a simple, easy-to-maintain alternative to donkey power, such as a motorized cart (Conservation Foundation, 1983:4).

Unfortunately, the Zabaline do not serve poor areas of Cairo; accordingly, large quantities of waste are thrown into the streets, because there is no alternative. Once in the streets, these wastes become the responsibility of the municipal sanitation department. Removal of household solid waste from the street is needlessly expensive and saddles an already overburdened public agency with this additional task.

For sewage and wastewater, Cairo is a municipal nightmare. The infrastructure system is totally undersized and totally broken down, simultaneously. In 1964 Cairo was threatened by overflow of the sewage system; various streets and basements of houses and other establishments were flooded by sewage water. An urgent project to deal with the situation was adopted to raise the capacity of the principal stations. Despite these measures, the vast increase of water consumption has caused flooding in various areas, and the sewage system has become a source of danger.

The total amount of waste water which the principal stations can handle is insufficient in terms of Cairo's water consumption which increases annually by a rate of 9 to 10 percent. The three major sewage networks of greater Cairo are currently functioning well beyond their means and are incapable of handling present flows. As a result, sewage washes into the streets. This is common enough in Cairo's poorer sections, but many of the city's affluent districts were suddenly introduced to the problem in December 1982. Two sewage mains collapsed right outside a pumping station, leaving much of Dokki and other upper-middle class neighborhoods flooded with sewage for two weeks.

Sewers are frequently misused. Solid wastes are shoved into manholes and plug up sewers. An inspection and cleanup project funded by the U.S. Agency for International Development (USAID) discovered that 90 percent of the pipes were clogged with logs, garbage, street litter, and even the cooking oil most Egyptian women use—it is very heavy and congeals like cement in the pipes. Human wastes from vaults in unsewered areas also are dumped into these inadequate sewers (see Conservation Foundation, 1983).

With the prevalence of human waste and other environmental deficiencies,

Egypt is plagued by countless illnesses. In addition to bacterial infections, viruses, and parasitic infections including typhoid, salmonellosis, amoebic dysentery, hepatitis, and poliomyelitis, there are serious problems associated with malnutrition. In 1982 infants and children (up to age four) accounted for roughly half of all deaths in Egypt. An estimated 104 of every 1,000 infants die before their first birthday; the mortality rate from ages one to four is believed to be 16 per 1,000 (World Bank, 1983).

Because of the fragility of the entire water network, water pressure levels are kept low to avoid ruptures and breaks within the system. Reservoirs are rarely filled to capacity causing inadequate and sporadic water service to many areas of Cairo.

In recent years, revenue from water tariffs has failed to cover even basic operating expenses, let alone generate necessary capital for future waste projects. Water tariffs, set at 50 percent below cost, are among the lowest in the world. As a result, the ability of sector organizations to provide quality service is frequently severely constrained. The government subsidy policy, designed to benefit lower-income people, has in reality had the opposite effects. Higher-income families are the beneficiaries of the low rate structures for piped water and sewage. We do not advocate abolishing subsidized water and sewers, but the system is unfair, and other measures are required to guarantee that such a subsidy will go to those who deserve it.

## Financial Problems

Cairo's budget in the last 30 years has confirmed three basic characteristics: (1) local revenues account annually for only a quarter to one-third of all city resources; (2) the budget as a whole is devoted overwhelmingly to current operational expenses with little left for investment; (3) the total size of the budget has been growing at best slowly and sometimes not at all (Waterbury, 1978:134).

In 1963 the city's (Cairo proper) approved budget stood at LE 29 million and rose gradually over a decade to LE 39 million; by 1977 it had reached 59 million (Waterbury, 1978:134; 1982:317)—an average growth rate of about 7.4 percent annually. Yet this growth was not evenly distributed over time. Between 1963 and 1973 the annual growth rate was 3.4 percent. If we accept a minimum of 2 percent inflation per year, then the budget was growing at a mere 1.4 percent annually during this period, significantly behind the population growth rate. Meanwhile, the budget grew 12.8 percent annually between 1973 and 1977. However, because of population increase and inflation which ran much higher during the second period—10 percent would be a conservative estimate—the actual growth rate was much slower. These figures work out to annual outlay of $18 per capita during the first period and $11.6 during the second. In reality, then, the budget has decreased while problems have magnified. This situation took place at a time when Cairo's budget allocation represented 13.7 percent of all recurrent outlays.

## Clash of Old and New

The old city of Cairo is having its full share of the malaise of an overcrowded city with an exhausted infrastructure. The cultural significance of historic Cairo has been accepted internationally. The old city was included in the World Heritage List by the World Heritage Convention in 1979. Consequently, historic Cairo ranks with the main Pharaonic monuments of Egypt and others of international significance in the world.

Although, traditionally, strong emphasis has been given in Egypt to the wealth of Pharaonic heritage, there is now a growing interest by many individuals and organizations in Islamic and Coptic heritage. However, attitudes of people who live and work in the old city toward the heritage around them are more difficult and complex to assess. To some, Cairo has an outdated urban fabric, but to others it is a precious and significant heritage.

In response to a request from the Egyptian government, UNESCO undertook in February 1980 a mission to prepare a report on a conservation strategy for the old city of Cairo. The study area, with an overall population of some 320 thousand (1976) consists of approximately 3.7 square kilometers. Within this area are 450 historic buildings of a total of 620 for the whole of Cairo. From 1966 to 1976, the residential population of the study area declined by some 8.6 percent (about 30 thousand people) in a period when Cairo as a whole expanded by about 3 percent per year.

Loss of residential population can be directly related to loss of dwellings in the area. A number of factors responsible include the impact of rent control leading to lack of maintenance and eventual destruction of dwellings; pressures from commercial interests seeking expansion space; voluntary building demolition by occupants to gain rights to new accommodations elsewhere; deterioration of the building fabric arising from general decay and rising ground water leading to building collapse; inadequate maintenance and use of inappropriate building technology.

The traditional pattern whereby craftsmen, merchants, and workers lived and worked in integrated communities is being rapidly replaced by a new pattern of bulk storage and manufacturing establishments which does not fit well into the old urban fabric. If these new developments are not checked, they will result in all the life being choked out of the old city over the next decade, with the irreparable loss of many crafts and skills, as well as the extinction of a way of life which might well survive and flourish in the future if only it received encouragement.

Historic monuments were well maintained until about 30 years ago, but the rising water table, combined with serious lack of maintenance, particularly of roofs, has led to rapid deterioration of the masonry and wooden roofs leading in some cases to total collapse of buildings. Many of the buildings are large, some enormous, so that repairing them is expensive.

The type of new housing and school construction now being built is usually alien to the traditional urban form of the old city, insensitive to its conditions, and interferes with the finely balanced pattern of traditional urban development (Lwecock, 1984).

Although the area has an extensive network of services (water supply, electricity, and drainage) for nearly all premises, these networks are overloaded and need maintenance. There is an accumulation of rubbish throughout the old city, with large rubbish dumps on any unused land, sometimes even around the mosques. The plan to build a massive underground sewer at this point, now being implemented, will make matters worse, as it is intended to serve the modern central area only; it will eventually be a dam to push up the high water table in the old city still further.

## Crime

Cairo, despite all the minuses it gets when considering the specifications of a sound and comfortable urban life, enjoys a definite plus when it comes to social conduct, security, and street safety, a fact easily verified by direct experience. Cairo's relative safety is also supported by criminal statistics, although in any country they are not very reliable; "dark figures" or unrecorded offenses, especially with regard to corruption, bribery, and immoral acts, are quite sizeable.

The volume of crimes of violence—murder and blows leading to death or infirmity—is much higher in rural than urban areas (49 percent of the total number of these crimes occur in upper Egypt, and 33.4 percent in lower Egypt, compared with 12.4 in the urban sectors). On the other hand, some types of crime are to a great extent urban types, such as embezzlement, bribery, forgery, and counterfeit.

There is a definite decrease in felonies and serious thefts (misdemeanors) in Cairo. Felonies decreased from 369 in 1973 to 265 in 1983; serious thefts from 7,146 to 3,525. Felonies in Cairo represented 12.4 percent of total felonies in Egypt in 1973, but only 16.3 in 1983. Juvenile delinquency also decreased from 2,884 in 1978 (25.3 percent of Egypt's total) to 2,494 in 1983 (15.4 percent of total). Juvenile vagrancy as well has sharply decreased year after year. From 1,869 cases in 1973 only 507 were recorded in Cairo in 1979 (Egypt, Ministry of Interior, 1984).

What explains the impressive decrease of criminality, juvenile delinquency, and vagrancy in the last ten years? Only juvenile vagrancy could be easily explained in light of the liberalization in this period which revitalized the economy and took care of basic needs of these young vagrants through better employment opportunities. This same liberalization of the economy opened the door wide for a wave of corruption, customs and tax evasions, bribery, and business fraud. Most of these offenses remain within the unrecorded dark figures.

One serious set of offenses seems to be constantly on the increase—drug crimes in line with the situation in many other countries. Drug offenses in Cairo, all felonies, increased from 1,812 in 1973 to 3,112 in 1983.

Cairo's dismal array of problems has been aggravated by its fiscal crisis. Solutions can be found but require large amounts of a scarce resource—money. Yet, spending more money on an already attractive city to save it, would probably make it more attractive to millions of poor peasants in the countryside, and hence enhance its domination and magnify its problems. Without an integrated development plan for the entire country, Cairo will increasingly suffer.

## Bailing the City Out of Its Misery: Can It Be Done?

In dealing with the serious problems of a city like Cairo, it is customary to hear of measures of a general character. Some have argued that Cairo should be declared off limits to any further industrial projects, require all entrants to the city to show proof of employment, limit all rentals to those who can show proof of employment, and decentralize government and public sector bureaucracies.[14] Such measures, one would expect, would be subject to considerable resistance on the part of public and private industries reluctant to sacrifice Cairo's proximity and government bureaucrats who definitely prefer Cairo. Finally, the migrants themselves over whom no one has control and the proof of work requirement would probably provoke a lucrative trade in false documents.

Raising the standard of living through faster growth of the economy would certainly improve living conditions and provide urban life with more amenities.[15] Advocacy of active programs of family planning and population control look better in theory. Even a successful family planning program will have no immediate significant effect in the near future. We now turn to more specific ways to provide help in alleviating urban pathological conditions.

### Rural Development

One important alternative is rural development. The uneven distribution of investment between urban and rural areas has had the effect of concentrating service and job opportunities in the big cities, particularly Cairo. Urban centralization acted as a magnet in drawing people away from the underprivileged countryside. The rural exodus grew considerably, exacerbated population pressure in the cities, and crowded the poor quarters.

Thus, one way of tackling this problem would be for government to pay greater attention to developing rural areas. This strategy would mainly be directed toward dealing with the cause of population movement to the cities. In a survey carried out among young people living in the countryside, 20.7 percent of those questioned said that they intend to emigrate to the towns (particularly Cairo) in search of suitable work (NCSCR, 1982).

Dispersal of the industrial infrastructure throughout the country would provide job opportunities for the unemployed, underemployed, and the skilled labor force, now drawn to large cities. Similarly, equalizing access to fundamental services, health care, good communication networks, and a movement for decentralization of technical and higher education—already underway—in all governorates under a plan that ensures equitable distribution of educational services would make it unnecessary for students to go elsewhere in search of higher or specialized education. Coupled with the dispersal of industry, this plan would consequently create job opportunities for new graduates in their home towns, thereby creating networks of services and employment away from the capital.

Finally, decentralizing administration, according more power for decision making to local authorities, should at least reduce the daily entry of people to Cairo, helping to release pressure on its services, especially transportation.

Similarly, promoting traditional craft and agricultural industries could lead to creation and consolidation of productive centers based primarily on human and material resources of the region. Small family enterprises of this kind would provide employment for men and women, young people, and children. This plan would revive industries of great economic and artistic value and provide an important source of income for villagers, whose agricultural work is very often seasonal. Success of projects of this kind in various parts of the country is clear proof of the profit to be derived by rural areas from craft industries.[16]

Thus, greater attention to rural areas and their development could curb the movement of population toward the capital, thereby alleviating the problems involved.

## Urban Development

The second step to help alleviate Cairo's problems is upgrading its urban physique. During the last few decades, the infrastructure has deteriorated steadily in Cairo, as described earlier. These physical conditions produce a quality of life not conducive to healthy development. In fact, the impoverished quality of life in all its aspects reinforces existing low standards of living. Besides being dysfunctional to any process of development, other serious negative effects of these physical conditions have detrimental effects on the socialization process of the new generation.

To do something about the deteriorating conditions in towns, which are increasingly becoming reservoirs of misery and pollution, three approaches are possible. The first, involving action by government organizations, would in theory seem the most radical and effective solution. In practice, however, projects of this kind come up against an awkward problem: the vicious circle of poverty, the passiveness of individuals inured to hardship by long suffering, their skepticism toward authorities and their promises. It is also very important to bear in mind that, while government action can be effective in providing basic facilities that require huge financial and technical resources, such action is fruitless unless it is continually kept going and followed up. The importance of support by the community concerned for official efforts is thus obvious.

The government of Egypt has recently launched an integrated program to upgrade its urban services. The program includes studies of the current water, wastewater, and solid waste collection systems operating in Cairo. Basic studies have greatly enhanced the governorate of Cairo's ability to make long-term policy decisions and organize new frameworks for improved urban service delivery in the city (Niemat-Allah, 1984).

Through many detailed analyses of existing systems, Egypt, in cooperation with several international agencies and firms, has designed two master plans to rehabilitate, improve, and expand Cairo's water and sewer system. A drastic program to provide piped water and build a sewage system for the whole of Cairo to cover its present needs could well cost over five billion dollars.

A second approach to improving deprived districts involves participation of those living in the areas concerned. Self-reliance in solving their problems would be one of the most effective solutions, provided the task was approached rationally and in a spirit of cooperation that sought to promote the general good of the community by tackling problems on the basis of purely local resources.

The third approach is an intermediate one, and involves joint action by the government and inhabitants of deprived areas with a view to solving their problems. Authorities would supply financial and technical assistance, and the community concerned would provide all possible forms of cooperation.

A number of pilot projects have been designed to alleviate the problems and inefficiencies inherent in the solid waste program. The pilot program for extending waste collection into presently unserviced, low-income areas now operates in five low-income communities with a total population of over 150 thousand. Costs are borne entirely by beneficiaries of the system.

## Deconcentration and Relocation

Finally, the third strategic pillar of solving Cairo's problems includes several deconcentration and relocation programs. Existing intermediate-sized cities are being developed as one way of accelerating decentralization and dispersion. Satellite settlements are created in the vicinity of Cairo to divert migration from the capital. One fears, however, that such cities might run the risk of becoming desolate dormitory cities or perhaps refuges for the better-off, leaving Cairo for the poor and the migrants. The telling example of Nasr City is exactly what should not be done. In addition, independent cities are promoted to absorb the Egyptian population from overcrowded large cities to the desert around the Delta as an alternative to alleviate the main cities, Cairo and Alexandria. Thus far, nine such cities have been proposed, and in 1979 about LE 170 million was budgeted for these new cities; about 3.2 percent (LE 321.6 million) of all investment in the current five-year plan is devoted for them as well.

Sadat City, the most recent of this third category, has been promoted by the government in view of diverting population from the main cities, to attract large heavy industries, mainly iron and steel and middle-sized industries located in Cairo and Alexandria.

The plan calls for relocating gradually to Sadat City the governmental agencies concentrated in the crowded cities and providing a complete range of civic, business, industrial, housing, educational, medical, and recreational facilities. The intention is to develop a balanced community where those relocating there will be able to live.

Location of the new city was chosen sufficiently far enough from Cairo, at about a hundred kilometers on the Cairo/Alexandria desert road. Choice of this location took into consideration availability and quality of ground water, easy access to electric power, and easy communication with other urban centers. Sadat City is planned to provide 165 thousand work opportunities: 60 thousand in the industry, 25 thousand in the main sevice sector, 60 thousand in the secondary service sector, and 20 thousand in construction.

Yet, such cities are not without their own problems. First, the idea of migrating toward the new cities is not very attractive to young people living in the countryside. A recent survey showed that 50.9 percent of those who intend to emigrate from their village reject the idea beforehand, while 46.7 percent find the prospect acceptable (NCSCR, 1982). The cities in question are not centers of attraction because of lack of services, transportation difficulties, and feelings of isolation associated with them. Second, government officials, even those working for the Ministry of Reconstruction, though admitting the importance of such measures and the role of the new cities in relieving pressure on large cities, would prefer a crowded Cairo to life in a provincial backwater.[17] Finally, the cost of relocating industry could prove to be too high and in most instances prohibitive; however, presence of an industrial nucleus around large cities represents an additional advantage to locating new plants where other companies are nearby to supply parts, markets, consultants, and ancillary services. Furthermore, the industrial nucleus is associated with a pool of relatively skilled labor.

It is the role of the government as the largest employer in the country to start establishing factories in such cities which then will attract people to work and live there. Until now, government policies have worked to concentrate people in large urban centers. Government's role in redistributing urban population would be probably put to the test against the task of making such cities a successful experience.

## Cairo of the Year 2000: Is There Hope?

It is becoming quite clear that the problems of Cairo, corresponding to those of Egypt to a greater extent, are not only directly related to demography but are in the final analysis socioeconomic.

Under the best expectation of fertility reduction, the population of Egypt will be around 60 million by the year 2000, and probably as high as 70 million, yet the crucial distinction is not between 60 or 70 million people. Rather, it is between 60 or 70 million nonproductive poor, dependent on the outside world to satisfy their needs, and a society of that many productive individuals who have the ability to develop their human potential to the fullest extent.

The Egyptian population will exhibit the following characteristics:

(1) *Age Structure*. About 50 percent will be below age 15. In addition to those aged 60 and over, they will represent about two-thirds of the total population, or about 40-46 million unproductive individuals in need of services, to be supported by an active age group of 20-23 million.

(2) *Population Density*. Population density will reach 2,000 per square kilometer. Effective development of new settlements requires an average of ten years. To keep density at its present level, new settlements for an average of one million per annum should now be in full swing, which is not the case.

How will the future of Cairo look by the year 2000? It is probably better to distinguish analytically between several aspects involved in this question—the demographic, the socioeconomic, or rather the political economy of the city, and its urban physical planning and development.

Demographically, assuming a current population of greater Cairo of 9.3 million in 1983, the Futures Group (1979?) calculates the number of Cairenes will double by the year 2000 and reach 25 million by 2010, if present rates of fertility and migration continue. Even with attainment of a two-child family average by the year 2000, the population would double to 18 million by 2010. At the most conservative estimates, the greater Cairo population will be around 15 million persons. This estimate will mean that the city, assuming the prevailing current rate of economic activity, would have to provide five million new jobs.

The number of Cairo inhabitants exceeding the available capacity of the major construction area as well as the urban and rural communities surrounding Cairo will then reach 5 million persons, and inner city density will reach an average of 46 thousand inhabitants per square kilometer. This expected population surplus must have a place either inside the province bloc or outside it. To absorb this number, requires constructing 50 thousand *feddans* (1.038 acres = 1 feddan) on the basis of an average density of 100 persons per feddan. Since it has been decided to maintain agricultural land without transgression by construction as much as possible, it would be necessary to resort to desert land to meet the extension required.

The last word in planning the future of Cairo goes to Hassan Fathy, a patrician gadfly attempting to inflict minimal damage on the city's stubborn course toward mediocrity. Fathy wants Cairo to be again what it once was, an organic entity growing through the spontaneous yet harmonized efforts of its citizens and citizen architects. All Cairenes were once involved in the construction and maintenance of their city, and had a perhaps unarticulated grasp of the implicit principles that gave the city its character. The implicit in medieval Cairo was always honored so that improvisations on the basic theme became masterpieces. Without the involvement of Cairenes, and citizen architects, the implicit principles of the city's future harmony will never emerge. Without this involvement, Fathy flatly declares, city planners should stick to cost accounting. The most a planner can do is help organize people's activities and movements within the city so that a maximum sense of community and self-sufficiency is fostered in each subcenter (Waterbury, 1978:143).

Fathy has noted two factors that have complicated the city's expansion. First, the original center of the city has continuously moved westward, leaving behind the center of gravity of the city's population in Azhar, Old Cairo area. Second, the city's northward and southward elongation forces an inordinately heavy flow through the new heart of the city. Fathy's argument is extremely simple. The center of the city can be moved eastward again by developing the Muqattum hills and tomb areas that lie between them and the built-up area of the city (Waterbury, 1978:143).

Indeed, one would hesitate to suggest so daring an operation if the case were not so desperate. No one goes lightly to a plastic surgeon to have his face remolded. . . . To reshape the Muqattum hills is an undertaking like shaping Hymettus—something never meant to be dreamt of unless to save the city's life. Then since it is the only solution, it must be entrusted to artists of surpassing genius. Only a person with the widest and most delicate visual sensibility can ensure that the new skyline will be as beautiful as the old.

Cairo can become a triumph of man's artistry, a work of cooperative art worthy to rank with the world's finest example of town-scape, with Verona. . . . But such a work as this is not to be undertaken lightly. Building a city for 5,000,000 new inhabitants on a 40-year plan establishing the character of Egypt capital for hundreds of years, recreating one of the greatest cities in the world, calls for the most responsible and humble approach (Hassan Fathy as translated in Waterbury, 1978:143).

Fathy's cry has gone, and probably will go, unanswered. The vision of breaking Cairo loose from its plight is obscured by the sheer welter of salvage operations needed to meet the short-term crises—paying for a place to sleep and getting to work—that are the lot of Cairenes' everyday lives.

What kind of urban phenomenon will Cairo represent in the year 2000? A United Nations study (1969:45) summed up the unpredictable future of Third World cities: "The process of urbanization may come to surpass itself, and give rise to geographic and social forms of human settlements to which current vocabulary can no longer be validly applied." In other words, no one knows.

Yet, in a more recent article, Professor Malcolm H. Kerr (1982:17) tried to assess Cairo of the year 2000, based on what is—or is not—happening today. He wrote:

a deepening gulf between a large, increasingly cosmopolitan and affluent business-based upper and upper-middle class, inhabiting a smart, new, well-paved, cleaned-up downtown Cairo with a skyline of hotels and office skyscrapers, and an ever-growing mass of underemployed and disaffected wretches inhabiting one of the world's largest slums, watched carefully by the police and the army.

There are of course other possible scenarios. However, judging by history, Cairo has survived more dramatic crises, and history proved the pessimistic shortsighted. If Cairo can make it successfully, it will provide both a model and a hope for other Third World cities fighting poverty and environmental decay.

# Notes

1. Our adjustments differ from hers in terms of subcity component districts. See Abu-Lughod 1971, chapters 11 and 12.

2. The district of Madinat Nasr, first established in 1972, is composed of the newly established Nasr City and some sections of the tracts of Eastern Za'afaran, and Eastern Sarayat from al-Wayli district and Eastern Montazah from Misr al-Jadidah district. Since much smaller divisions of tracts do not appear in the 1960 census, adjustment could not be made for it.

3. The district of al-Tebeen was annexed in 1969 from al-Sacf province—Giza Governorate—under the jurisdiction of Helwan district. In 1972, it was established as an independent district encompassing five tracts: Northern Tebeen, Southern Tebeen, Tebeen public housing, Steel City, and Hikr al-Tebeen. The 1960 Census tables do not show any of the details except for the then al Tebeen village without any further breakdown. Therefore, attempts to adjust for boundaries and population suffer from the same deficiency as Madinat Nasr.

4. Except for the district of Bulaq, the industrial arsenal during the era of Mohamed Ali, the entire northern city is a product of the 20th century. (See Abu-Lughod, 1971:171.)

5. Figures are quoted after Waterbury (1982), calculated from official census data. (See table 10.1.2, p. 308.)

6. Qasr al Nile was not established as an independent district until the 1960 census. In 1947 it was divided between the district of Abdin (five shyakhas) and Sayidah Zaynab (one shyakha). (See CAPMAS, *Detailed Census Results*, 1960. 1976.)

7. See Waterbury (1982:329), who gives a population figure of 322,480 for the entire district—which he questions.

8. A similar trend took place in the southern and northern fringes of Cairo. (See Fakhouri, 1972.)

9. Two shyakhas are called byth commons and, officially, Ma'adi Al-Saraiyat Al-SharQiyah, and Ma'adi Al-Saraiyat al-Gharbiyah. It is no coincidence that the word *Saraiyat* (singular *Saraiyah*) means "palace" in Arabic. The adjective is intended to distinguish the modern residential quarters from the original village of Ma'adi, which is a slum.

10. This conclusion is confirmed and applicable to other major Egyptian cities as well. Using B. S. Morgan's index of residential differentiation and data from the 1960 and 1976 censuses to analyze patterns of residential distribution by occupation in 12 major Egyptian cities, Abdalah (1983) found that the highest occupational group is quite segregated from the other groups. Separation of the lowest groups from the middle was, however, found to be less marked than was expected in all 12 cities.

11. In terms of annual household income, classes are categorized as follows (in Egyptian pounds): the destitute, less than 300; the poor, 300-500; the borderline, 500-1,000; the upwardly mobile, 1,000-2,000; the secure, 2,000-3,000; the rich, over 5,000. For a more detailed picture, see Ibrahim (1982b).

12. *The Economist*, special correspondent, 1975. For a splendid description of the life and culture of bus riding in Cairo, see Waterbury (1978:147-150).

13. See the heated debate on the issue in *Al-Ahram al-Iqtisadi* (18 July 1983):757; (15 Aug. 1983):761; (29 Aug. 1983):763; (5 Sept. 1983):764; (3 Oct. 1983):768; and (31 Oct. 1983):772.

14. This was the opinion of most members of the Greater Cairo Planning Commission. (See Waterbury, 1978:128.)

15. Evidence drawn from Sri Lanka suggests that reduced economic growth has been conducive to reducing migration to Colombo. But this is a price Egypt cannot afford. See Vining (1985) and National Center for Social and Criminological Research and Institute of Social Studies (1982).

16. Taylor has alluded to the importance of cottage industry on rates of emigration to urban centers and oil-producing countries in the two neighboring villages of Dahshur with cottage textile industry Zawiyyat Dahshur with no such industry. In a field visit to the village of Salamoon Al-Qumash (Daqahliya Governorate, northeast of the Delta, near Port Said) which has traditionally maintained a famous cottage industry, several heads of households operating the industry complained about the increasing difficulty of selling their products because of stiff competition from imported products from Port Said free zone and decreasing returns from their enterprise. They associated this diminishing return with an influx of emigration from the village to the oil-producing countries and urban centers (Mohamed M. Mohiedden, personal field notes, 21 June 1983).

17. This information came from Professor Magdi Hijazi of Cairo University, during a conversation on a research project he was conducting for the Ministry of Construction in the fall of 1983.

# Bibliography

Abdalah, Ahmed A.M.M. 1984. "The Residential Segregation of Occupational Groups in Some Egyptian Cities: 1960-1976." In *Studies in African and Asian Demography: Cairo Demographic Center Annual Seminar, 1983*. CDC Research Monograph Series, no. 12. Cairo: Cairo Demographic Center.

Abdel-Fadil, Mahmoud. 1980. *The Political Economy of Nasserism: A Study in Employment and Income Distribution Policies in Urban Egypt, 1952-72*. Occasional Paper no. 52, University of Cambridge Department of Applied Economics. Cambridge: Cambridge University Press.

Abu-Lughod, Janet L. 1961. "Migrant Adjustments to City Life: The Egyptian Case." *American Journal of Sociology* 67:22-32.

———. 1971. *Cairo: 1001 Years of the City Victorious*. Princeton Studies on the Near East. Princeton, NJ: Princeton University Press.

Al-Kholi, Lutfi. 1975. "These Are the Bayyumiyun." *al-Tali'a* 1:81-85 (in Arabic).

Al-Neklawi, Ahmed. 1973. *Cairo: A Study in Urban Sociology*. Cairo: Dar-al-Nahda al-Arabia (in Arabic).

Arab Republic of Egypt. Central Agency for Public Mobilisation and Statistics. 1978. *Detailed Results of the 1976 Population and Housing Census: Cairo Governorate*. No. 93-15111-1978. Cairo (in Arabic).

———. 1981. *Statistical Yearbook. Arab Republic of Egypt, 1952-1980*. Cairo (in Arabic).

Arab Republic of Egypt. Ministry of Interior. 1984. *Public Security Reports: 1974-1984*. Cairo (in Arabic).

Arab Republic of Egypt. Ministry of Transport/Louis Berger International, Inc. 1977. *Egypt National Transport Study Interim Report*. Vol. 1. Cairo (in Arabic).

ARE. *See* Arab Republic of Egypt.

CAPMAS. *See* Arab Republic of Egypt. Central Agency for Public Mobilisation and Statistics.

Conservation Foundation. 1983. "Cairo: A Third World City in Growing Trouble." *Conservation Foundation Letter. A Monthly Report on Environmental Issues* (Sept.).

Copetas, A. Craig. "Cairo's City of the Dead." *New York Times* (21 Mar. 1981):23.

The Economist, Special Correspondent. 1975. "On the Rooftops, in the Cemeteries." *The Economist* 256 (27 Sept.):59.

El-Baradei, Mohammad Mustafa. 1985. "3 Million and 459 Thousand Cars in the Year 2000 Will Need to Increase the Road Surface Three Times." *Al-Ahram* (12 Jan): 13.

El-Khorazaty, Mohammed Nabil Ezzat. 1984. "Lessons of 'What-If' Thirty Years Ago in Population Policy." *Population Studies* (Cairo) 11(69):3-18.

El-Safty, Madiha. 1984. "Housing in Cairo: The Social Perspective." Paper presented to the Conference of the Aga Khan Award for Architecture, National Center for Social and Criminological Research, Cairo, November (in Arabic).

Fakhouri, Hani. 1972. *Kafr el-Elow: An Egyptian Village in Transition*. New York: Holt, Rinehart & Winston.

Friedrich, Otto. 1984. "And if Mexico City Seems Bad . . . A Trio of Third World Megacities Face Similar Problems." *Time* 124(6 Aug.):36+.

The Futures Group. 1979? *Egypt: The Effects of Population Factors on Social and Economic Development*. Cairo: Egyptian State Information Service.

Haynes, Kingsley E., and Sherif M. El-Hakim. 1979. "Appropriate Technology and Public Policy: The Urban Waste Management System in Cairo." *Geographical Review* 69:100-08.

Hijazi, Izzat. 1971. *Cairo: A Study of an Urban Phenomenon*. Cairo: National Center for Social and Criminological Research (in Arabic).

Ibrahim, Saad Eddin. 1982a. *Internal Migration in Egypt: A Critical Review*. Research Monographs Series no. 5. Cairo: Research Office, Population and Family Planning Board.

———. 1982b. "Social Mobility and Income Distribution in Egypt, 1952-1977." In *The Political Economy of Income Distribution in Egypt*, edited by Gouda Abdel-Khalek and Robert Tignor. The Political Economy of Income Distribution in Developing Countries, vol. 3. New York: Holmes & Meier.

Kerr, Malcolm. H. 1981. "Rich and Poor in the New Arab Order." *Journal of Arab Affairs* 1(1):1-26.

———. 1982. "Egypt and the Arabs in the Future: Some Scenarios." In *Rich and Poor States in the Middle East: Egypt and the New Arab Order*, edited by Malcolm H. Kerr and El Sayed Yassin. Boulder, CO: Westview Press; and Cairo: American University in Cairo Press.

Lwecock, R. 1984. "The Conservation of the Old City of Cairo." *See* El-Safty, 1984.

*Middle East* (Oct. 1981):45-47. "Mosaic. A Myth for the Rubbish Tip."

Mohi-el-Din, Amr. 1975. "Employment Problems and Policies in Egypt." Paper presented to the International Labour Organisation/Economic Commission for Western Africa Seminar on Manpower and Employment Planning in Arab Countries, Beirut, May.

Morcos, Wedad Soliman. 1980. "Trends and Patterns of Urbanization in Egypt during the 1966-1976 Intercensual Period." *Population Studies Quarterly Review* 7(52):38-71 (in Arabic).

NCSCR. *See* National Center for Social and Criminological Research.

National Center for Social and Criminological Research and the Institute of Social Studies. 1982. *Development Potential at Low Levels of Living: A Pilot Study*. The Hague and Cairo.

Niemat-Allah, M. 1984. "Urban Service Delivery in Cairo." *See* El-Safty, 1984.

Ouda, Mahmoud. 1971. *Migration to the City of Cairo, Its Patterns and Motives*. National Review of Social Sciences, vol. 2. Cairo: National Center for Sociological and Criminological Research (in Arabic).

Stewart, Desmond Stirling. 1965. *Cairo*. Cities of the World, no. 1. London: Phoenix House.

Taylor, Elizabeth. 1984. "Egyptian Migration and Peasant Wives." *MERIP Reports* 14(5):3-10.

United Arab Republic. The Authority of Statistics and Census. 1962. *The General Population Census of 1960. Cairo Governorate*. Cairo.

United Nations. Bureau of Social Affairs. Population Division. 1969. "World Urbanization Trends, 1920-1960." In *The City in Newly Developing Countries. Readings in Urbanism and Urbanization*, edited by Gerald William Breese. Modernization of Traditional Societies Series. Englewood Cliffs, NJ: Prentice-Hall.

Vining, Daniel R., Jr. 1985. "The Growth of Core Regions in the Third World." *Scientific American* 252(4):42-49.

Waterbury, John. 1978. *Egypt: Burdens of the Past, Options for the Future*. Bloomington: Indiana University Press in association with American Universities Field Staff.

———. 1982. "Patterns of Urban Growth and Income Distribution in Egypt." In *The Political Economy of Income Distribution in Egypt. See* Ibrahim, 1982b.

———. 1983. *The Egypt of Nasser and Sadat: The Political Economy of Two Regimes*. Princeton, NJ: Princeton University Press.

Wikan, Unni. 1980. *Life Among the Poor in Cairo*. Translated by Ann Henning. London: Tavistock.

Wirth, Louis. 1938. "Urbanism as a Way of Life." *American Journal of Sociology* 44:1-24.

The World Bank. 1983. *World Development Report 1983*. New York: Oxford University Press for the World Bank.

———. 1985. *The World Bank Atlas 1985*. 18th ed. Washington, DC.

Wren, Christopher S. 1978. "For Thousands in Cairo, Home Is a Tomb with the Dead." *New York Times* (13 Dec.):A-2.

# 9

# Mexico City

## Martha Schteingart

Mexico City has been described repeatedly as an example of calamitous and pathological urban development. Such catastrophic visions of social reality in the city often use partial and misleading information and sometimes apply biased interpretation of that urban reality. My principal aim is to analyze a wide range of aspects and problems that accompany rapid metropolitan expansion to give a more objective and balanced interpretation of the positive and negative qualities of urban growth.

I emphasize the demographic and economic characteristics of Mexico City, referring at the same time to its role in the national economy and to the problems of hyperconcentration. I present an overview of the internal organization of metropolitan space, including historical processes, and provide some explanations of observed phenomena. Within this context lies a series of urban problems which technical experts and wide sectors of the population consider the most serious. Conclusions from my own research supplement contributions of other researchers on various aspects of Mexico City's development and emerging problems.

## Twentieth-Century Population and Economic Growth

Development of an industrial city depends to a great extent on the already established standard of the preindustrial infrastructure. Mexico City, as Mexico's principal economic and political center for centuries, had the more favored situation for industrial development. The city offered the most sophisticated infrastructure in the country, the largest consumer market, a concentration of the few existing industries, and a relatively well-trained labor force. These factors, combined with the city's being the seat of the federal government, which facilitated commercial transactions, significantly spurred later dynamic growth (Garza and Schteingart, 1978a).

Mexico City's rapid growth during this century is well known. In 1900 there were 344 thousand inhabitants and an annual growth rate of 3.1 percent—indicating the already important amount of migration toward the city which intensified in the 1920s to reach a 5 percent annual growth rate (while the nation grew at only 1.6 percent). These migratory processes were in part the outcome of the Revolution of 1910. Although the Agrarian Reform tended to keep the rural population on their lands, it was insufficient to overcome the lack of balance between the agricultural and nonagricultural sectors of the economy, which resulted in migration of many peasants. The main center of attraction was Mexico City (Garza and Schteingart, 1978a).

From 1940 to 1970 the city experienced annual growth rates of over 5 percent, almost doubling its population with each decade. In the 1970s, growth dropped slightly to an annual 4.8 percent, still very high for a metropolis the size of Mexico City. In 1940, the population was 1.7 million; by 1980 it had soared to over 14 million, making Mexico City at this moment the world's most populated metropolis (see table 9.1).

## Economic Base and Employment

Along with its increase in population, the city's industrial development was accompanied by the proliferation of a great variety of services, territorial expansion of the urban area, and a process of metropolitanization. Since the 1940s, the nation's industrialization process, a logical outcome of the "substitution of imports" project, has fostered further growth and industrial concentration in the capital city. Established industrial units increased from 3,180 in 1930 to 34,543 in 1975, climbing from 3.8 percent to 29 percent of the national total. Consequently, Mexico City had an increasing share of national production, rising from 29 percent in 1930 to 45.4 in 1975 (Garza, 1981). But this enormous industrial concentration occurred with characteristics that limited the sector's capacity to absorb great quantities of the labor force, especially since the 1960s. Thus, the growing urban population has been increasingly employed in the alternative third (service) sector which has absorbed most migrants displaced from agriculture.

The economically active population in the industrial sector grew from 31 percent in 1940 to 38 percent in 30 years, despite a small decline between 1960 and 1970. Employment in commerce fell sharply from 27 to 14 percent in the same three decades while service employment grew from 30 to 36 percent (Bataillon and d'Arc, 1973). Currently more than 58 percent of the economically active are estimated to be working in the third sector of the economy.

TABLE 9.1
Concentration of Population and Index of Primacy—
Mexico City, 1900-2000 (in thousands)

| | Population | | | Primacy |
| Year | Nation | MZMC[a] | % | Index[b] |
|------|----------|---------|---|----------|
| 1900 | 13,607.3 | — | — | 3.4 |
| 1910 | 15,160.4 | — | — | 3.9 |
| 1921 | 14,334.1 | — | — | 4.3 |
| 1930 | 16,552.6 | 1,263.6 | 7.6 | 5.7 |
| 1940 | 19,649.2 | 1,670.3 | 8.5 | 6.5 |
| 1950 | 25,779.3 | 2,952.2 | 11.4 | 7.2 |
| 1960 | 34,923.1 | 5,125.4 | 14.7 | 6.1 |
| 1970 | 48,381.1 | 8,882.8 | 18.4 | 6.1 |
| 1980 | 71,387.0 | 14,445.0 | 20.3 | 6.0 |
| 1990 | 99,669.0 | 21,841.7 | 21.9 | 5.6 |
| 2000 | 135,089.0 | 30,860.7 | 22.8 | 4.9 |

SOURCE: Garza and Schteingart (1978b); Unikel et al. (1976); MZMC data for 1980 from Centro de Estudios Economicos y Demográficos (1975).
a. Metropolitan Zone of Mexico City.
b. Index for Mexico City and Guadalajara.

The largest contribution to the increase in employment in services is the expansion of public administration, which contributed substantially to growth of the middle classes. However, expansion of services is also a sign of growth of a population employed in low-productivity activities (such as repairing and cleaning services) with low incomes representative of urban underemployment.

Mexico City became the seat of new production, exchange, and consumption activities, generating new social groups and expansion or modification of others already there. Apart from new sectors of the middle class appearing on the urban scene, the proletariat and subproletariat grew. These groups have had different influences on the state apparatus and its strategies concerning industrial and economic processes as well as distribution of those groups in the urban space and production of the city's built environment.

During the 1940s and 1950s, industry produced the need for a large work force. Later its capacity for absorption began to slow because it was consolidated by a pattern of accumulation which promoted concentration of income and increased unemployment.

Thus, a surplus work force developed which helped maintain salaries at a low level. Between 1938 and 1965 minimum salaries increased by 315.9 percent while the cost of living soared by 744.8 percent (COPEVI, 1977). On the other hand, open unemployment (people actively seeking jobs) appears to have doubled each decade since 1940 to reach 100 thousand in 1970, jumping to 220 thousand in 1974.

At least 24 percent of the economically active in the Federal District were estimated to be involved in fringe occupations (those remunerated by less than minimum wage). Were it possible to include not only those who received very low salaries but were also those who received larger salaries but were not permanently employed (and whose income was consequently reduced by distributing it over the whole year), we could have increased the group of underemployed persons to 35 percent of the economically active in the Federal District in 1970 (Hewitt de Alcántara, 1977). By 1978 the proportion of underemployed was estimated at 47.6 percent of the economically active.

## Migration to Mexico City

Rural-urban migration has been an important cause of the growth and multiplication of Mexican cities. From 1940 to 1970, 6.2 million persons moved from the country or small towns to large urban centers, half of them making their way to Mexico City which grew very rapidly during this period (Unikel, 1976).

The importance of Mexico City's migratory growth compared with natural increase (produced by fertility minus mortality) has been reduced in recent years and declined even further in the last few years. Nevertheless, about a thousand persons arrive daily in search of work from other parts of Mexico (Federal District Offices estimates). This mass of migrants is forced to leave the rural areas by dissolution of the peasant economy, modernizing agriculture, or social stagnation of small population centers. They come seeking the new opportunities Mexico City can offer and continue to have an important impact on its development.

The indirect impact of migration on growth is important because of the predominance of young adult migrants who have low mortality and high fertility. Female migration predominates because of the high supply of work for women, principally domestic employment.

More migrants than urban natives occupy fringe jobs, but the difference diminishes with their permanence in the city. Likewise, male migration allowed consolidation of an important sector of industrial workers; relatively fewer are concentrated in manual service activities. In contrast, more than half of female migrants are in manual service compared with 20 percent of the natives. Consequently, women who arrive in the city and recent male migrants confront the greatest difficulties of all groups in the process of urban assimilation (Oliveira and García, 1984). Marginality and poverty can occur as temporary phenomena for the newly arrived migrant. However, these phenomena have become part of the social structure because of the constant refueling of the labor market by new migrants (Muñoz et al., 1977).

## Mexico City in the National and International Context

Industrialization and modernization, which had begun in Mexico by the end of the last century, accelerated notably after 1940. Facilitated by the

Second World War, the substitution-of-imports policy was introduced during this period. The policy's success is basically a result of prohibiting traditional imports and constructing an infrastructure for industrial production. This situation, combined with maintaining low salaries, stimulated private, national, and foreign investment.

From the 1950s and into the 1970s, Mexico's economic growth was sustained. Between 1940 and 1974 the gross national product grew to an annual rate of 6.4 percent, but this growth always presented some negative features: external dependence and the great concentration of income growth in the capital at the expense of other regions of Mexico, which have remained very backward (Garza and Schteingart, 1978a). As a result of declining oil prices and huge foreign debts, the Mexican economy has entered a period of deep crisis which has accentuated even more some of the negative consequences of Mexico City's growth.

In 1940 Mexico City contained 7.9 percent of the Mexican population, but its economic importance was much greater. During 1950-70 the city's contribution to the gross national product rose from 30.3 to 37.4 percent. The huge step forward in the 1940s occurred because of increases in industrial activities. Although Mexico City's population grew, its commercial dominance grew even faster. About half of all economic nonagricultural activity in the nation is concentrated in Mexico City. Similarly, in 1972 the city was absorbing 50.8 percent of the national demand for industrial products and 42.5 percent of that for durable consumer goods. This consumption concentration explains in part the tendency for industrial enterprises to locate in Mexico City.

Approximately 68 percent of the banking capital stock and reserves are concentrated in Mexico City, as well as 42 percent of the short-term deposits and 93 percent of the long-term ones. The city also has a large concentration of educational, cultural, and recreational activities. Above all is the great concentration of political power and public administration, adding to the city's dominance over the rest of Mexico. In 1975, 191 thousand of the 516 thousand public employees at the national level lived in the capital. This figure would rise to 500 thousand if employees of decentralized public businesses were added (Garza and Schteingart, 1978a). At present, about 35 percent of federal public employees are concentrated in Mexico City. The distribution of higher-order service activities between the principal urban center and the rest of the country reenforces Mexico City's primacy, thus making any attempt at decentralization difficult.

As for Mexico City's international role, although its demographic size would place it on top of the world hierarchy, the city cannot be classed as a "world city" (Friedmann and Wolff, 1982) largely because it is not a global decision-making center. While a third of Mexico City's industrial production is under the control of transnational capital, practically all its production is directed toward the national market. Consequently, Mexico City plays an insignificant role in the international division of labor. Neither is the city headquarters for international capital, which for many reasons remains in the principal cities of the core capitalist countries. As a producer of world market commodities, Mexico City's importance is relatively low since the country's main export items (e.g., petroleum and other minerals) are neither produced nor exported from there.

Large cities of underdeveloped countries like Mexico tend to be general receivers rather than generators of capital, industrial products, culture, and so on. Because of this phenomenon, they cannot really be considered world cities. Mexico City, nonetheless, is the dominant center of Central America and fulfills an important function at a regional level (Garza and Schteingart, 1984).

## Internal Structure of the Metropolitan Zone of Mexico City

At present, the Metropolitan Zone of Mexico City[1] (MZMC) has 17 million inhabitants and its urbanized area covers about 1,200 square kilometers. The MZMC is located in two federal entities: the Federal District and the State of Mexico.

Of the administrative units forming part of the metropolitan zone, there are 16 "delegations" in the Federal District and 16 municipalities in the State of Mexico. In 1970, there were already 10 metropolitan municipalities and the central city was divided into 12 *cuarteles* which after 1970 were converted into 4 delegations. These delegations added to the remaining 12 constitute the Federal District's total administrative units.

Mexico City's basic spatial structure was originally patterned after the Aztec City of Tenochtitlan. Few changes were made during the colonial period and the greater part of the 19th century. Only after the expropriation of the Catholic Church in 1856 did significant transformations take place. As a result, a real estate market developed and soon lands in the area surrounding the city center were subdivided and sold. This situation caused the first movement of the economic elite away from the center to urbanized zones in the western fringe. Changes made during the dictatorship of Porfirio Díaz (1876-1911), especially in constructing infrastructure (avenues, sewage systems, drainage canals, etc.) and the formation of districts for the upper and lower classes, gave extra impetus to the tendency of the elite to move.

By the end of the Porfirian period, spatial segregation was already established on the city's map. The lower income sector moved north and east, the higher group toward the west. During the 1940s, new industries were established, mainly in the north of the city. Migrants arriving to supplement the work force essential for rapid industrialization also settled in the delegations in the north of the Federal District. At the same time there was new construction in the central areas because of expansion of the state government bureaucracy, commerce, and service activities. These changes in central land use resulted in large numbers of middle-class residents moving to the periphery west and south of the city. Migrants arriving from rural zones were absorbed into the center and saturated the old *vecindades* (slums).

In the 1950s, the city began to grow beyond the Federal District's northern boundaries absorbing large amounts of territory in the state of Mexico. While middle-class families settled in fully urbanized subdivisions such as Ciudad Satelite (established in 1957), the poor began to occupy the unimproved lands of the Texcoco lake bed. This area lies northeast of the city and is the most inhospitable part of the Valley of Mexico, plagued by constant flooding, dust storms in the dry season, and an almost complete lack of urban services.

In the 1960s, the flight from the central city continued to accelerate. During that decade the delegations corresponding to this area lost some 800 thousand persons. At the same time the spread of the urban area into the State of Mexico was more notable particularly with establishment of new industries and proliferations of irregular settlements. An eloquent example of this accelerated urban expansion is Nezahualcoyotl, a clandestine subdivision whose population grew in this decade from 65 thousand to 650 thousand (an unprecedented rate in Mexico and perhaps in all Latin America).

In 1970, the urban area covered 633 square kilometers compared to the 117 of 1940. During the 1920s the Federal District's urban growth was concentrated in the southeastern zone. In the State of Mexico, lower income communities proliferated in Ecatepec (also toward the northeast) and continued multiplying in Nezahualcoyotl, which by 1975 had 1.3 million inhabitants.

According to COPEVI (1977), popular settlements occupy 65 percent of the metropolitan urbanized area. These communities generally consist of illegally acquired lands without services and with limited and deficient facilities and self-constructed housing.

Middle-class suburbanization continued as illustrated by the residential subdivisions which multiplied on both sides of the Queretaro highway in the northwest of the metropolitan zone following the initial establishment of Ciudad Satelite. Between 1960 and 1977 different municipalities in the State of Mexico approved some 190 residential subdivisions with 292,500 lots—300 percent more approved in 15 years than had been approved in the Federal District over a 35-year period. The subdivisions account for approximately 36 percent of the urban area growth in this entity between 1970 and 1977 (Schteingart, 1983).

The urban area expanded by nine times its size between 1940 and 1983. It comprises the growing decentralization of productive, commercial, and service activities, as well as the suburbanization of different social strata resulting in a social division of space.

## Population Distribution and
## Intraurban Sociospatial Differences

We have already referred to out-migration from the city's center and the phenomenon of suburbanization. Yet, from table 9.2, we can have a clearer idea of the changes that have taken place in the metropolitan zone with regard to population distribution.

TABLE 9.2

Population Growth and Its Distribution—Metropolitan Zone of Mexico City, 1940-1980
(in thousands, with percentages in parentheses)

| Territorial Subdivision | Population | | | | | Annual Growth Rate | | | |
|---|---|---|---|---|---|---|---|---|---|
| | 1940 | 1950 | 1960 | 1970 | 1980 | 1940-50 | 1950-60 | 1960-70 | 1970-80 |
| Central delegation[b] | 1,448 (86.7) | 2,235 (75.7) | 2,832 (55.2) | 2,903 (32.7) | 2,812 (19.5) | 4.3 | 2.4 | 0.25 | -0.3 |
| Other Federal District delegations[c] | 222 (13.3) | 688 (23.3) | 1,984 (38.7) | 4,056 (45.7) | 6,561 (45.5) | 10.25 | 9.7 | 6.9 | 4.7 |
| State of Mexico[d] | — | 29 (1.0) | 309 (6.1) | 1,924 (21.6) | 5,072 (35.1) | — | 16.6 | 14.5 | 9.0 |
| Metropolitan zone | 1,670 (100) | 2,952 (100) | 5,125 (100) | 8,883 (100) | 14,445 (100) | 5.5 | 5.3 | 5.4 | 4.8 |

SOURCE: Unikel (1976); 1980 data from Mexico, Secretaria de Programación y Presupuesto-Consejo Nacional de Población (1982).
a. Calculated using the formula AGR = 2 (P1 − Po/P1 + Po) x 1/N x 100, where P1 and Po = population at the beginning and end of the period; N = number of years.
b. Central delegations of the Federal District are Benito Juárez, Cuauhtémoc, Miguel Hidalgo, Venustiano Carranza.
c. Other delegations numbered 7 in 1940 and 1950, 10 in 1960, 11 in 1970, 12 in 1980.
d. Municipalities in the state of Mexico which integrate the MZMC were 1 in 1950, 2 in 1960, 10 in 1970, 16 in 1980.

Between 1940 and 1980 the central delegations and Federal District continued to lose weight in the metropolitan zone while the State of Mexico's municipalities grew in population and relative importance. Although in 1950 only 1 percent of the population lived in this entity, by 1980 it had reached 35 percent of the total. In contrast, in 1950, the central delegations totalled 75.7 percent of the metropolitan population and by 1980 only 19.5 percent. The sharp plunge in the center's population growth from 1950 and the large increase in the growth rate in the State of Mexico, particularly in the 1950s and 1960s, underline the tendency.

To gain a more systematic picture of the changes taking place in the last few decades, we present some conclusions of a recently completed work (Rubalcava and Schteingart, 1985). Using census information for 1950, 1960, and 1970 (1980 information was unavailable) and applying factor analysis, the study describes the most pertinent characteristics of sociospatial differentiation and its evolution in time. Census variables used were socioeconomic characteristics (economic activity, occupation, education, houshold head's income) and those related to housing (crowding index, tenure, water supply).

In the factor analysis, correlations of each variable with the remainders serve as a departure point for determining the factors. The factors appearing for three years under consideration are related to the phenomenon of *urban consolidation*, an essential characteristic of intraurban differentiation. Through construction of these factors it was possible to consider over a period of 30 years six zones indicating a gradation of sociospatial situations from the center to the periphery. These zones tend to be concentric in that the distance from the original city center is connected to the urban expansion and to their consolidation through time. This analysis has also shown that urban decentralization to peripheral areas in general leads to a greater access to educational and basic living services for the population (despite the consumption deficits occurring in urban growth where lower income groups predominate).

Through analysis of the internal variability of the consolidation index in the six zones (the difference among the political administrative units they comprise), we were able to observe that changes in the population distribution which we saw above accompany an increase in the internal heterogeneity of the upper zones and a growing homogeneity in those with inferior conditions. This situation indicates a differentiated consolidation process—more heterogeneous (with greater social mixture) in the central zones and more homogeneous (with greater segregation of poor groups) in the periphery.[2]

## Factors Influencing the City's Expansion and Internal Structure

The forms of production of the built environment which depend on real estate agents' and private constructors' practices, state policies, and the low-income sectors' initiatives are responsible in large part for the spatial distribution of the social groups and their different access to collective consumption. In this distribution, ground rent also plays an important role.

On the other hand, the city not only manifests current social structure and recent urban processes but also reflects various social structures and urban processes which have occurred historically. For this reason, I shall refer to the city's origins and previous evolution in my analysis.

In general, state policies were important for shaping Mexico City's internal space. For example, fiscal incentives and local infrastructure stimulated industry location in the north of the Federal District and afterwards in the State of Mexico. Federal District Department (FDD) investment was directed with preference to public works which tended to satisfy more the demands of economic activity than of the working population (Perló, 1981). Thus, for the proletariat and subproletariat a long time passed before there was a state housing program. Perhaps the rent control law dictated in the 1940s, which affected an important part of the housing stock in that period, constituted a policy calculated to mitigate the housing conditions of those sectors aggravated by rapid urbanization. In contrast, for the middle classes, the state implemented some housing programs, which, although still not substantial in the 1950s, signified an important step.

In the 1960s, the state put a new housing financing program into practice which had strong repercussions in the real estate market. This program supported private developers and the expansion of large residential subdivisions. The program was directed basically toward the middle classes, while the lower strata had no alternative but to invade private lands, occupy *ejidal* and communal land,[3] or become victims of unscrupulous speculators or buying lots in irregular developments (this is how Nezahualcoyotl grew). People's response intensified toward the end of the 1960s and beginning of the 1970s when invasions of urban lands notably increased, as did social movements struggling for improvement of lower income settlements (Navarro and Moctezuma, 1980).

After 1970, the state assumed a position of more active intervention in the economy. Public investment was not only targeted for infrastructure development but also for development of health and educational services and implementation of new housing programs for workers. The incorporation of the Fondos de la Vivienda, in particular, produced an important leap forward for housing production with state support. Furthermore, some public projects have opened new residential areas in the north and south fringes of the city. Taking advantage of existing infrastructure, private firms have produced new developments in neighboring areas and influenced peripheral urban growth.

In spite of the existence of some residential subdivisions developed directly by the state, most were built by private enterprises which began to develop real estate businesses (particularly based on land speculation) toward the end of the 1960s. The promotional private sector continued to expand and transform in the 1970s with an increase in the importance of large enterprises and financial capital, with growing state support not only for providing housing financing but also, in many cases, for urbanized lands at below market prices.

This trend indicates the articulation of public and private sectors in production of the urban built environment.

Another important program which affected living conditions of the lower income sector was creation of a vast apparatus for the regularization of land ownership in the popular settlements—that is, selling agricultural land (*ejidos*) for self-constructed housing (Schteingart, 1983a). This program has contradictory results in that while it has served to introduce services and better living conditions in the lower income districts, it has also resulted in price increases within these same areas and in some cases ejection of the poorest families who cannot bear the added cost of regularization.

In the last few years, urban policies in the MZMC have been oriented toward a more efficient city with better facilities (at the consequent cost of a strong and growing external debt) without land invasions or clandestine housing developments. At the same time lower income districts are being regularized and integrated into the city. With the oil crisis, which has strongly affected the country's economy, state housing programs have been more limited and alternatives for the lower income sectors have continued to narrow;[4] also the inflationary process has strongly restricted access of a wide range of people to a minimum standard dwelling (Schteingart, 1983b).

Production of a built environment has resulted in an enormous extension of the urban area. The area has grown at low densities (leaving empty zones close to nuclear centers in the hope of their increasing in value). A strong sociospatial division and a permanent increase in the distances between living and working areas with the consequent increase in transportation needs are other results of development.

## Urban Development Problems

In dealing with the previous points, I have already referred to some of Mexico City's basic problems. We have seen that the city's high primacy problems within the system of cities contributes to the country's uneven development. To this I should add the inefficient allocation of financial resources, inefficient use of natural resources, and inequitable distribution of income in favor of the urban sector.

Unemployment and underemployment affect wide sectors of the population as does extended metropolitan growth accompanied by the multiplication of the urban poor. Although one cannot deny that these problems have been aggravated by an annual addition of 600-700 thousand new inhabitants (each year the equivalent of a medium-sized city), they cannot be explained simply in terms of population growth. The problems may also be seen in the context of a model of dependent and unbalanced national development which presents problems of structural unemployment and working force exploitation, and works for the benefit of the land-owning, industrial, financial, and real estate sectors (who also profit from the concentration of population and economic activities).

Mexico City's growth problems further aggravate related dilemmas such as lack of services and adequate housing; expansion of low-income settlements; transportation; ecological and environmental deterioration in the city and its hinterland; and growing difficulties in confronting the greater demands for water, essential for developing urban activities.

## Lack of Housing and Services

By 1970 an estimated 577 thousand new dewllings would be required to eliminate Mexico City's housing shortage. This figure corresponds to 242 thousand units for homeless families, 148 thousand to replace deteriorated housing, and 187 thousand to eliminate overcrowding (Garza and Schteingart, 1978a). According to FDD estimates, the actual deficit affects more than 800 thousand families in the district.

Tap water reaches 80 percent of homes with a primary network of 540 square kilometers and a secondary network of 12 thousand. Nevertheless, in the delegations in the worst condition, from 40 to 46 percent of the population lack water service. Although a sewage system serves 70 percent of the population, its present deficit affects some three million people (FDD data). Most lacking in services are the lower-income settlements in the State of Mexico where water and sewer installation is more problematic and urban growth is most accelerated.

## Expansion of Lower-Income Settlements

I have already referred to the location of lower-income settlements in the MZMC and to the proportion of population they absorb. These areas are characterized basically by irregular land tenure, lack of services, and precarious housing, in general, self-constructed by the *colonos*. Although a gradation of situations exists implying different degrees of consolidation and consumption, most of these settlements present very poor conditions, now aggravated by the economic crisis, increased unemployment, and limited possibilities for housing improvement (Schteingart, 1983b).

Regarding land tenure, a large part of these districts is illegally settled on ejidal and communal lands or on state-owned property, derived from the Texcoco's drying lake bed (in the State of Mexico). That Mexican cities grow to a great extent on nonprivate property[5] constitutes a specific situation which sets them apart from other Latin American cities. Nevertheless, this situation has not benefited the low-income sectors with no access to the capitalist real estate market. An important part of communal or ejidal lands was occupied through legal and illegal means by the well-to-do. Some of those also served as support for the lower income settlements. Illegal land purchase by the colonos, tolerated by the state, has served to mitigate the conflicts produced by irregular settlements. These conflicts tend to be much greater in countries where urban expansion takes place almost exclusively on private property and where confrontations occur between colonos and powerful owners of the peripheral land. Yet the legal situation in force does not allow the low-income sector to settle without their going through a long series of repeated payments for regularization of developments (Schteingart, 1983a).

## Urban Transportation Problems

Mexico City has a pressing need for new transportation policies. A transportation scheme has been developed based on the proliferation of private automobiles with little coordination and support of public transport. This system is deficient and expensive for low-income workers. On buses there is an obvious increase in the saturation index, a longer waiting period in stations and at bus stops, and the percentage of wages destined for transport rose in the average family from 9.4 percent to 13.4 between 1968 and 1978 (Ibarra, 1981).

In the MZMC at present some 23 million person-trips are registered daily; of this total only 16 percent are in private automobiles and the rest are by public means (bus, taxi, trolleybus, subway). The subway concentrates some four million trips with a network which as a result of recent extensions covers around 100 thousand square kilometers with 6 lines and 85 stations. Nevertheless, on some lines and during rush hours, a 25 percent overload of pasengers has been calculated (Mexico, Secretaría de Programación, 1984).

Almost a third of the person-trips take place on the city's periphery, particularly in metropolitan municipalities where there are no subway lines with subsidized fares. To get to work, those living in this zone have to crowd into increasingly more saturated buses or take "collective taxis" whose prices have increased considerably in the last few years.

The transportation scheme implies an annual increase of 10 percent in the number of private cars. At present the number of cars in circulation is 2.2 million—97 percent of the total transportation units but only making 16 percent of the total trips. Automobiles promote inefficient energy consumption; they use 66 percent of the gasoline total and public transportation less than 22 percent (Campbell, 1981). The lack of road space and parking places for this growing mass of automobiles increases traffic jams in the city center and on freeways leading to the periphery. Road systems developed in the last few years created enormous expenses for the public administration and encouraged greater private car use without solving the problem of lack of capacity which continues to worsen.

## Biological and Environmental Degradation

The MZMC is located in the southwest of the Valley of Mexico, in the country's southern region at some 23 hundred meters above sea level. The valley covers an area of approximately 96 hundred square kilometers and is open only toward the north and southwest; in other directions mountain chains rise more than 600 meters above the valley area.

The Valley of Mexico has undergone drastic transformation. Much of it was covered by a series of lakes of which only the Texcoco salty lake bed remains. This salt base creates a barrier to any kind of construction in the northwest MZMC. Probably 73 percent of the forest area has been lost and 99 percent of the lake area, resulting in erosion of 71 percent of the soil. In addition, seven hundred hectares of agricultural land are lost annually (Mexico, Comisión de Desarollo, 1981).

The valley's geographical location in the intertropical zone, as well as its altitude, conditions its climate which is also influenced by winds. There is little rain and clear skies in winter and spring; in summer and autumn, the valley has clouds and heavy rain. Vegetation, also important in climate transformation, has been seriously eroded which affects temperature changes—now registering more extreme maximums and minimums. These geographic, climatic, and environmental change factors play an important role in the city's ecological and environmental degradation and, as we shall soon see, in providing water necessary for urban activities.

## Air Pollution

The MZMC has frequently been called one of the most polluted areas in the world. Altitude, rains, and limited winds, together with industrialization and the pattern of urban growth (the transportation scheme, diffusion of popular settlements, and land use in general), have aggravated the situation.

In the Valley of Mexico, the atmosphere is badly affected by the thermal inversion produced when a mass of cold air, at a certain height, does not allow the renewal of warmer air. During the rainy season, when a cold air mass penetrates, the thermal inversion is upset. Thus, the six-month dry season is most favorable for pollution.

The heavy circulation of automobiles, their faulty maintenance, and traffic jams, combined with the city's altitude,[6] result in heavy air pollution. Added to this are some 60 polluting industries such as oil refineries, thermoelectric plants, and cement factories, emitting vast amounts of air particulates (Mexico, Secretaría de Programación, 1984).

Other polluting sources are the Texcoco lake bed with its sandy, loose, salty soil and scant vegetation; arid eroded zones (20 percent of the MZMC is in the process of advanced erosion and another 17 percent suffers moderate erosion); the 14 thousand tons of waste matter dumped in the open air and 20 thousand put in clandestine dumps. Excrement deposited in open places is dehydrated by the sun and its residues distributed by the winds. The highest concentrations of these dust particles, according to measurements by the Secretary of Health and Assistance, were found in the southeast and northeast MZMC where lower income people live.

The highest monthly mean averages of sulphur dioxide are in the southwest and northwest of the MZMC. Concentrations in the northwest can be blamed on emissions from combustion processes from a major oil refinery and thermoelectric industries. The highest levels of carbon monoxide occur during the hours of the most intensive vehicle transit, particularly in the center and north of the urban area. The pollution level varies according to the city's zones and the season. The lower income population is, of course, exposed to heavier pollution since they live in areas close to industry or in districts east of the MZMC.

In table 9.3, I present estimates of emissions of dust particles, sulphur dioxide, hydrocarbons, carbon monoxide, and nitrogen oxide. Here we can observe the increase in total emissions (that of dust particles that have fallen annually) and the great amount of carbon monoxide coming from vehicles (moving sources).

There has been very limited use of emission control devices in industry and transportation units; consequently in 1983 total emissions had increased by 45 percent compared to 1972. Pollution in the MZMC is six times higher than accepted limits for humans—a likely cause of 90 percent of the respiratory illnesses and infections the capital's population suffers (Marcó del Pont, 1984).

## Water Problems

In the 16th century the extensive system of lakes in the Valley of Mexico began to disappear. Springs were used indiscriminantly and this natural source dried up since the water was absorbed at a rate greater than its natural rate of replenishment. Overexploitation of subterranean water caused a sinking of up to nine meters in the city (Ramirez, 1981). Since the 1960s, Mexico City's traditional water supply has been in a state of crisis. Great increases in demand, limitations imposed by the natural environment, and techniques used for gathering and consuming water made it essential to create new schemes for its supply, which led to greater dependence on water imported from external basins. The great majority of these water sources are at considerable distances, making the necessary infrastructure expensive to construct and maintain.

In 1980 consumers in the Valley of Mexico used 2.4 million cubic meters of water daily. Of this total, 78 percent came from subterranean waters, 6.3 percent from surface waters, and 15.4 percent was imported. Water use included 53 percent for domestic purposes, 11.6 percent for commerce and services, 18.9 percent for industry, 23.5 percent for agriculture, and 16.4 percent for public purposes (data from Mexico, Comisión de Conurbación, 1981, and a 1982 Valley of Mexico Water Commission report).

Domestic water use, the most important, varies greatly according to social group. Residents of the high-income Chapultepec zone were consuming some 450 liters per person daily while daily per capita consumption in low-income Nezahualcoyotl was only 50 liters. Nine percent of the consumers use 75 percent of the water total and more than two million have very limited access to it. Policies concerning utility rates appear to have reenforced this pattern (Mexico, Secretaría de Programación, 1984). In 1980, demand for water exceeded supply by 284 million cubic meters per day.

### TABLE 9.3
Total Emissions from Fixed and Moving Sources—Mexico City, 1972-1983
(in thousands and percentages)

| Year | From Fixed Sources | | | | | | From Fixed and Moving Sources | | From Moving Sources | | Total |
| | Dust Particles | | Sulphur Dioxide | | Nitrogen Oxide | | Hydrocarbons | | Carbon Monoxide | | |
| | Tons | % | Tons | % | Tons | % | Tons | % | Tons | % | |
|---|---|---|---|---|---|---|---|---|---|---|---|
| 1972 | 153 | 5.76 | 2C7 | 7.61 | 51 | 1.92 | 314 | 11.82 | 1933 | 72.89 | 2653 |
| 1973 | 158 | 5.52 | 217 | 7.57 | 53 | 1.84 | 338 | 11.80 | 2101 | 73.27 | 2868 |
| 1974 | 152 | 4.68 | 236 | 7.27 | 57 | 1.75 | 371 | 11.43 | 2429 | 74.87 | 3244 |
| 1975 | 129 | 3.79 | 278 | 8.20 | 64 | 1.90 | 381 | 11.23 | 2541 | 74.88 | 3394 |
| 1976 | 117 | 3.41 | 305 | 8.89 | 70 | 2.04 | 386 | 11.26 | 2553 | 74.40 | 3431 |
| 1977 | 117 | 3.48 | 284 | 8.26 | 100 | 2.90 | 391 | 11.37 | 2544 | 73.99 | 3438 |
| 1978 | 124 | 3.59 | 307 | 8.89 | 74 | 2.15 | 398 | 11.54 | 2547 | 73.84 | 3449 |
| 1979 | 128 | 3.64 | 343 | 9.74 | 81 | 2.31 | 407 | 11.54 | 2562 | 72.74 | 3521 |
| 1980 | 233 | 3.68 | 374 | 10.37 | 87 | 2.42 | 418 | 11.61 | 2589 | 71.89 | 3600 |
| 1981 | 136 | 3.71 | 384 | 10.44 | 90 | 2.38 | 435 | 11.83 | 2628 | 71.56 | 3672 |
| 1982 | 140 | 3.73 | 393 | 10.46 | 91 | 2.43 | 452 | 12.03 | 2680 | 71.33 | 3757 |
| 1983 | 147 | 3.66 | 400 | 10.44 | 93 | 2.41 | 470 | 12.20 | 2747 | 71.31 | 3851 |

SOURCE: Mexico. Subsecretaría de Mejoramiento del Ambiente (1980).

Residual and sewage waters are produced by diverse urban activities. The need to remove them from the city resulted in construction of hydraulic works which allowed their extraction from the Valley of Mexico toward the River Tula where half the water is used for irrigation. A variety of chemicals and microorganisms (many of them toxic) pollutes these waters and affects the population by exposing it to diseases either through direct consumption or from eating foods that have been in contact with the polluted water. This situation occurs when these waters are used for irrigation or pollute the subterranean waters used later. Faced with this critical situation which affects the entire water provision system, the population's health, and agricultural production, it has been considered indispensable to develop residual water and sewage treatment plants.

## State Responses to Mexico City's Problems

To analyze urban policies for Mexico City we should start by presenting the political and administrative context within which they have been proposed or implemented. The FDD Office is the principal public agency charged with solving metropolitan problems within the Federal District. It is governed by a *regente* (mayor) appointed by the president of the republic; the mayor appoints the delegates who are responsible for the delegations within the Federal District. These authorities are not elected, as are other state and municipal authorities.

The Organic Law of the Federal District (1941) introduced an Advisory Council composed of representatives of chambers of commerce and industry, landowners, tenants, peasants, employees, and so on. With reforms introduced to this law in 1970, neighborhood councils were established to assist the Advisory Council and the delegates acquired greater responsibility through a process of administrative decentralization. In 1980, local committee participation became legal and a representative's election mechanism was established. These representatives formed neighborhood councils and their presidents participated in the Advisory Council (Rodríguez Araujo, 1980). Nevertheless, citizen participation in local government is still very limited and even more so is their decision-making power regarding urban development, undoubtedly strongly linked to the Mexican political structure, dominated by one large party.

The 1975 Urban Development Law of the Federal District established the need for a master plan to guide land, housing, communications, and transport development. In 1976, the Master Plan for the Federal District Development was passed. With these laws, city planning became formally recognized. Although the need for coherent and comprehensive state-controlled planning of urban growth and spatial organization is accepted, Mexico City is still far from realizing these objectives. Some public policies contribute to spatial structuring—which has nothing to do with aims of the master plan. This situation is aggravated by these laws and plans being applicable only in that part of the MZMC corresponding to the Federal District and excluding the

State of Mexico municipalities which form an important portion of the metropolitan territory. Municipal authorities are elected and also depend on the state of Mexico's governor. Therefore there has never been any consistency or compatibility between actions undertaken in the Federal District and those enforced by municipal and central authorities in the State of Mexico.

The 1976 Federal Law of Human Settlements attempts to deal with such problems. The Commission for the Conurbation of Central Mexico and Limits between the Federal District and the State of Mexico Commission were created; however, they have not solved the coordination problems between the two entities. It would appear that only a metropolitan government could overcome the existing division between the district and state.

These planning and institution-creation initiatives for the MZMC are part of a wider attempt at state intervention on territorial aspects appearing in the 1970s. For example, in 1977 a federal ministry was established to coordinate policies referring to human settlements and an important set of national, regional, state, and municipal plans was formulated.

In the first two years of the present government (1982-83), seven programs have already appeared dealing totally or partially with urban and regional problems. Two pertain to Mexico City (particularly the need to reduce economic and population concentration); a third proposes measures to solve the most acute problems of the metropolitan zone and central region. This last program, formulated by the Ministry of Planning and the Budget, tries to overcome limitations of FDD plans. The aim is to coordinate this agency's actions and those of the State of Mexico and the federal government, thereby defining the MZMC and central region as its field of action. Another important aim of the program is reducing the excessive location of economic activities in the MZMC. To this end, the ministry will attempt selective decentralization, avoiding heavy concentrations of industries in a few locations by encouraging their distribution throughout the entire six-state territory included in the central region defined by the program. This strategy, rather than decentralizing, seems to stimulate a megalopolis embracing the MZMC and nearest cities. However, the measures proposed to achieve this objective are so imprecise and inadequate that it is unlikely they could ever be carried out.

Recently, the FDD has again presented a plan for Mexico City (Plan of Urban Restructuring and Ecological Protection) which repeats some of the proposals of previous plans regarding organization of metropolitan space and adds some new aspects, especially with regard to the zone's environmental degradation. The plan comes within the New National System of Democratic Planning which, to resist the crisis, uses the National Development Plan 1983-88 as its basic instrument. Strategies presented underline the urgency to restrict expansion of the urban area which has not reached the limit beyond which ecological degradation would be irreversible. Here the proposal is to declare as a territorial reservation some 77 thousand hectares of forest and uninhabited areas and prohibit new settlements. With regard to urban restructuring, the plan is to develop eight sectors of the city in almost complete self-sufficiency and achieve a complementary internal diversification which could avoid problematic displacements. Urban development would be supported by eight urban centers functioning as development poles, and having a concentration of commercial, service, and housing activities. To promote this scheme, public investment would be coordinated (in housing, services, transport, infrastructure); private investment would be stimulated; and there would be relocation of activities and land.

These ideas appear attractive in making spatial organization more balanced and rational. Nevertheless, their shortcoming is that they are not the outcome of a serious study of the social conditions producing that organization. Moreover, they would have to overcome powerful counterforces among the private sector, low-income groups, and interests within the state apparatus.

Although plans also mention basic economic aspects of urban development (heavy centralization, financial imbalance, unemployment), their treatment is very superficial, apart from being based on some false suppositions. Once again we confront an impractical proposal with an emphasis only on physical and spatial aspects. It appears to be merely a formality to place the plan titularly within a new system of democratic planning, while in fact it repeats previous schemes and proposals whose ideological function immediately springs to mind. Many of the concrete practices of public institutions continue flagrantly to contradict them (Ward, 1981).

## Land Policies

To explain internal urban space in the MZMC, it seems helpful to consider land policies (particularly those directed to ejidal and communal lands) and their regularization in lower income settlements.

Government actions to regulate land use and transfer within the MZMC have taken a variety of forms. Perhaps most important is expropriation of ejidal and communal land for public use or social benefit. These expropriations are governed by the Agrarian Reform Law. In general, successive administrations have failed to limit expropriations of land to cases of obvious social benefit, because of the ambiguity of the relevant laws as well as pressure from powerful private-sector interests.

While intervention of some state units has resulted in use of ejidal and communal land for speculative subdivisions designed for the well-to-do, some public housing organizations have had recourse to private agents to acquire land for their projects. The process of converting ejidal and communal lands into private property has contributed to the increase in land prices, especially in the State of Mexico, thereby displacing the lower income sectors who have less and less access to urbanized lots (Schteingart, 1983a).

Beginning in the 1970s the government established a number of institutions charged with regularizing land tenure within the MZMC. For illegally occupied ejidal land, the official policy is to expropriate it, compensate the affected *ejidatarios*, and sell it to its current occupants at reduced prices.

In recent years, particularly in the Federal District, the policy has been to regularize the majority of settlements to integrate them into the city. Regularization nevertheless does not mean economic improvement for all settlers. Many must leave the district since they cannot meet the regularization payments. Locating settlements within the city is also relevant since regularization does not occur when a settlement is situated in an area where land use has been designed for other social sectors. Already-established relations between colonos and the official party or relevant authorities are also factors. Thus legislative change and policies regarding collectively held land are urgently needed.

## Urban Transport Policies

Efficient urban structure is often derived from routes of public transport, either subway or bus. However, in Mexico, state actions have allowed development of a scheme based on the extensive proliferation of automobiles. This action could be explained by the impetus given to the car industry, construction of major highways, low-priced gasoline, and so on. Public transport, particularly buses, was until recently in private hands; in 1981 it was taken over by the local government which improved routes, focused greater attention on maintenance, and increased the number of units in circulation. Nevertheless there is still no detailed evaluation of results of these changes. Subway construction began only 15 years ago. Because of recent extensions and creation of new lines, only lately has the subway begun to cover a large area of the metropolis and the system is still very limited. The huge cost of these transport projects, particularly for an underdeveloped country experiencing a strong economic crisis, makes it problematic whether extending Mexico City's transportation system will continue.

The development program for the metropolitan zone emphasizes the need to modify the investments distribution scheme in favor of public transport and recommends that the subway be set up as the structuring element of the multinodal system with bus, trolleybus, and tram services being organized as feeder elements or substitutes. Improving the organization of existing systems and not spending money on increasingly large works is also considered necessary. To help integrate State of Mexico municipalities, the program proposes building a suburban railway and coordinating municipality road systems with that of the Federal District through a hierarchical structure of arterial roads. The plan presented by the FDD only refers to transport appropriate to the proposed eight centers. When this scheme is completed, the hope is that transportation will be more efficient and economical because of reduced person-trip hours.

## Environmental Policies

Up to the present, very little has been done to stop increasing pollution and environmental deterioration. Seriousness of the situation has led the present administration to give priority to this problem. Although there were already some public bodies treating these problems, their importance increased with creation of the Department of Ecology within the Ministry of Urban Development and Ecology. In addition, a series of plans was proposed, a campaign to heighten public awareness of these issues was developed through the mass media, and a reenforcement of university programs was designed to train professionals in this field. In the FDD, new offices have also been established to deal with these problems.

Among recent proposals are those included in the Ministry of Urban Development and Ecology plan (Mexico, Secretaría de Desarollo, 1984). They emphasize the need to establish monitoring centers to determine air quality levels and maintain control over vehicular emissions. They oversee necessary changes in the automobile industry to lower smoke and noise emissions and eliminate the lead content in gasoline. The FDD plan proposes establishment of an industry relocation program, prohibition of new industries with high pollution potential, and creation of ecological reservations to avoid continued environmental deterioration.

For the moment, despite all the institutional and media promotion, very little is being done to implement these policies. For example, vehicle emission control is hardly applied, even with public vehicles, because there are insufficient resources to implement it at this time, according to authorities. Of course, it is even more difficult to establish industry controls; during economic downturns, entrepreneurs are in the worst possible condition to increase production costs through installation of antipollution devices.

Despite the urgency of these problems, we do not find influential groups in Mexico organized to fight for improved environmental conditions. Nor have popular urban movements, more active in recent years, given their support to this pressing problem.

## Water Provision Policies

Policies for providing water to Mexico City have been directed toward increasing the proportion of clean water coming from external sources. In the 1950s water began to be imported from the Lerma Valley, but this source was insufficient to meet growing demands resulting from urban growth. In recent years, a series of projected works to bring water from external basins has involved three water-gathering systems: the Cutzumala, Tecolutla, and Amacuzal Systems. Intended to be completed by the end of the 1980s, these projects involve enormous investment because of the need to transport water from great distances and lower altitudes. Such projects require massive pumping installations and resulting heavy energy use.

The first stage of the Cutzumala project is completed; the second is scheduled to end in 1987. With these works it will be possible to increase the water supply by 24.5 cubic meters per second. Prospects for further expansion of these very large works are doubtful because of the enormous debt it would create for the FDD whose financial situation is critical. Thus there are proposals insisting on the need to improve water gathering from local sources by avoiding, for example, piped water loss which affects 40 percent of the total water supply being carried to the MZMC. Also proposed is extending the gathering capacity and use of rain water.

To limit the city's water consumption, it is recommended that size of the contiguous urban areas be controlled to fit the availability of this vital resource and that large water-consuming industries be prohibited. A new rate policy reenforces the aim of rationalizing water use (Plan of the Federal District Department and Program for the Development of the MZMC and the Central Region).

The particular characteristics of the Valley of Mexico established that, from the beginning, with the drying up of the lakes, city expansion would affect the environment. This impact is much greater because of the metropolitan process in recent decades. It is ironic that a place once covered by lakes, where water excess resulted in constant flooding, has been transformed into an area where it becomes increasingly difficult for a large part of the population to have access to this basic resource.

## Mexico City in the Year 2000

Even assuming that the future mean annual increase of the MZMC population will slow down, it could reach 30.8 million in the year 2000 (see table 9-1). At this rate of expansion, it is estimated that from 1990 to 2000, the MZMC will probably absorb nine more municipalities from the State of Mexico and one each from the States of Tlaxcala and Hidalgo (CEED, 1975). Thus, by the end of the century, that zone will be physically located within four federal entities, forming a megalopolis including the metropolitan areas of Mexico City, Toluca, and very likely, Cuernavaca.

Mexico City's projected industrial growth has been calculated to be 440 percent over the next 20 years. Contradictory[7] past and current government programs for industrial decentralization have had negligible results and have at times been self-defeating. Emergence of this vast megalopolis will further accentuate the country's great spatial disparities in economic development, as well as the already overwhelming urban problems of the capital itself, requiring the federal government to devote an ever-larger proportion of its revenues to its maintenance. For example, if the anticipated growth occurs, in the year 2000 the Federal District will require about 40-45 percent of the federal public investment—it now absorbs 33 percent.

To accommodate urban growth during the next 15 years, it will be necessary to develop another urbanized area with infrastructure and services similar to the present one, to increase enormously the capacity of highways converging on the city, and to provide substantially more water, energy, and food. In these circumstances, such an endeavor would not contribute positively to Mexico's development.

Internal problems of the metropolis will greatly worsen if measures are not enforced that drastically change the policies or lack of controls now in existence. For example, the daily demand for person-trips could grow to more than 40 million; the vehicle total could reach some 6 million units and access to the city could be permanently clogged by nearly 100 thousand freight carriers. The current enormous emissions of polluting agents could triple, thus increasing respiratory illness and lowering still more mean visibility in the city (already diminished from 10-15 to 2-4 kilometers between 1937 and the 1980s). There would be demands for 100 million cubic meters of water per day—water that to a great extent would have to be brought from very distant sources at extremely high cost and enormous energy outlay (at present the demand is for 36.6 million cubic meters per day). The cost of providing water, its transportation to the city, the necessary investment to widen the water network, meter installation, and so on will all increase water costs so much that in the year 2000 Mexico City's population could be drinking some of the world's most expensive water.

In the Federal District alone, an estimated five million people would not have tap water service, seven million would lack a sewage system, and it would be necessary to improve or build at least 2.5 million housing units by the end of the century. These social deficits would be produced in spite of maintaining a public investment per capita of twice the national average.

The projections by the FDD give us a very troubled vision of Mexico City's future in the year 2000. Unfortunately, for the moment, many of the initiatives and policies proposed to address those problems and to stem present trends are not being implemented.

## Concluding Remarks

Probably efforts will be made to address certain problems, but there are strong structural limitations to application of many of these proposals. Such policies must be viewed in the context of governmental decision making in a country with an extremely dependent capitalist economy and high inegalitarian distribution of resources.

Because of this situation, there will likely be a significant transformation of state policies (and of metropolitan development perspectives) only if basic changes are made in Mexico's socioeconomic structure. My view is that there must be a democratization process at national and local levels to open the way for greater participation of the masses and popular organizations in decision making about the future of Mexico. Otherwise, technocratic and antipopular urban development plans, rather than modifying present tendencies, may accentuate Mexico City's growth only to benefit its already powerful minorities. Better mobilization of independent popular organizations and more activism by those who suffer most intensely the negative consequences of metropolitan growth can create progressive new options. Such developments, of course, will not be encouraged by the city's privileged groups, nor by firmly established government leaders.

# Notes

1. Here the words *Mexico City* and *Metropolitan Zone of Mexico City* are used interchangeably. The latter describes more accurately the territorial extension of the central city and the political administrative units contiguous to it which have metropolitan characteristics (as work places or living areas for workers employed in nonagricultural activities) and maintain a direct and constant interrelation with the central city. As the sum of political-administrative units, the metropolitan area also contains rural population and activities. The urban area embraces only the occupied and urbanized zones which, spreading out from the central nucleus, present physical continuity (see Unikel, 1974).

2. Here zones were defined from a combination of correlations among the variables and not from individual values. It seems relevant to mention some values of the most important variables in the highest and lowest zones. For example, in 1970 the highest zone had an average of 84.3 percent of dwelling units with tap water, a crowding factor of 1.4 persons per room, and 69.8 percent of persons with basic education; corresponding figures for the lowest zones were an average of 33.3 percent, 2.8, and 31.5 percent. These figures underline the wide differences between the extreme zones. But if we observe the evolution undergone by the values, in general, improved conditions have been achieved in each zone. (Now we are completing this research by introducing data from the last census.)

3. Communal lands have their origin in the colonial period when the Spanish rulers gave them to the indigenous population to be used collectively by a village. The *ejido* was established by the Agrarian Reform in the 1915 Code. Ejidal-held land allows the peasants only the usufruct obtained from land produce; sale is prohibited. This type of property can only be expropriated for public use when there is evidence of this being superior to the social use of the ejido.

4. Difficulties in finding land to settle and shortages in self-construction because of the great price increases in materials (much greater than a worker's salary increases) have limited this form of housing production. Consequently room rentals in modest housing of low-income settlements have proliferated.

5. In the Federal District between 1940 and 1975, growth of the urban area covered some 42.8 percent of private property, 26.5 percent of communal lands, and 20.7 percent of ejidal lands. In the State of Mexico, the urban area has expanded by 21.9 percent on ejidal lands, 27 percent on communal lands, and 27.8 percent on state-owned lands (resulting from the drying up of Texcoco Lake), and by only 22.8 percent on private lands (Schteingart, 1981). The figures demonstrate clearly that the city grew on nonprivate lands to a much greater extent than in the Federal District.

6. At this altitude there is an increase of 30 percent in hydrocarbons and 100 percent in carbon monoxide produced compared to what is produced at sea level.

7. I find contradictions between the aims of the governments of the Federal District and State of Mexico on one hand and the national objective of decentralization on the other. Both the FDD and State of Mexico have continued to promote the city's growth by issuing new industrial and residential development permits and developing very costly urban projects.

# Bibliography

Bataillon, Claude, and Hélène Riviere d'Arc. 1973. *La Ciudad de México,* translated by Carlos Montemayor and Josefina Anaya. Sep-Setentas no. 99. Mexico City: Secretaría de Educatión Pública (in Spanish).

Campbell, Timothy. 1981. "Resource Flows in the Urban Ecosystem—Fuel, Water, and Food in Mexico City." Working Paper, Institute of Urban and Regional Development, University of California, Berkeley, no. 360. Berkeley.

CEED. See Centro de Estudios Económicos y Demográficos.

Centro de Estudios Económicos y Demográficos. 1975. *Estudio Demográfico del Distrito Federal.* Mexico City: CEED, El Colegio de México (in Spanish).

Centro Operacional de Vivienda y Poblamiento. 1977. *Investigacion sobre vivienda.* Vol. 2. Mexico City (in Spanish).

COPEVI. See Centro Operacional de Vivienda y Poblamiento.

Departamento del Distrito Federal. 1984. *Programa de reordenación urbana y protección ecológica del Distrito Federal.* Mexico City (in Spanish).

Friedmann, John F., and Goetz Wolff. 1982. "World City Formation: An Agenda for Research and Action." *International Journal of Urban and Regional Research* 6:309-44.

Garza, Gustavo. 1981. "El proceso de industrialización de la Ciudad de México: 1845-2000." *Lecturas del CEESTEM* (Centro de Estudios Económicos y Sociales der Tercer Mundo) 1(3):103-11 (in Spanish).

Garza, Gustavo, and Martha Schteingart. 1978a. *La acción habitacional del Estado en México.* Collección-Centro de Estudios Económicos y Demográficos no. 6. Mexico City: El Colegio de México (in Spanish).

———. 1978b. "Mexico City: The Emerging Megalopolis." In *Metropolitan Latin America: The Challenge and the Response,* edited by Wayne A. Cornelius, Jr., and Robert Van Kemper. Latin American Urban Research, vol. 6. Newbury Park, CA: Sage.

———. 1984. "Ciudad de México. Dinámica industrial y estructuracion del espacio en una metropólis semi-peiférica." *Demografía y Economía* 18:518-604 (in Spanish).

Hewitt de Alcántara, Cynthia. 1977. "Ensayos sobre la satisfacción de necesidades básicas del pueblo mexicano entre 1940 y 1970." *Cuadernos del CES,* no. 21. Mexico City: Centro de Estudios Sociológicos, El Colegio de Mexico (in Spanish).

Ibarra, Valentin. 1981. "El papel económico del transporte de personas en la Ciudad de México." *Lecturas del CEESTEM* (Centro de Estudios Económicos y Sociales del Tercer Mundo) 1(3):77-83 (in Spanish).

Marcó del Pont, Louis. 1984. *El crimen de la contaminación.* Atzcapotzaldo, Mexico: Universidad Autonoma Metropolitana-Atzcapotzaldo (in Spanish).

Mexico. Comisión de Conurbación del Centro del Pais. 1981. *Balance hidráulico del Valle de México.* Mexico City (in Spanish).

Mexico. Comisión de Desarrollo Agropecuario del Distrito Federal. 1981. *Estudio de espacios abiertos en al Distrito Federal.* Mexico City (in Spanish).

Mexico. Secretaría de Desarrollo Urbana y Ecología. 1984. *Plan de ordenamiento ecológico e impacto ambiental, México.* Mexico City (in Spanish).

Mexico. Secretaría de Programación y Presupuesto. 1979. *Basic Information on the Structure and Characteristics of Employment and Unemployment in the Metropolitan Areas of Mexico City, Guadalajara and Monterrey.* Mexico City.

———. 1984. *Programa de desarrollo de la Zona Metropolitana de la Ciudad de México y de la Región Centro.* Mexico City (in Spanish).

Mexico. Secretaría de Programación y Presupuesto—Consejo Nacional de Población. 1982. *Datos básicos sobre la población de México, 1980-2000.* Mexico City (in Spanish).

Mexico. Subsecretaría de Mejoramiento del Ambiente. 1980. *Situación actual de la contaminación atmosférica en al Area Metropolitana de la Ciudad de México.* Mexico City (in Spanish).

Muñoz, Humberto, Orlandina de Oliveira, and Claudio Stern. 1977. *Migración y desigualdad social en la Ciudad de México.* Mexico City: El Colegio de México (in Spanish).

Navarro, Bernardo, and Pedro Moctezuma. 1980. "Ejército industrial de reserva y movimientos sociales urbanos en México, 1971-1976." *Revista Teoría y Política* no. 2 (Oct.-Dec.) (in Spanish).

Oliveira, Orlandina de, and Brigida Garcia. 1984. "Migración a grandes ciudades del Tercer Mundo: Algunas implicaciones sociodemográficas." *Revista Estudios Sociológicos* 2(4): 77-103.

Perló, Manuel. 1981. "Apuntes para una interpretación en torno al proceso de acumulación capitalista y las políticas urbanas del Distrito Federal: 1920-1980." *Revista Vivienda. Infonavit* 6(6).

Ramírez, Pablo. 1981. "Problemas del abastecimiento de agua potable a la Ciudad de México hasta el año 2000." *Lecturas del CEESTEM* (Centro do Estudios Económicos y Sociales del Tercer Mundo) 1(3):94-97 (in Spanish).

Rodríguez Araujo, Octavio. 1980. "Gobierno y organización en el Distrito Federal." In *Atlas de la Ciudad de Mexico*. Mexico City: Departamento del Distrito Federal (in Spanish).

Rubalcava, Rosa María, and Martha Schteingart. 1985. "Diferenciación socioespacial intrurbana en el Area Metropolitana de la Ciudad de Mexico." *Revista Estudios Sociológicos* 3:481-514 (in Spanish).

Schteingart, Martha. 1981. "Crecimiento urbano y tenencia de la tierra. El caso de la Ciudad de México." *Revista Interamericana de Planificatión* no. 60 (in Spanish).

———. 1983a. "La incorporación de la tierra rural de propiedad social a la lógica capitalista del desarrollo urbano: El caso de México." In *Relacion campo-ciudad: La tierra, recurso estratégico para del desarrollo y la transformación social*. Mexico City: Sociedad Interamericana de Planification (in Spanish).

———. 1983b. "La promoción inmobiliaria en el área metropolitana de la ciudad de México, 1960-1980." *Demografía y Economía* 17(53):83-105 (in Spanish).

———. 1984. "El sector inmobiliario y la vivienda en la crisis." *Comercio Exterior* 34:739-50 (in Spanish).

Unikel, Luis. 1974. "La dinámica de crecimiento de la Ciudad de México." In *Ensayos sobre el desarrollo urbano de México*, edited by Edward E. Calnek et al. Mexico City: Secretaría de Educatión Pública (in Spanish).

Unikel, Luis, with Crescencio Ruiz Chiapetto and Gustavo Garza Villareal. 1976. *El Desarrollo urbano de México: Diagnóstico e implicaciones futuras*. Mexico City: Centro de Estudios Economicos y Demográficos, El Colegio de México (in Spanish).

Ward, Peter M. 1981. "Mexico City." In *Problems and Planning in Third World Cities*, edited by Michael Pacione. New York: St. Martin's Press.

# 10

# São Paulo

## Vilmar Evangelista Faria

The Metropolitan Area of São Paulo (MASP), encompassing over 13 million people, is one of the largest and fastest growing urban agglomerations of the world. In 1982, the MASP had one of the highest per capita incomes among Third World metropolises. At the same time, however, about 30 percent of its labor force earned monthly salaries barely above the physical survival level, and in several of its poorest areas, infant mortality was well above 100 per thousand.

For comparative purposes, it is important to realize that this metropolitan area is the central growth pole of the Brazilian industrial economy, the tenth world industrial power measured in terms of gross domestic industrial output.

The industrial economy centered around the MASP is structured to serve, mainly although not exclusively, the Brazilian internal market, thus differing from many other Third World cities which perform international exporting functions.

## The Metropolitan Area of São Paulo: Brief Historical Background

From an administrative point of view, the Metropolitan Area of São Paulo—legally created in 1969—comprises 37 autonomous municipalities

AUTHOR'S NOTE: I would like to thank Pedro Luiz Silva, Eduardo Fagnani, Celso Lamparelli, Guido Lopez, and Sonia Maria Begueldo. Without their active help it would have been impossible to write this chapter.

(*municipios*) including São Paulo. The metropolitan agglomeration covers 7,967 square kilometers, containing almost 10 percent of the total Brazilian population and nearly 20 percent of its urban people. Together with the metropolitan area of Rio de Janeiro, the MASP heads the Brazilian system of cities, organized in a complex hierarchy of more than 480 *urban centers* (urban areas of 20,000 or more inhabitants) where more than 60 million persons lived in 1980.

The core of Brazilian modern urban-industrial activities is located in the MASP: 45 percent of the total value added by the Brazilian transformative industry was generated there (1975 data), and here the more important modern urban activities are concentrated, such as banking and finance, the largest publicity agencies, two of the four national newspapers, and the largest and most important Brazilian university.

Growth of the MASP, however, is a relatively recent phenomenon. Even though some of its municipalities are among the country's oldest (the city of São Paulo, for instance, was settled in 1554), until the first half of the 19th century they performed only local functions. The city of São Paulo itself performed only regional central-place functions until the end of the last century since Rio de Janeiro was, indisputably, the national metropolis. In fact, by 1890, while Rio already had more than 500 thousand inhabitants, the city of São Paulo barely surpassed 60 thousand.

The development, first of São Paulo, and later (after 1940) of the MASP as a whole is the result of the three-phase development process which has been taking place in Brazil during the last hundred years. The first phase, characterized by expansion of the coffee-export sector in the State of São Paulo, took place in the last decades of the 19th century and first decades of the 20th century. During this period, the city of São Paulo benefited from locational advantages as well as an import-export commercial center and, increasingly, as an industrial area supporting the expanding coffee economy. In this initial period, the area benefited from growing waves of European migrants which provided it with a diversified pool of human resources, from an expanding network of communications with its hinterland, and from an important internal urban infrastructure and urban network.

The second phase, starting in the 1920s, was dominated by consumer goods import-substitution industrialization. The city of São Paulo gained from its proximity to the growing internal market created by the money-based coffee economy and offered some advantages in terms of hydroelectrical energy, good transport connections with the hinterland, and a diversified pool of human resources and entrepreneurial talents. By 1940, the State of São Paulo had already surpassed Rio de Janeiro in its value added contribution to the industrial sector, and the city of São Paulo, as its state capital, got the lion's share of this industrial expansion.

After the Second World War, a third development phase took place, dominated by expansion of the capital and durable goods sectors of industry. This new wave of growth in Brazil in the last 30 years transformed the country's structural conditions. The city of São Paulo and urban areas surrounding it have been the spatial focus of these changes and, as a consequence, urbanization of the area and its later metropolization gained momentum.

To analyze the demographic component of metropolization after 1940, it is useful to distinguish three subregions within the MASP: the core, formed by the municipio of São Paulo and two adjacent municipalities; the inner periphery formed by 17 other municipalities surrounding the core; and the outer periphery of the MASP, formed by the remaining municipalities.

Until 1940, urban growth was concentrated in the core region of the MASP. After that, population started to accelerate at the inner core, and after 1960 also at the outer periphery (see table 10-1). From 1970-80, population at the core grew at an annual rate of 3.8 percent, in the inner periphery at 6.6 percent, and at the outer periphery at 5.9 percent yearly. As a result of these differential growth rates, by 1980 about 72 percent of the MASP population lived at the core, 24 percent in the inner periphery, and 4 percent in the outer periphery.

Migration from other regions in the State of São Paulo and other parts of the country (particularly from the northeast) has been the main source of metropolitan population growth. Even in the last decade, when the pre-existing population base was large enough to feed the growth process, internal inmigration contributed 77.2 percent to the decennial population increment although a significant amount of outmigration also occurred (see table 10.2).

By 1980, among MASP inhabitants, 57 percent were immigrants; over a third came from other Brazilian urban areas; more than a fourth had moved to the MASP in the last decade; and fewer than 5 percent had arrived in the final year of the decade (see table 10.3). An expanding supply of cheap labor over a long period of intense growth has been one of the main factors accounting for two characteristics of the MASP: a dramatic growth of its urban base and persistence of high levels of poverty together with the expansion of wealth. In fact, the pace of industrial and service expansion of the MASP, fueled by an almost unlimited supply of cheap labor, created a metropolitan area where social contrasts are quite intense.

The MASP exemplifies the enigma of an industrially complex, urban, modern, and poor society. As a sociologically intriguing phenomenon, a substantial amount of fresh research must be done to understand this new reality better. A brief description of the economic structure, income distribution, and urban services situation of the MASP follows.

## Economic Structure and Income Distribution

The economic base of the MASP, compared to Brazil as a whole, is highly specialized in performing industrial and modern central-place service functions (see table 10.4). The percentage of persons employed in the secondary

TABLE 10.1

Population Growth, Core and Periphery—Metropolitan Area
of São Paulo (MASP), 1950-1980 (in thousands and percentages)

| Year | | | MASP Areas | | | | | |
| | | | Periphery | | | | | |
| | Core | | Inner | | Outer | | Total | |
| | N | % | N | % | N | % | N | % |
|---|---|---|---|---|---|---|---|---|
| 1950 | 2,198 | 82.5 | 346 | 13.0 | 119 | 4.5 | 2,663 | 100.0 |
| 1960 | 3,331 | 80.0 | 808 | 16.9 | 152 | 3.1 | 4,791 | 100.0 |
| 1970 | 6,305 | 76.8 | 1,638 | 20.0 | 263 | 3.2 | 8,206 | 100.0 |
| 1980 | 9,163 | 72.0 | 3,090 | 24.3 | 465 | 3.7 | 12,719 | 100.0 |
| Annual growth rates | | | | | | | | |
| 1950/1960 | | 5.7 | | 8.9 | | 2.5 | | 6.1 |
| 1960/1970 | | 5.1 | | 7.3 | | 5.6 | | 5.5 |
| 1970/1980 | | 3.8 | | 6.6 | | 5.9 | | 4.5 |
| 1950/1980 | | | | | | | | |

SOURCE: FIBGE (1981).

TABLE 10.2

Population Growth Components—Metropolitan Area
of São Paulo, 1970-1980 (in thousands and percentages)

| Components | N | % |
|---|---|---|
| Natural increase | +2095 | 48 |
| (2.3% annual average) | | |
| In-migration | +3839 | 77 |
| Out-migration | −1096 | 25 |
| Total | 4383 | 100 |

SOURCE: Census of Brazil (1980).

sector of the economy was almost twice as high in the MASP as in all of Brazil.
The manufacturing industry subsector had the lion's share of the MASP labor
force. Nearly a third of Brazil's industrial labor force is located in the MASP.
The percentage of labor force participation in distributive and personal
services also reaches higher levels in the MASP than in the country as a whole.
Together, the manufacturing industry and distributive and personal services
in 1980 accounted for more than 80 percent of the MASP employment
compared to slightly over half the national employment.

The MASP is, therefore, Brazil's most industrialized area and rough
estimates indicate that its contribution to the total value added by the
manufacturing industry in Brazil by 1980 had been around 40 percent.
Analysis of the internal structure of MASP manufacturing shows a highly
modern and diversified industrial sector. The capital and intermediate goods
industry in 1980 generated 77.2 percent of the total value added by the
manufacturing sector as a whole in the MASP and employed more than 60
percent of total industrial employment. The MASP's main manufacturing

TABLE 10.3

Natives and Migrants—Metropolitan Area of São Paulo,
1980 (in thousands and percentages)

| Population | A City of São Paulo | | B MASP | | A/B |
|---|---|---|---|---|---|
| | N | % | N | % | % |
| Migrants | | | | | |
| Total | 3,646 | 51.2 | 7,169 | 57.0 | 51 |
| From urban area | 2,176 | 30.6 | 4,566 | 36.0 | 48 |
| Less than 10 years | 1,408 | 19.8 | 3,384 | 27.0 | 48 |
| From outside state | NA | — | 1,920 | 15.0 | — |
| Less than 1 year | 206 | 2.9 | 562 | 4.5 | 37 |
| Total | 7,114 | 100.0 | 12,589 | 100.0 | 56 |

SOURCE: Census of Brazil (1980).

TABLE 10.4

Economically Active Population by Economic Sector—Brazil,
MASP, and São Paulo, 1980
(in thousands and percentages)

| Economic Sector | Brazil | | MASP | | São Paulo | | MASP as Percentage of Brazil | São Paulo as Percentage of MASP |
|---|---|---|---|---|---|---|---|---|
| | N | % | N | % | N | % | | |
| Primary | 12,661 | 30 | 41 | 1 | 11 | — | — | 28 |
| Secondary | 10,772 | 26 | 2,426 | 46 | 1,556 | 42 | 23 | 64 |
| Manufacturing | 6,939 | 16 | 1,991 | 38 | 1,271 | 34 | 29 | 64 |
| Construction | 3,171 | 8 | 372 | 7 | 242 | 7 | 12 | 65 |
| Other industry | 662 | 2 | 62 | 1 | 43 | 1 | 9 | 69 |
| Tertiary | 18,838 | 45 | 2,838 | 54 | 2,147 | 58 | 15 | 76 |
| Distributive services | 5,838 | 14 | 887 | 17 | 652 | 18 | 15 | 74 |
| Personal services | 7,032 | 17 | 1,105 | 21 | 844 | 23 | 16 | 76 |
| Social services | 4,693 | 11 | 549 | 10 | 412 | 11 | 12 | 75 |
| Other services | 1,274 | 3 | 298 | 6 | 239 | 7 | 23 | 80 |
| Total | 42,271 | | 5,305 | | 3,714 | | 13 | 70 |

SOURCE: Census of Brazil (1980).
NOTE: Percentages do not total 100 because of rounding.

sectors are metallurgical, automotive, electrical, and mechanical industries.
Together, they account for nearly half the manufacturing employment in the
metropolitan area (see table 10.5).

As for distributive and personal services, modern branches of these
subsectors are located in the MASP such as the financial, information-
processing, and research and development subsectors. It is important to
emphasize that a large number of traditional and informal tertiary activities
still remain in the MASP.

Economic development of the MASP in the last 30 years, although based in
expansion of a highly capital-intensive industrial sector and on growth of a

TABLE 10.5

Occupied Persons by Manufacturing Industry Subsectors—
Metropolitan Area of São Paulo, 1980
(in thousands and percentages)

| Industrial Subsectors | N | % |
|---|---|---|
| Metalurgical | 263 | 15 |
| Automotive | 215 | 13 |
| Electrical and communications | 169 | 10 |
| Mechanical | 161 | 9 |
| Textile | 135 | 8 |
| Garments and shoes | 103 | 6 |
| Chemical | 81 | 5 |
| Food processing | 76 | 4 |
| Miscellaneous | 62 | 4 |
| Nonmetallic minerals | 60 | 3 |
| Editorial | 57 | 3 |
| Pharmaceutical | 28 | 2 |
| Subtotal | 1410 | 82 |
| Other subsectors | 301 | 18 |
| Total | 1711 | 100 |

SOURCE: Census of Brazil (1980).

modern tertiary and quaternary economy, has not been able to produce a more equitable income distribution. Income for the MASP working population has not risen commensurate to the gains in economic productivity occurring during the period. An elastic supply of labor, deep regional and sectoral heterogeneity, and persistence for a long period of an authoritarian regime that curtailed political and labor union mobilization and participation are some of the factors contributing to the persistence of widespread poverty in the MASP.

Although wages are better in the MASP than in Brazil as a whole, in this more developed and industrialized region of the country, nearly half the residents hold jobs earning less than twice the minimum per capita subsistence wage (see table 10.6). Purchasing power at these salaries is not sufficient to buy minimum food for a young family of three to survive.

Economic activities and different income classes are not evenly distributed throughout the several municipalities and subregions within each municipality. On the contrary, economic activities are highly segregated spatially as are the different economic segments of the population. There is an intra-metropolitan territorial division of labor: in some municipalities and districts within them, for better or worse, there is a high concentration of modern and high-productivity industries (some districts of the city of São Paulo and São Bernardo, for instance) while other municipios specialize in low-productivity industrial and service endeavors; still others perform the function of labor force reproduction (municipalities and districts known as *cidades dormitorios*, such as Barueri and Diadema). This spatial segregation of classes of activities and income groups, resulting from the development process, is closely related

TABLE 10.6
Monthly Income Classes, Urban Labor Force—Brazil and
Metropolitan Area of São Paulo, 1982

| Income Classes | Brazil | Southeast | State of São Paulo | MASP |
|---|---|---|---|---|
| $0-$45 + $45-$90* | 35.0 | 28.9 | 22.2 | 15.2 |
| $90-$180 | 28.3 | 28.6 | 29.1 | 29.6 |
| $180-$450 | 24.5 | 27.9 | 31.9 | 35.4 |
| $450-$900 + over $900 | 12.2 | 14.5 | 16.7 | 19.7 |
| Total N | 33,681,851 | 18,373,543 | 9,899,853 | 5,534,185 |

SOURCE: FIBGE (1982).
*Minimum subsistance wage = $90 U.S. (1982).

to an array of urban problems (such as shortage of housing, transportation, and public services) facing the MASP. We now turn to these aspects.

# Growth and Poverty:
# Urban Problems Facing the MASP

The major problems facing the São Paulo metropolitan area are:

(1) Urban growth at annual rates exceeding 5 percent, which means the population will double in less than 15 years.
(2) A pattern of capital accumulation increasingly based on foreign multinational corporations (even though producing for the internal market).
(3) Long periods of political demobilization and repression.
(4) Highly segregated patterns of spatial organization.

Most of these urban problems, which have accumulated over the years to an astonishing amount, are linked to structural characteristics of the development pattern and the persistence of widespread urban poverty. Others are linked to the nature of state power and the organizational pattern of public services; still others have links to the nature of the relationship among different levels of government and to distribution of governmental responsibilities. See table 10.7 for a set of indicators summarizing the main problems the MASP was facing at the beginning of the 1980s.

The size and complexity of the problems underlying the indicators provided in table 10.7 are easy to grasp in terms of social and economic attributes and governability. Let us briefly examine the most important ones.

## Housing and Urban Infrastructure

Under the pressure of rapid population growth as well as of rapid increase in economic activities, the process of urban expansion in the MASP has been chaotic, especially if we consider the economic and political influence of

TABLE 10.7
Main Deficits—Metropolitan Area of São Paulo, c. 1982

| | |
|---|---|
| Percentage of solid wastes collected but not adequately treated | 78 |
| Percentage of persons living in households not served by sewage system | 64 |
| Percentage of persons living in areas without solid waste collection | 33 |
| Percentage of households not served by treated water systems | 17 |
| Percentage of persons living in substandard houses | 11 |
| Modal class of hours daily spent on urban transportation | 3-4 hours |
| Coefficient of infant mortality (%) | 55 |
| Percentage of persons over age 10 functionally illiterate | 22 |
| Percentage of employed persons not covered by Social Security | 21 |
| Percentage of families earning less than twice the minimum per capita subsistence wage | 34 |
| Percentage of persons holding irregular jobs | 23 |
| Percentage of open unemployment (people seeking jobs while currently unemployed) | 10-15 |

NOTE: Compiled by the author from various sources.

interests linked to urban land development and the construction industry. Together with the lack of adequate public programs for housing and urban infrastructure—especially for the poor—these factors have operated to produce an array of urban problems difficult to solve. It is estimated that the housing deficit, in terms of shortage and quality, in the MASP is now around 300 thousand units: about 500 thousand persons live in slums (*favelas*) and about a million live in deteriorated multiple-family housing (*corticos*).

Another aspect of this problem is the large number of families living in areas opened for urban settlement that lack even minimum residential necessities and urban infrastructure. At the several urban peripheries in the MASP are land developments (*loteamentos*) where the working class people have been building substandard houses by themselves, without having adequate rights to the land, even though this land was purchased at market prices. It is estimated that in the city of São Paulo alone there is an area of 311,474.774 square meters of land illegally occupied and developed.

The rapid, chaotic, illegal, and highly segregated process of urban expansion has placed a severe burden on the city to offer basic infrastructure, particularly waste collection and treatment and urban transportation.

Regarding the latter, two factors have been contributing to urban transportation problems. First is the spatial segregation of activities and housing by income groups. Most of the working class population has to travel daily across long distances to reach places of work. Second, the development model of the last 30 years has been strategically linked to automobile transport. As a consequence, emphasis has been given to individual transportation at the expense of collective mass transportation and, for the latter, preference has been given to buses instead of developing subways and railways, at least until recently. (The city of São Paulo subway system started to operate in 1975.)

The result (table 10.8) has been that about 40 percent of daily trips are made by individual cars and almost 90 percent by either cars or buses. The impact of this mode of urban transportation in terms of traffic congestion, cost, and

TABLE 10.8
Daily Trips by Type—Metropolitan Area of São Paulo, 1982
(in thousands and percentages)

| Type | N | % |
|---|---|---|
| Collective | 11,800 | 61 |
| Bus | 9,800 | (83) |
| Subway | 1,180 | (10) |
| Railway | 820 | (7) |
| Individual (autos and cabs) | 7,500 | 39 |
| Total | 19,300 | 100 |

SOURCE: Fagnani (1985: Tables 9 and 11).

environmental pollution is easy to assess. Moving around the MASP to work, study, or even to shop is costly, dangerous, and time consuming; modally, a person spends between three and four hours daily in travel.

## Health and Education

The state of health and educational services in the MASP, in aggregate terms and compared with the country as a whole, is in better shape than the services previously discussed. In terms of desirable levels, however, the situation still deserves close attention from the government.

Even when the coefficient of infant mortality was 55 deaths per thousand at the MASP level, in some areas within the MASP this rate was still well above 80. The same occurs with availability and geographical distribution of hospital beds. By 1982, 55,558 hospital beds were available in the MASP—less than 5 per thousand inhabitants. Moreover, they were extremely poorly distributed and in at least eight MASP municipios no hospital beds were available.

As for basic education, about 90 percent of the population aged 5-14 were enrolled in school. The main problems once again are linked to widespread poverty and deficient urban services: schools are unevenly distributed, forcing students to travel long distances and once enrolled, students have difficulty remaining in school, mainly for reasons connected to household poverty. The latest available information indicates that for each ten students entering first grade only four would remain students by the eighth grade. One strategic way to keep students in school is to provide food all year round; otherwise most of them will leave school or have an inadequate academic performance because of factors associated with malnutrition.

However, the main problem associated with MASP health and educational services concerns daily work and the wages received. Most schools and health centers are ill equipped and personnel receive very low wages. As a result, the quality of daily services is bad.

## Water and Environmental Pollution

Although about 60 percent of the fresh water used in the MASP comes from outside the region, the water supply system functions at satisfactory levels. The main problem here is cost: it is not uncommon to have numerous households in an area the water supply systems serves not connected to the system simply because their household budgets do not permit such an expenditure.

By 1980, only 36 percent of the population of the MASP were living in areas provided with sewer systems. Worse yet, the proportion of sewage collected that was receiving adequate treatment was negligible. Most residential and industrial sewage flows directly into rivers and other water reservoirs in the MASP. As a result the level of water pollution is quite high. The two main rivers are completely dead, a situation similar to that of the Thames River in the 1950s.

About nine thousand tons of solid wastes are produced daily in the MASP. Hospital solid waste represents about a hundred tons, industrial solid residues about 29 hundred tons, of which 130 tons are highly lethal. Only a small fraction of this solid waste is collected and treated—about 10 percent. A large proportion remains in open-air deposits of different sizes and locations—some dangerously near water supply sources and most in areas completely uncontrolled by public authorities. Existing plans to improve the situation, if executed, would only postpone a part of the problem for a few years.

A further problem affecting various areas of the MASP is periodical flooding. Because of problems related to the absence of drainage systems and to inadequate control of rivers crossing the region, some localities are routinely inundated. Damage to property, particularly housing for the poor, can be quite extensive.

Air pollution is another serious threat facing the MASP. The topographical and meteorological situation of the MASP, particularly during the winter, is unfavorable to pollutant dispersion and air quality reaches undesirable levels several times during the year. There is no adequate air pollution control legislation and even the minimal existing law is difficult to enforce.

The main sources of air pollution are automobiles and buses (about a million vehicles) and industries, as well as open-air combustion of solid wastes. Some of these environmental problems are highly correlated spatially and attributed to social segregation patterns in lower class neighborhoods where most of them occur simultaneously.

It is difficult to estimate the amount of resources necessary to solve this cumulative array of problems affecting the quality of life of the MASP and, particularly, its poorest segment. A quote from a recent press conference given by São Paulo's Mayor, Dr. Mario Covas, Jr., gives a rough idea of the magnitude of the problems:

If, by any special circumstance, the city government could get the $13 trillion *cuzeiros* needed to eliminate in a year the problems the city has in terms of urban infrastructure, housing, and urban services, São Paulo would gain six thousand kilometers of good streets, one thousand kilometers of adequate waterways, four thousand nurseries and about a million new houses for the poor; the city would also gain hundreds of health care centers, tens of hospitals and green spaces (my translation from *Folha de São Paulo*, 1985:21).

In other words, the city of São Paulo would need 12 times its present annual budget to address its present problems (about 30 billion dollars, or a third of the Brazilian external debt).

## Impact of the Present Crisis

The situation in the Metropolitan Area of São Paulo has been the structural result of several decades of sustained growth and development, during which only a few periods of stagnation and recession of short duration occurred.

As is well known, a debt crisis hit the Brazilian economy very deeply and the cumulative effects of the crisis, particularly after 1981, immersed the country in a recession that continued through 1986. The recession has been affecting, particularly, some important segments of the industrial sector as well as the construction industry. While this is not the place to analyze the causes and characteristics of the recession, it is relevant to point out that the MASP has been suffering severely with the crisis since an important proportion of the Brazilian manufacturing industry is located in the MASP.

The slowing down of economic activities has been affecting the quality of life in the MASP in different but cumulative ways. For one, rates of open unemployment which in the last decade oscillated around 2 or 3 percent of the MASP labor force, reached nearly 10 percent in the first quarter of 1985, even by underestimated official figures. In an economy like Brazil's where an important proportion of the economically active population makes a living by holding temporary, ill-paid, and informal jobs, distinguishing *open* employment from various forms of unemployment is important.

If we include disguised forms of employment (such as those not looking for jobs anymore and those holding very unstable and ill-paid jobs), the rates of total unemployment reached almost 15 percent of the MASP labor force in 1985 (see table 10.9). These unemployment figures are even higher for specific segments of the population of the metropolitan area, such as youths.

Paralleling the rise of the open unemployment rate was the increase of informal economic activities located in marginal sectors of the urban economy. A rough indication of this phenomenon is given by the proportion of employed persons working without legal registration, one of the guarantees workers have that they will be covered by existing social benefits. For example, in construction, the percentage of formal employment decreased by 12 points from 80.5 to 67.8 between 1976 and 1983.

Another consequence of the present crisis has been acceleration of inflation which had reached 200 percent annually by mid-1984, with the correspondent impact on the cost of living, particularly food and urban transportation prices. Food costs were also aggravated by the agricultural policy subsidizing

TABLE 10.9
Specific Unemployment Rates—Metropolitan Area of São Paulo,
February-April, 1985 (in percentage of labor force)

| Specific Groups | Open | Disguised | Total |
|---|---|---|---|
| | | Unemployment | |
| Total economically active population | 9.2 | 5.0 | 14.2 |
| Women | 12.4 | 5.7 | 18.1 |
| Youths (less than 18) | 29.2 | 7.2 | 36.4 |
| Recent migrants | | | 17.9 |
| Nonhead of households | 13.8 | 6.6 | 20.4 |
| Nonwhite | 10.4 | 6.2 | 16.6 |

SOURCE: São Paulo (1985).

sugar cane used for alcohol production to help solve the energy crisis and production of export staples to help Brazil's balance of payment problems. Both resulted in a decrease in food supply. Transportation costs were further hurt by the oil crisis, but this has not been devastating because of the relative success of the alcohol alternative. The impact of food and transportation price increases on the household budget of the lower segments of the MASP population is heavy since more than 60 percent of lower class household budgets is spent on food and transportation.

The crisis hit more deeply those firms facing problems of financing as a result of increases in interest rates as well as those technologically less advanced. Once again this impact, given different locational industrial patterns prevailing in the MASP, has been spatially concentrated, aggravating unemployment in some subregions.

Finally, the decline of economic activities affected the fiscal base of state and municipal governments and the performance of public services. Resources for these sectors are based directly on funds related to the pace of employment expansion, as is the case of housing (Banco Nacional da Habitação-Sistema Financeiro da Habitação) and social security and health provision (Instituto Nacional de Assistencia Medica de Previdencia Social) which nearly collapsed.

Growing unemployment and the deep financial crisis of the housing, social security, and health provision of federal programs in a situation of widespread structural urban poverty put a severe burden on state and local governments. It has also led to an increased amount of urban violence—the most tragic indicator of the depth of the present social crisis in the MASP.

Violence in an urban area of the size, complexity, heterogeneity, and social inequality as the MASP is an endemic phenomenon. Although ethnically heterogeneous, and presenting a certain degree of ethnic spatial segregation, there is no visible violence along ethnic cleavages. Migrants of Italian, Japanese, Portuguese, and Arab origin live relatively peacefully with Brazilians of European and African background, despite existing prejudices and discrimination. Urban violence in São Paulo, besides being related to several aspects of modern urban life, is anchored in the persistence of widespread structural poverty.

With the worsening socioeconomic situation resulting from the recession, urban violence has been increasing since 1981, particularly criminal offenses against property. Studies in this area indicate high longitudinal correlations between indicators of crimes against property and level of industrial employment (negative correlations of about .65) on one hand, and indicators of the cost of living for the working class (positive correlations of about .60) on the other hand. Also occurring in the MASP, mostly at the beginning of 1983, were forms of social violence such as rioting, although these could be attributed to the organizing efforts of certain political groups.

In summary, the debt crisis and recession have been aggravating the social situation of the area which is facing, despite some signs of economic recuperation, severe problems of unemployment, rising cost of living, and increasing urban violence.

There is, however, another side of this coin of poverty and social turmoil. Also in the MASP are the most well-organized and autonomous trade unions, such as the metallurgical workers' union; a growing array of grass roots social movements; and an active and progressive intelligentsia. In São Paulo, two of the most important and modern political parties—Partido do Movimento Democratico Brasilerio (PMDB) and the Partido dos Trabalhadores—find strong active support, providing the country with a new leadership coming from several sectors of society. The MASP, has, therefore, an enormous potential for progressive change.

The Metropolitan Area of São Paulo can be characterized as polarized by the contrast between its impressive performance as an economic center and the levels of poverty and inequality still prevailing in the area. One is struck by the modernity and complexity of its industrial base, strength of its financial institutions and commercial network, sophistication of some segments of its personal service subsectors, richness of its cultural life, and its potential for sustained economic growth.

At the same time one observes the sprawling of favelas and corticos, lack of adequate urban infrastructure, pollution of its environment, violence of its streets. The situation in the MASP is one of concern for the hugeness of its problems and its growing ungovernability. By way of conclusion I offer some policy measures at the macrolevel that are necessary requisites to addressing a multitude of more specific problems São Paulo faces.

## Setting an Agenda for the Near Future

First and foremost, it is necessary to take measures to end the present recession and stop rampant inflation. If the Brazilian economy returns to its historical rates of growth, present unemployment rates will decrease and control of inflation will reduce the cost of living problem. To put the economy in a growth pattern, the country—and the international economic system—will have to give a structural and long-term solution to the problem of the external debt, in terms of long-term rescheduling of payments (interests and principal) and of in-flow of new investments. According to Brazilian economists, a major restructuring of the Brazilian financial system is urgently needed. These measures, even if they have a national impact and focus, are crucial for the metropolitan economy since it is in the MASP where most of the more dynamic and modern sectors are located which will benefit from a national economic recovery.

Second, it is important to reform the Brazilian fiscal system in terms of tax structure, changes, and distribution, as well as in terms of control by different levels of government. State and local governments should have control over larger portions of fiscal income as well as an autonomous—even though democratically controlled—and more flexible taxing capacity. These reforms would increase the financial resources of local power and the more democratic control would allow planning of local and state government expenditures to take local demands into due consideration. Decentralization in fiscal policy and administration would benefit the MASP insofar as a large portion of this fiscal income is generated in the area and also because the income distribution of the MASP could support a more progressive and differentiated tax structure.

Third, it is necessary to change the prevailing structure of the public sector operating in the area of basic social services (particularly housing, public health, transportation, energy and water supply, and environmental control). Agencies in charge of these sectors need to have more of their resources free of links to financial returns and should be planned to operate efficiently but oriented toward goals of social equality rather than market-oriented economic success. This orientation would increase public services for those who cannot pay market prices for them and would benefit the poorest segments of the MASP population, particularly if local and metropolitan governments have some autonomy for allocation of public resources into these services.

Fourth, political, judicial, and administrative reform should be considered; its main objective should be a new organization of the municipal and metropolitan political-administrative structure. At the municipal level, different measures facilitating differentiation of local powers according to size, economic base, and function within the metropolitan division of labor should be taken. Increasing executive decentralization and improving existing mechanisms and creating new ones of popular control over public decisions are urgent needs. At the metropolitan level, coordination authorities supported by adequate amounts of financial resources, political power, and juridicial legitimacy have to be created to deal with problems requiring structural metropolitan consideration.

Fifth, it is necessary to plan ahead for technological innovations to come. There is some risk that much industry located in the MASP will become obsolete in the near future. Without an adequate long-term policy to attract new industry and facilitate the restructuring of existing ones, there is a risk of economic decay and urban blight. This aspect is crucial for the city of São Paulo with its older industrial base. Measures to attract new, dynamic industries are essential. Otherwise, the city of São Paulo could suffer the same type of economic decay which has occurred in large metropolitan areas of developed countries.

Economic recovery, although necessary, is not a sufficient condition for improvement of the situation in the MASP. The resources generated by long-term economic development, under conditions of democratic control of public expenditures at the local and metropolitan levels, should be channeled primarily to improve housing, transportation, and basic infrastructure (sewage and solid waste collection and treatment, environmental pollution, and flooding).

Another major problem is related to population growth. By the year 2000 the Metropolitan Area of São Paulo will have around 25 million inhabitants. Natural increase is already declining so the main problem in the future will be migration, not in relative terms but mainly in the absolute number of people migrating to the area. The solutions again lie outside the metropolitan area: economic development of outlying regions should take place to create other poles of attraction.

Without an adequate mix of national and local policies prospects for the Metropolitan Area of São Paulo are bleak. More than 20 million people living under the existing conditions is a political powder keg, nor is their situation morally acceptable.

## Bibliography

Brazil. Fundação de Instituto Brasileiro de Geografia e Estatística. Secretarìa de Planejamento da Presidência da República. 1976. *Pesquisa Nacional for Amostra Domicílios*. Rio de Janeiro.

———. 1981. *Sinopse Preliminar do Censo Demografico. IX Recenseamento Geral do Brasil-1980*. Rio de Janeiro.

———. 1982. *Pesquisa Nacional for Amostra Domicílios*. Rio de Janeiro.

———. 1983a. *Censo Demográfico. IX Recenseamento Geral do Brasil-1982*. Rio de Janeiro.

———. 1983b. *Pesquisa Nacional for Amostra Domicílios*. Rio de Janeiro.

Fagnani, Eduardo. 1985. "Pobres Viajantes. Estado e Transporte Coletivo Urbano Brasil—Grande São Paulo 1964/84, São Paulo." Master's thesis, State University of Campinas, São Paulo.

Faria, Vilmar E. 1983a. "Crescimento Econômico, Urbanização e Pobreza: O Caso de São Paulo." *Estudos FUNDAP. A Questão Urbana e os Serviços Públicos*. No. 1. São Paulo.

———. 1983b. "Desenvolvimento, Urbanização e Mudancas na Estrutura do Emprego: A Experiência Brasileira dos Ultimos Trinta Annos." In *Sociedade e Política no Brasil pós-64*, edited by Bernardo Sorj and Maria Hermínia Tavares Almeida. São Paulo: Brasiliense.

FIBGE. *See* Brazil.

Folha de São Paulo, Staff. 1985. "Idéias para cobrir o déficit." *Folha de Sao Paulo* (27 Feb.):21.

Greenfield, Gerald Michael. 1982. "Privatism and Urban Development in Latin America: The Case of São Paulo, Brazil." *Journal of Urban History* 8:397-426.

Langenbuch, Juergen Richard. 1971. *A Estruturação da Grande São Paulo; Estudio de Geografia Urbana*. Bibliotea geográfica brasileira ser. A, Books, no. 26. Rio de Janeiro: Departamento de Documentação e Divulgação, Instituto Brasileiro de Geografia.

Morse, Richard McGee. 1974. *From Community to Metropolis. A Biography of São Paulo, Brazil*. New & enlarged ed. New York: Octagon Books.

São Paulo (state). Fundação Sistema Estadual de Análise de Dados, and Departamento

Intersindical de Estatística e Estudos Socio-econômicos. 1985. *Pesquisa de Emprego e Desemprego na Grande São Paulo*. São Paulo.

São Paulo Justice and Peace Commission. 1978. *São Paulo: Growth and Poverty: A Report from the São Paulo Justice and Peace Commission*. [Translation of *São Paulo, 1975*.] London: Bowerdean.

Warren, Dean. 1969. *The Industrialization of São Paulo, 1880-1945*. Austin: University of Texas Press for the Institute of Latin American Studies.

# Index

# About the Editors

MATTEI DOGAN is scientific director at the National Center for Scientific Research, Paris, and professor of political science at the University of California at Los Angeles. Author, coauthor, editor, or coeditor of 15 books and a contributor to 38 others, his most recent publications are *How to Compare Nations*, *Le Moloch en Europe*, and *Comparing Pluralist Democracies*.

JOHN D. KASARDA, Kenan professor of sociology and chairman of his department at the University of North Carolina at Chapel Hill, is also a Fellow of the Carolina Population Center. His research focus is urban economic restructuring and public policy. He has published extensively on urbanization processes in the United States and abroad.

# About the Contributors

ISAAC AYINDE ADALEMO, professor of geography and deputy vice-chancellor of the University of Lagos, Nigeria, holds a Ph.D. in geography from the University of Michigan. Coeditor of *Population Issues in Planning Metropolitan Lagos*, he served as associate consultant to the Lagos Metropolitan Master Plan Project.

VILMAR EVANGELISTA FARIA holds a Ph.D. in sociology from Harvard University. He is on the faculty of the Instituto de Ciencias Humanas da Universidade Estadual de Campinas in São Paulo and editor of *Bahia de Todos os Pobres*.

JAMES W. HUGHES, chairman and graduate director of the Department of Urban Planning and Policy Development, Rutgers University, has been a Woodrow Wilson and a Ford Foundation Fellow. He has served on a number of federal and state housing and planning task forces and is a contributing editor to *American Demographics*.

EMRYS JONES, professor emeritus of geography at the London School of Economics, is the author of several books on cities and social geography, including an atlas of London. He has been a consultant on several new towns and cities in England.

AHMED M. KHALIFA is Director General, Regional Arab Center for Social Science Research and Documentation (UNESCO) in Cairo. Recipient of the

1984 State Award of Highest Merit in Social Science in Egypt, he is editor of the *National Review of Social Sciences*, Cairo.

IVAN LIGHT is professor of sociology at the University of California, Los Angeles. He is the author of *Ethnic Enterprise in America, Cities in World Perspective*, and coauthor of *Immigrant Entrepreneurs: Koreans in Los Angeles*.

MICHAEL L. McNULTY is professor of geography and director, Center for International and Comparative Studies, University of Iowa. His publications have dealt with problems of urban and regional development in the Third World. He has been active with the National Science Foundation.

MOHAMED M. MOHIEDDIN holds a Ph.D. in sociology from the University of North Carolina at Chapel Hill. He has been a researcher with the National Center for Social and Criminological Research and the Supreme Council for Family Planning in Cairo.

RHOADS MURPHEY is professor of history at the University of Michigan where he formerly taught Asian studies and geography. A writer and scholar focusing on China and South Asia, his most recent book is *The Fading of the Maoist Vision*.

HANS NAGPAUL is associate professor of sociology at Cleveland State University and formerly taught at the University of Delhi. He has worked with the Indian National Planning Commission and has published four books and numerous articles.

HACHIRO NAKAMURA is professor of sociology in the Institute of Social Sciences, University of Tsukuba, Japan. He has published widely in the field of urban sociology, specializing in Japanese urban neighborhood associations, community power, the urban community, urban populations, and urbanization in developing countries.

MARTHA SCHTEINGART, an Argentinian architect and urban sociologist, has been a professor and researcher at el Colegio de Mexico since 1975. Her area of specialization includes urban and housing problems in Mexico and other Latin American countries, particularly Argentina and Chile.

GEORGE STERNLIEB, who holds a doctorate from the Harvard Business School, is director of the Center for Urban Policy Research and professor of urban planning and policy development at Rutgers University. He is a member of the Census Advisory Committee on Population Statistics, a trustee of the Urban Land Institute, and has served on several presidential task forces on urban development.

JAMES W. WHITE is professor of political science at the University of North Carolina at Chapel Hill and a Fellow of the Carolina Population Center. He is the author of *Political Implications of Cityward Migration* and *Migration in Metropolitan Japan*.